THIN ICE

Unlocking the Secrets of Climate in the World's Highest Mountains

MARK BOWEN

A JOHN MACRAE BOOK
HENRY HOLT AND COMPANY NEW YORK

Henry Holt and Company, LLC
Publishers since 1866
175 Fifth Avenue
New York, New York 10010
www.henryholt.com

Distributed in Canada by H. B. Fenn and Company Ltd.

Portions of the text have appeared in different form in *Natural History* and *Climbing.*

Library of Congress Cataloging-in-Publication Data
Bowen, Mark (Mark Stander)
Thin ice : unlocking the secrets of climate in the world's highest mountains / Mark Bowen—1st ed.
p. cm.
Includes bibliographical references and index.
ISBN-13: 978-0-8050-6443-8
ISBN-10: 0-8050-6443-5
1. Atmosphere, Upper—Observations. 2. Climatic changes—Research. 3. Tropics—Climate—Research. 4. Climatology—Research. I. Title.

QC879.59.B69 2005
551.51'4'072—dc22 2005040426

Henry Holt books are available for special promotions and premiums. For details contact: Director, Special Markets.

First Edition 2005

Designed by Kelly Too

Printed in the United States of America
1 3 5 7 9 10 8 6 4 2

for Andrew and Anna

I do believe that we ought to pay more attention to the opinion of philosophers, that "nothing but nature can qualify a man for knowledge."

-HENRY WADSWORTH LONGFELLOW

Strive to interpret what really exists.

-LOUIS AGASSIZ

It's coming to grips with reality. The mountain doesn't yield to any dialogue or self-appraisal. It's what you are in one way or the other.

-LONNIE THOMPSON

CONTENTS

THIN ICE

Greenland Sea
GREENLAND
Camp Century
GISP2/GRIP
Dye 3
ICELAND
Bona-Churchill
WRANGELL MOUNTAIN RANGE
Hudson Bay
Labrador Sea
Norwegian Sea
NEWFOUNDLAND
Gulf of St. Lawrence
Great Salt Lake
ROCKY MOUNTAINS
SIERRA NEVADA
GREAT BASIN
NORTH AMERICA
ATLANTIC OCEAN
Tropic of Cancer
PACIFIC OCEAN
Equator
Chimborazo
SOUTH AMERICA
Moche
THE ANDES
Huascarán
Lima
Quelccaya
Cuzco
Lake Titicaca
Arequipa
Sajama
Salar de Uyuni
The Altiplano
Tahiti
Tropic of Capricorn
75°
60°
45°
30°
15°
180°
165°
150°
135°
120°
105°
90°
75°
60°
45°
30°
Patagonia
Tierra del Fuego
Drake Passage
Antarctic Circle
ANTARCTICA
Antarctic Circle
Antarctic Peninsula
Dyer Plateau
Plateau Remote
Pole of Relative Inaccessibility
Siple Station
South Pole
Vostok
Byrd Station
Ross Ice Shelf
Dome C
0
1000 Miles
0
1000 Km

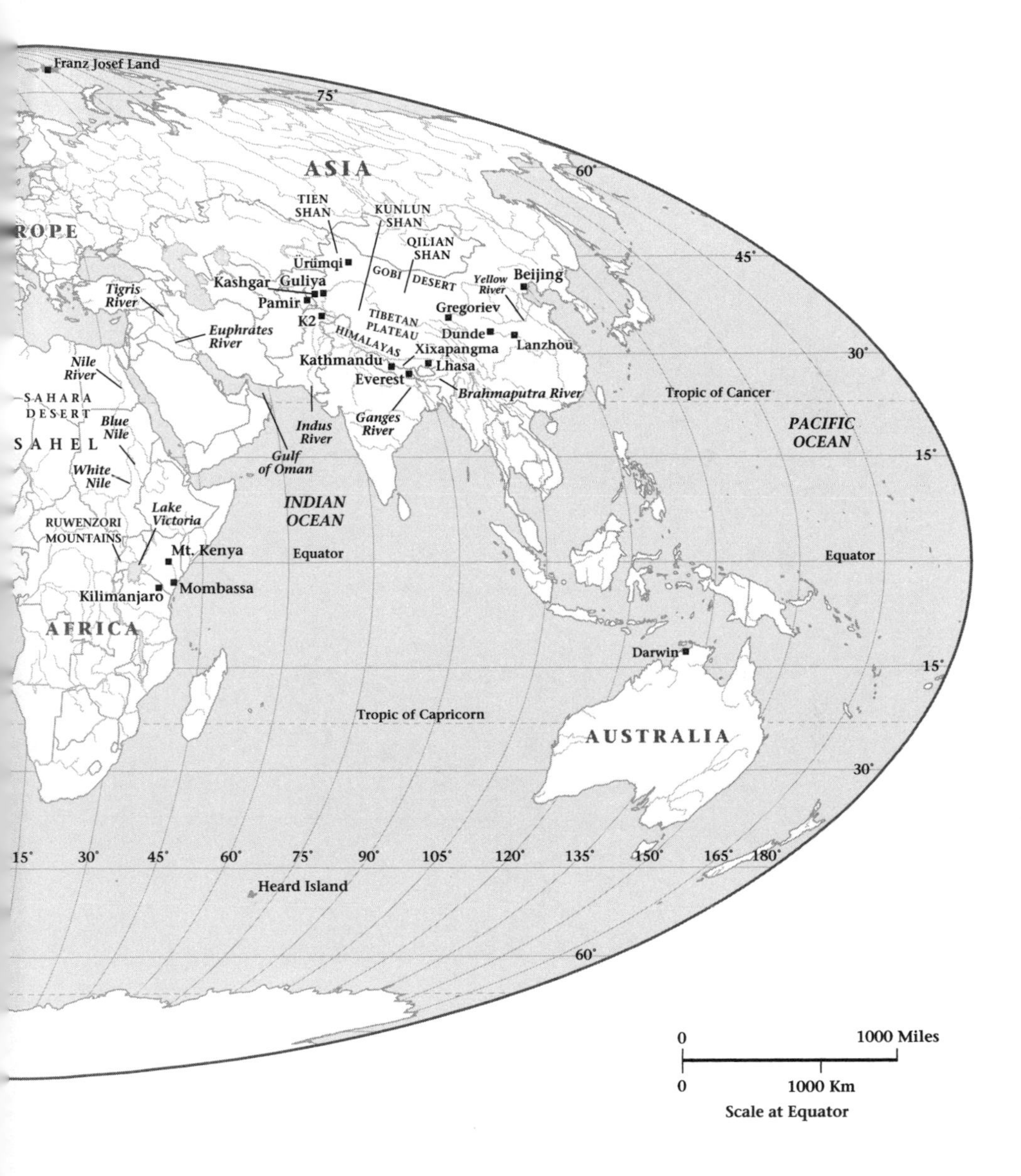

0 1000 Miles

0 1000 Km

Scale at Equator

PROLOGUE

A CALL FROM THE BLUE

My life took a turn in the summer of 1997 with an unexpected phone call from a magazine editor in New York.

She explained that they were rather deep into a story, and they needed some help. A climatologist named Lonnie Thompson, from Ohio State University, was drilling ice cores on the summit of Nevado Sajama, the highest mountain in Bolivia. The magazine had hired a writer to join Thompson's expedition, but she had backed out just that morning, the day before she was due to fly south, for she had no experience in high mountains and was wary of going to altitude—wisely so, as it would later turn out.

The editor, Máire, had gotten my name from a climbing magazine, for which I was just completing an article about a trip I had taken to the island of New Guinea to climb Carstensz Pyramid, the highest mountain between the Andes and the Himalaya. Anyway, somewhere between her desperation and the fact that I had never seen the Andes, we worked out a deal. She sent me a stack of scientific papers by overnight mail and put me in touch with Lonnie's wife and collaborator, Ellen Mosley-Thompson, who sent me another stack of papers and gave me the information I would need to get to the mountain. (I didn't read any papers until I got back.)

Both women also pointed out that Lonnie had taken a unique hot-air balloon along to Bolivia, designed by an accomplished balloonist named Bruce Comstock, who had joined the expedition as well. The idea was to float the ice directly from the summit of the mountain to a freezer located on the back of a flatbed truck, which would be parked on the surrounding plain. If the plan worked, it would be the highest balloon launch in history.

I quickly decided that I wanted to be standing on the summit when the balloon took off, but that there was no way I was going to ride in that thing. For, apart from the high likelihood of having an accident while you're up there, the deadliest risk in going to the top of a high mountain is high altitude edema, abnormal fluid retention in the lungs or brain, and the surest way to bring it on is to ascend too rapidly. Sajama, just shy of 21,500 feet, reaches a fair distance into what climbers call the "death zone," which begins, some say, at about 18,000 feet; and, according to Ellen, Lonnie expected the balloon to fly on July 4, a scant ten days after my plane eventually touched off from a runway at Logan Airport, a few hundred yards from the waters of Boston Harbor.

LA PAZ IS DRAPED FROM THIRTEEN TO ELEVEN THOUSAND FEET DOWN A HUGE GASH IN the side of the world's second-highest plateau (to the Tibetan), South America's Altiplano: dun-colored desert above; dark green rain forest below. It is one of the highest cities in the world and also, thanks to its setting, one of the most beautiful. Five-summited Illimani, the second-highest mountain in the country, can be seen floating like a cloud in the sky from nearly every vantage point in the city. La Paz's airport is located on the Altiplano above the city, and flights from the United States tend to land at about dawn. Through the exceptionally clear morning air, the sharp profiles of surreal white peaks jut from the desert's edge into a cobalt, starry sky.

Visitors have been known to collapse as they walk from their plane to the terminal building, and the advice I had gotten was to spend the first day lying down. But as my cab dropped from the edge of the Altiplano into the canyons that hold the city, and as the sun rose on Illimani before me, I decided to ignore that advice. What would the accidents of travel bring in this vivid land?

Ellen had put me in touch with Bernard Francou, a French glaciologist at the Instituto de Hidraulica e Hidrologia, a collaborative set up by France and Bolivia, and connected with the University of La Paz. Bernard was helping Lonnie with field logistics and ostensibly collaborating on the science. One thing he told me when I dropped in on him that first afternoon was that climatologists monitor El Niño with something they call the Southern Oscillation Index, a measure of the difference in air pressure between the central and western tropical Pacific. The index begins to drop as an El Niño begins to build. He held up a graph to demonstrate that the index had begun to drop a few months earlier: in other words, an El Niño was on the way. Bernard figured it would bring high easterly winds to Sajama, which might spell trouble for the balloon.

So I waited in La Paz for a few days, mostly with a group of British expatriate climbers, following their advice on how to adjust to the altitude, which was to join them in the bars every night. Their ringleader, and the vital force in Bolivian mountaineering at the time, was Yossi Brain, a guide and guidebook author who had climbed Sajama many times. He told me it was just a "walk-up" but advised me to treat it with respect.

"Leave High Camp for the summit at three A.M.," he said. "Rope up on the glacier. Be prepared for very cold weather and high winds." Yossi also pointed out that there was at least one dead body up there: in 1946, the second successful ascent team was caught in a blizzard, and one of its members lost sight of his companions and was never seen again.

JUST BEFORE DARK ON MY FIFTH EVENING IN BOLIVIA, I LEFT FOR THE MOUNTAIN IN THE truck that would carry the freezer. Its owner, Javier, was driving. He had worked a full day, so I was afraid he might fall asleep at the wheel. But soon the flat road and the waste and darkness around us had their effect not on him but on me. I drifted off to sleep . . . to awake suddenly a few hours later, a few inches above my seat, as our truck crashed across a giant pothole in the road that rings Sajama.

Javier and his technician, Manuel, were chattering away in Spanish in the front seat, knocking back gulps from a two-liter Coca-Cola bottle and making frequent forays into a large bag of coca leaves on the console between them. (Javier also had a thermos of coffee stuffed between his seat and the door.) Manuel passed the bag of leaves back to me, along with a small bag of *legía,* a catalyst made of potato ash, quinoa ash, and powdered limestone. He showed me how to stuff a pinch of the ash in with the leaves, chew the mixture to a pulp, prod it into my cheek, and squeeze the bitter, grassy juice into my mouth. My cheek and tongue went numb, a soft charge lit up the neurons in my head, a wave of peace and well-being washed through me—and then I became aware of the dark presence of Sajama: an enormous black hump obscuring a quarter of the sparkling, star-filled sky. The brightest Milky Way I had ever seen arched overhead from horizon to horizon.

Javier had that bug-eyed momentum that six hours' driving with the aid of three stimulants tends to produce, and I don't believe he had noticed our company yet. I pointed and said, "Sajama." He and Manuel grunted and nodded, and we proceeded to lurch and bounce right past the mountain and off into the desert for the better part of an hour, quite possibly across the border into Chile. I finally tapped Javier on the shoulder and pointed through the rear window at our destination, a good distance behind us now. We jolted around in a K-turn and raced back. We were looking for a hacienda somewhere near the base of the mountain.

I now realized that I had our only map and neither of them had ever been there before. We stopped at every collection of huts we found by the side of the road, waked the stunned inhabitants, and entered their single rooms to ask where we were and if they knew anything about a hacienda. The skinned limbs and furry skulls, feet, and tails of dead game hung from the rafters, dangling before our eyes as we gathered around the map, exhaling frost into the flashlight beam.

These were native Aymara people; Javier and Manuel spoke only Spanish; I speak only English. Our hosts would twist the map around sleepily . . . right side up, upside down, sideways . . . draw a finger across a road or a river in a glimmer of comprehension, squint, shake their heads, twist the map again. . . . We would thank them and leave. After a few fruitless attempts, we pulled into an abandoned churchyard and promptly fell asleep in our seats.

PREDAWN LIGHT REVEALED A FLAT, FREEZE-DRIED LANDSCAPE, ALL ORANGE, BROWN, and gray, dotted with pointed tufts of the wild grass called *paja brava* and streaked with free-ranging herds of the gentle and curiously regal llama. Sajama towered in the crisp sky to the east, its summit crowned by an inverted dish of cloud. That meant Lonnie was sitting in a hurricane up there.

The sun cleared the horizon to bleach Las Payachatas, classic twin volcanoes floating pink above the plains to the southwest. The hacienda stood out sharply in the wide emptiness of the desert: a cluster of adobe huts a mile up the road. As we pulled in, a herd of silent, milk-heavy llama waited outside a pen of bushy, rust-colored tree limbs stuck in the sand. They were soon joined by the hacienda owners, Don Luis Alvarez and his tiny wife, Teodora, and five disgruntled balloonists.

Lonnie and his drilling team had set up camp on the summit two weeks before, and most of the balloon was up there, but the balloonists had not yet succeeded in getting themselves to the top. Nearly one hundred meters of ice had been drilled, and it was time to start bringing it down; but it would take four or five balloon trips to transport all the ice from this effort, and there was no time for that now. The best we could hope for was one.

So I waited at that altitude for two days before moving up to base camp, a tiny cluster of tents on the vast orange and brown flatland between two long ridges, below Sajama's brooding and monumental six-thousand-foot west face.

PART I

THE SAJAMA EXPEDITION

·1·

THE MOUNTAIN GOD

The Aymara are descended from the Tiwanaku, a mysterious pre-Incan people who once ruled an empire encompassing most of present-day Chile, Peru, and Bolivia from a city on the shores of Lake Titicaca, an inland sea on the high Altiplano. Like their ancestors, many of today's Aymara revere the snow-covered mountains as gods, for they produce the rare trickles of water that make life possible in this inhospitable land. The Aymara believe water is to the land as blood is to the body—the sacred essence of life and fecundity—and that mountain spirits show their wrath through ferocious storms or by holding the water back, in the sky. In the eerie, looming presence of Sajama, one can understand and even share in this belief.

Modern society also treats Sajama with respect. Commercial jets are forbidden from flying overhead, for the mountain's core contains an enormous magnet that may twist the needle of a compass more than twenty degrees away from true magnetic north.

When Lonnie Thompson arrived in small Sajama village, a few miles west of the mountain, with twenty-odd scientists, a fleet of outfitters from La Paz, the five Peruvian mountaineers who have supported his high-altitude work for decades, and six tons of equipment, the villagers decided they did not want him disturbing the local deity. Similar visitors from the University of Massachusetts, Lonnie's collaborators, had set up a satellite-linked weather station on the summit the previous year, and the subsequent growing season had been poor. The villagers were afraid a larger expedition might induce more severe retribution. Impassioned negotiations ensued.

Bernard Francou remembers the bargaining as the most engaging cultural experience of his six years in Bolivia. It fell to Lonnie to convince the people of Sajama—more baleful an assemblage than any grant review committee back in the States—that his work was relevant to their own lives.

He told them that the ice on their mountain could give him information not only about Bolivia's changing climate but also about El Niño, the mysterious change in the weather that visits the Altiplano every two to seven years. Their mountain and their dry vegetable plots, which provide a marginal existence even in the best of years, feel El Niño's effects quite directly.

Since it generally appears at around Christmastime, the phenomenon was named for the Christ child by anchovy fishermen living less than a thousand miles northwest of here, in the coastal villages of Peru and Ecuador. They noticed that the cold waters that they fished, normally one of the most productive fisheries on earth, rose in temperature with El Niño, and the anchovy disappeared. This disrupts the entire aquatic food chain and the economy based upon it. Seabirds, which feed off the anchovy and produce the mountains of guano that the local farmers use as fertilizer and export all over the world, disappear as well. And while the sea withholds its bounty, it drives huge thunderheads into the skies above the equatorial coastal plain, normally one of the driest places on earth, producing storms, flash floods, and the occasional deadly landslide.

Simultaneously, more than nine thousand miles to the west, New Guinea, one of the wettest places on earth, experiences drought, as do Australia and the Indonesian archipelago. The Asian monsoon generally fails, while, farther west, the Horn of Africa receives torrential rain.

El Niño's sister, La Niña, who usually visits on his heels, makes Peru even drier than usual and New Guinea even wetter. This climatic seesaw, which scientists have named the "El Niño/Southern Oscillation," or ENSO, spans the Pacific and has immediate collateral effects nearly everywhere on earth.

And ENSO is just one—though perhaps the most notorious—of the myriad oscillations and seesaws known to climatologists. Seesaws and oscillations, either in time or in space—some real, some perhaps imagined and the subject of intense debate—are the basic stuff of climatology. A cycle may span years, as with El Niño, decades, several centuries, or hundreds of millennia. Only a few have El Niño's global reach.

To find a particular cycle, a climatologist must generally tease it out from among dozens of others. Consider the motion of the sun across the sky, which can change temperature by as much as sixty degrees in a single day; the hourly or weekly changes we call weather; the seasons; subtle changes in our orbit around the sun, having periods of tens or hundreds of thousands of years; sunspots and other cycles in solar output—and all of these have different strengths at different latitudes and longitudes. It's like a kid playing with a yo-yo, bouncing on a pogo stick, in a car on springs, zooming along on a roller coaster—worse, actually.

The devastation wrought by the last few El Niños has made the phenomenon a household name in recent years. Its east–west, transpacific seesaw has become relatively well known; but Sajama, in Bolivia, stands at the southern end of a second, lesser-known, north–south seesaw with equatorial Peru. When El Niño brings storms to Peru, Sajama experiences drought; when La Niña brings drought, Sajama receives rain.

AT THE NEGOTIATING TABLE, LONNIE EXPLAINED THAT IF HE COULD FIND BETTER WAYS to predict the onset and strength of these climatic siblings he might help the people of Sajama choose appropriate crops for the corresponding wet and dry years. For some inscrutable reason, however—perhaps out of caution, for they are not yet entirely confident in their predictions—he and Bernard did not reveal that the Southern Oscillation Index was dropping and an El Niño was on the way.

But they did point out that a global change of another sort was taking place, that Sajama's glacier was retreating (as the locals had already observed), and that every glacier in the tropical zone will probably disappear within fifty years. Sajama lies just inside the southern edge of the tropics—the main reason Lonnie wanted to drill there. Its high glacier should outlast the others, which are generally lower and nearer the equator, but the odds are that it, too, will disappear while the children, playing as he spoke in the streets of that village, are still alive. And since Sajama's snow is their primary source of water, its disappearance stands to have a dramatic effect upon their lives.

Lonnie happened to have some idea just how dramatic that effect might be. Two ice cores he had recovered fifteen years earlier, on the summit of the Quelccaya massif on the northern Altiplano, in Peru, had shown that climate change had played a major role in the births and deaths of numerous Andean civilizations five hundred and more years ago—the ancestors of these very people among them.

Archeologists have long known that the Tiwanaku abandoned their city by the lake shortly after A.D. 1000. They had no written language, and there was no explanation for their demise until the Quelccaya ice cores revealed that a severe drought had set in on the Altiplano just before the Tiwanaku disappeared. The drought persisted for about four hundred years, and when it withdrew, the first Aymara city-states emerged. Furthermore, the drought commenced about a century after a warm spell began on the Altiplano, and its span coincides with an event that is sometimes called the Medieval Warm Period, in which temperatures in many parts of the world, including northern Europe, rose for about four centuries. Based on this evidence, one might expect the present warming to cause another devastating drought in this arid land.

You can tell just by looking that the people of the Altiplano are tough—not threatening, strong in the face of adversity. The Aymara of today grow the same crops, tuned to the Altiplano's cold, dry climate, that the Tiwanaku farmed more than a thousand years ago: tubers like potato, *oca, olluco,* and *mashwa,* and the unique, cold-adapted grains quinoa and *caniwa.* Their way of life and animist belief system have withstood the Incas, the Spanish conquistadors, and—at least until now—the lure of the city. But a few bad years might change that. Already El Alto, the slum by the airport on the Altiplano above La Paz, is swelling from a steady trickle of migrants across the high, windswept plain.

WHEN THE SCIENTISTS FINALLY WORKED OUT A DEAL WITH THE PEOPLE OF SAJAMA, A delegation from the village joined them at base camp to conduct a ceremony similar to those that had taken place ten and more centuries earlier in the city of Tiwanaku. They gathered together in a circle around a bonfire and passed around an endless supply of coca and legía, chasing them with shots of grain alcohol. (One scientist testifies that his head went completely numb.) Two *yatiri,* or holy men, offered songs and prayers to Sajama and Pacha Mama ("Lady Earth"), seeking fertility for their families and fields and forgiveness for the transgressions they planned. Lastly, the *yatiri* laid a white llama on a smooth stone, used before for this purpose, and slit its throat. They spilled blood, the magic elixir, into a ritual ceramic cup. Singing songs and scattering figurines of humans and animals in the fire, they sprinkled the animal's blood on the sand.

Lonnie reached the summit two days later.

·2·

AN ISLAND IN THE SKY

About three weeks after the llama sacrifice, I met a geochemist from Penn State named Todd Sowers as he passed through base camp on his descent from the summit. He had left some gear at the intermediate, nineteen-thousand-foot High Camp, so he invited me to join him as he climbed back up to retrieve it. This gave me a chance to take another step toward acclimatization and get an introduction to climatology at the same time.

High, healthy Andean glaciers receive 80 to 90 percent of their snow in the austral summer, from November to April. During the dry season in between, the wind lays a thin layer of dust and pollen on the snow. Viewed edgewise—that is, down the walls of a crevasse or along a recently exposed margin—an Andean glacier resembles layer cake. Puffy layers of clean, white snow alternate with dark, paper-thin dust bands. Annual layers can be counted by eye, and years of low snowfall are obvious. Thus, a high, healthy glacier archives climate in the manner of tree rings.

There is some art to the estimation of annual snowfall, however, because compression and the lateral flow of the glacier will cause layers to thin as they recede into the depths, buried by layers from subsequent years. At a certain depth, the compressed snow, called firn, packs to ice. The depth of this firn–ice transition varies from a few inches to hundreds of yards.

An ice core is harvested in cylindrical segments up to eight inches in diameter and one meter (or three feet) long. If there are no discontinuities caused by flow (easily identified), and the drill doesn't break down or stick in the hole (not an uncommon occurrence), a core will sample progressively older layers of ice from the surface of a glacier all the way to the bedrock or sand at its base. Detailed information about climatic conditions at the time each layer fell may be inferred by looking closely at the chemistry and structure of the ice and at the impurities it contains, either chemical or particulate—dust and pollen, for instance, sticks or leaves, even frozen bugs. Todd Sowers's specialty is analyzing the air bubbles that are trapped as the airy firn packs to ice.

These bubbles serve as samples of the overlying atmosphere from a time *soon after* the enclosing ice fell as snow. The reason for the time lag is that air from above the surface will diffuse into cavities in the firn until it is compressed enough to seal those cavities. They are usually sealed within one or two hundred years, but in areas of low accumulation and extreme cold, such as Antarctica, the process may take a few thousand years. Estimates of the close-off time can be made, but now that climatology is beginning to focus on fine details, such as where past changes began and how they spread geographically, this subject can lead to strenuous (and sometimes quite personal) disagreement. In the subtler points of climatology, as in other aspects of life, timing is turning out to be everything. As Ellen Thompson puts it, "Without a timescale, it's like trying to build a body without a skeleton."

One, perhaps obvious, use of the bubbles is to estimate the atmospheric levels of greenhouse gases in climates of the past. And, as Todd tossed out matter-of-factly on our hike up Sajama, ice cores from the polar regions, particularly those from Russia's Vostok Station in Antarctica, show that atmospheric temperature has risen and fallen in concert with atmospheric carbon dioxide for the past half-million years. To Todd and his colleagues, the greenhouse effect is not a theory, it is an experimental fact. And to the overwhelming majority of them, the debate as to whether fossil fuel burning stands to make the earth noticeably warmer in the near future ended years ago. This is not one of those fine details.

More interesting to Todd is the fact that he can use the air bubbles to estimate the age of the ice. If the average amount of atmospheric methane, a powerful greenhouse gas, were to change, for instance, it would take a year or two for the new level to mix completely in the air all over the earth—the blink of an eye in paleoclimatic terms. In the mid-nineties, Todd was involved in measuring methane levels in the two-mile-long GISP2 (Greenland Ice Sheet Project 2) core from the summit of the Greenland ice cap, in which visible annual layers date back more than forty thousand years (and a fleet of dedicated graduate students has been on hand to count them). Subject to the uncertainty in bubble close-off time and layer counting—and since layers do sometimes merge with one another or sublime or blow away, the count can be off by a few percentage points—the curve of methane concentration against time in the air above Greenland can be used as a standard to construct timescales for ice cores from other regions. One simply stretches or squeezes the timescale for the methane curve from the unknown core until its peaks and valleys match up with GISP2.

I didn't catch everything Todd said as we ventured briefly above the eighteen-thousand-foot mark and into the death zone that day. He grabbed my notebook once or twice and made a few graphs and scrawls, which I studied later. In the methane graph, he drew a large dip at about twelve thousand years ago and labeled it "Younger Dryas," the first time I'd seen or heard that term. The Younger Dryas was a cold snap that occurred a few thousand years after the most recent ice age began to loosen its grip. The atmosphere began to heat up about eighteen thousand years ago and then, during the Younger Dryas, dropped again for almost two thousand years. I see the name "Lake Agassiz" also in my notes and recall that Todd was the first to tell me a now-familiar story about the huge lake that lay on the vast, receding ice sheet that once covered North America.

As this climatological tale goes, Lake Agassiz surged through a depression in the wall of the melting continental ice sheet near what is now the St. Lawrence River Valley, on Canada's eastern coast; when the resulting pulse of cold, freshwater entered the northern Atlantic Ocean, it supposedly shut off a massive ocean current, called the great conveyor (another phrase scrawled in my notebook), which carries warm equatorial water north along the surface of the Atlantic and was thought until very recently to be the source of the excess heat that makes northern Europe and Scandinavia so warm relative to similar latitudes in North America.* The shutting off of the conveyor, so this line of conjecture goes, sent the whole earth into a cold spell. When Lake Agassiz had sufficiently drained, the conveyor picked up again, and temperatures gradually rose to the levels they had reached before the Younger Dryas. This marked the beginning of the present climatic epoch, which is known as the Holocene.

This story is probably the most popular in the folklore of modern climatology and has seen quite a bit of play in the popular press. It has even generated a feature-length movie, *The Day After Tomorrow*. Unfortunately, however, the work that Lonnie was doing just then on Sajama would show that the story probably isn't true.

The story of Lake Agassiz and much else in the field of ice core climatology, including the techniques Todd has helped develop to understand

*This bit of received wisdom was still very strongly in place as Todd and I walked that day. It was not disproven until 2003.

the gas bubbles, have arisen mainly from the huge drilling efforts, involving small armies of scientists and specialized support crews, that have been undertaken in Greenland and Antarctica. The polar ice sheets contain 99 percent of the permanent ice on the earth's surface and 95 percent of all the freshwater.

Yet here I was risking my life to meet a very small team working on the other puny 1 percent, and in the tropics no less—an unlikely place even to be thinking about ice. I would soon learn that the very implausibility of drilling ice in the tropics is the reason Lonnie's discoveries have surprised and changed the thinking of his colleagues. He collects ice from low latitudes but high altitudes. The great mountain ranges near the equator still hold quite a few high glaciers, but at this unique moment in our planet's history every one that has been studied is melting away.

In contrast to the drilling sites on the flat polar ice sheets, where it is possible to shuttle crews and ice in and out by plane, Lonnie's sites are so high—and often so precariously situated—that even helicopters can't land nearby. Not only must he and his teammates lug their own bodies and equipment to the summits, but their frozen quarry must often be carried for days, through rain forests or across deserts, before reaching safe haven in the nearest freezer. Hence the balloon.

Sajama, a circular volcano rising from a flat plain, encircled by a road, was the perfect place to test this unique scientific tool for its potential use in Tibet, for example, where it is not uncommon for the trail from drilling site to road head to span the better part of a hundred miles. The plan was for the balloon to fly just before dawn, to float the ice through the warm lower altitudes not only swiftly but also at the coldest time of day.

And Todd had a special interest in this, because the main reason it was important to keep the ice in its pristine frozen state for every step of the journey, from Sajama's summit to Lonnie's cold rooms half a world away, was to preserve the air in the trapped bubbles. Lonnie had retrieved ice from the tropics many times before, but he had never succeeded in making sense of the trapped air in a tropical core.

AT MID-MORNING ON THE DAY AFTER OUR WALK—TODD HAVING HIGHTAILED IT OUT TO the comforts of civilization—a group of Bolivian porters strolled in to base camp, led by a lively, ponytailed rock climber named Gonzalo. They began packing provisions for High Camp and the summit. I decided to move up with them. When we had packed our various loads, we sat in the traditional circle and passed around the traditional plastic bags of coca and legía and a small Nalgene bottle of somewhat less traditional pure ethanol.

I passed on the booze but partook of the coca, and it seemed to work wonders on the trek that day, supplying energy and easing the nausea that is one of the hallmarks of acute mountain sickness. The leaf is legal and ubiquitous in Bolivia and continues to play a prominent role in the sacred and medical traditions of peoples all through the Andes. It's too bad that poor farmers have become dependent on the cash they can earn by supplying it to the cocaine trade, for in its natural form it seems less pernicious than coffee. It produces neither anxiety nor tension, and its effects disappear the moment it leaves the mouth. I'd call it an herb rather than a drug.

It was a laid-back group that floated up the hill that day. My companions had a knack for acclimatization. Whenever they caught me grunting or gasping for air, they'd pat the air gently and say, "Slowly, slowly . . . Take your time . . ."

HIGH CAMP WAS LAID OUT ALONG A RIDGE NEAR A LARGE MOLAR OF ROCK AT 18,700 feet, still 3,000 feet below the summit. The true high-altitude porters lived there, sometimes five to a two-man tent. As it is somewhat inconvenient to take a bath on a knife-edged ridge at an altitude where water comes in chunks, the balloonists in particular found the lack of hygiene at High Camp to be the greatest hurdle of the ascent. They had figured out a way to avoid sleeping there by driving partway up the ridge leading directly to High Camp, thereby avoiding base camp (and, incidentally, breaking the rules of Sajama National Park). This allowed the three of them who ever got any higher than High Camp to walk all the way from Don Luis's hacienda to the summit in a single day. Not only did this seriously compromise their ability to get themselves and their equipment to the top, it also stood as a prime example of the contrary posture they took with respect to the expedition as a whole.

But the porters were an engaged and enthusiastic bunch. The evening I arrived, a fiery twenty-one-year-old named Genaro invited me to his tent for dinner, where he informed me that he had climbed to the summit twenty-three times in the previous eighteen days—a feat that would have made headlines had he come from Europe or the United States, but here, as in the other great mountain ranges, all in a day's work for a native.

My plan was to reach the summit the next day—if only to breathe the thin air and return to High Camp—for the balloon was nearly in place at the top; but I couldn't sleep that night. My mind didn't race exactly; it flowed: over hill, over dale, around the next bend . . . smoothly enough, but on and on and on. At first I thought it was the coca tea I had enjoyed with dinner; then I realized it was the altitude. Somewhere in the night I decided to stay put the next day.

I nearly had a stroke when I emerged in the morning light and got a look at the spot I had chosen for relieving myself of said tea before going to bed the previous night. It was less than two feet from the edge of an icy shoot that dropped a few thousand feet to base camp. One slip on the hard snow in my smooth-soled sneakers and I would have died almost certainly. People *have* died in the mountains doing precisely that. It may have been the closest call I've had in more than twenty years of mountaineering. I still shudder to think of it.

After breakfast, I stood at the door of the small cook tent and watched Genaro and his friends shoulder pack frames loaded with unwieldy snow stakes or the large insulated boxes Lonnie uses to store his ice cores, and then stumble off and gradually shrink to black dots on the giant white dome silhouetted against a spanking blue sky. To continue the work of acclimatization, I decided to take a short walk myself, so I headed upward.

The path led directly to a labyrinth of *nieve penitente,** which in a few hundred yards mercifully gave way to a smooth but steep slope of snow. To my right was a chute slightly longer than the one I had nearly pitched myself into the previous night, leading down, as before, to the tiny tents at base camp. Holding very still, I braced my left hand against the steep slope at shoulder height and let my head spin with vertigo. When equilibrium returned, I surveyed the route above. The porters had fixed ropes up the steep section, but the sun had melted a workable layer of slush on the surface. I was wearing stiff-soled boots now, and the porters' boots had cut out a series of steps over many trips, so I decided to ignore the ropes and enjoy the freedom of climbing unfettered so high in the air.

I stopped to rest on a rocky outcrop at the top of the slope, and soon began debating with myself—probably out loud, vague as I was from the altitude. It didn't take long for the inexplicable climbing urge to win out: I strapped a pair of crampons to my boots and took a step onto the white summit dome. My sleeping bag and pad were back in High Camp, but I had extra clothing with me, and I figured Lonnie had erected a small village on the summit. If he and his friends had no spare gear, I planned to bivouac as I had many times before, with my legs stuffed in my backpack.

GRADUALLY LAS PAYACHATAS, WHICH HAD PREVIOUSLY DOMINATED THE VIEW TO THE west, seemed to bow at the feet of Sajama. To the north, Illimani and Huayna Potosí, the other white gods of the Altiplano, emerged above the sharp edge of the snow.

*A Spanish phrase meaning, literally, "penitent snow." Hans Ertl, a German who climbed Sajama in 1951, described the *penitentes* as "knife-sharp ice-stalagmites, some over a yard high, which are distributed in close array up to the summit like armored protection of a modern fortification."

Inching up the dome to the pounding of my heart, I counted steps and breaths to moderate my pace—thirty lurching steps, stop; forty or fifty desperate gasps, start. My thoughts growing dim, I began to rely on intuition to judge the state of my mind, the pain in my lungs, and the pressure in my head, ready to turn back at any moment. One sometimes gambles in a bid to reach a summit, but as the British mountaineer and barroom philosopher Don Whillans once observed, "The mountains will always be there. The trick is for you to be there, too."

As afternoon sunlight turned the snow to gold, the porters loped by on their way down. The angle slowly eased. Finally, I saw the pole from the drilling rig sticking up from the summit like a single palm tree on a tiny desert isle. A frigid bunker came into view: a few mountain tents on sunken platforms of ice, battened against the wind; a snow cave for the scientists; a small tent by the cave's entrance, where the solitary cook melted snow for water and handed down one-pot meals; a pit for the ice cores; random piles of gear; a stiff blue snowsuit covered with hoarfrost and frozen to the ground; a wall of snow blocks and white cardboard boxes shielding the rig and its small down-clad crew from the wind; dozens of solar panels connected by cables to a silent electromechanical drill.

The summit's circular aspect made for the perception of immense space. Past the curving white surface, which dropped off quickly in every direction, was desert the color of canvas, several miles below. To the west, snowcapped peaks dotted the spine of the Andes, which stretched for hundreds of miles, north and south. Due north, far across the Altiplano, a few solitary mountains hovered above the horizon like clouds of carved crystal. To the south, across the border in Chile, stood snowcapped Guallatiri, the highest of the three Nevados de Quimsachata, puffing smoke into a dazzling sky.

A hooded figure in a quilted red parka broke from the work and shuffled over to greet me. It was Lonnie. So that's where I met him: on top of the highest mountain in Bolivia. He had a gray beard and sky-blue eyes. His voice was weak, and cracked when he spoke. After introducing himself (and breaking into a hacking cough more than once), he led me over and introduced me to the other walking mattresses toiling at the drill: Bruce Koci from Alaska; Victor Zagorodnov and Vladimir Mikhalenko from Russia; Benjamín Vicencio M., the leader of the Peruvian mountaineers; and Patrick Ginot from France. I watched them work.

Soon the sun sank too low to provide power. The pyramidal shadow of the mountain stretched across the desert to meet the horizon. The sky turned purple, then gray. We shoehorned ourselves into the snow cave, a rectangular hole fifteen feet long and five feet wide—too low for standing—which served as dining room, living room, and lounge. We sat along opposite walls on benches cut in the snow, knee to knee, thigh to thigh, backs and seats against cold white surfaces, sipping tea and soup in near silence.

After dinner I decided to point out that I had no gear. Bruce Koci then forced himself slowly to his feet and climbed into the twilight. A few minutes later he called from above, holding four of the six-inch foam pads they use to insulate the ice and two of the fluffiest down sleeping bags I had ever seen—much warmer than my own back at High Camp. He helped me lay out the pads in a tent and pile the bags on top.

"There," he said. "Wrap yourself in these like a fox in his tail."

A fine gesture. I have now spent months in harsh conditions with Bruce and learned that he's like that all the time.

AFTERWARD, LONNIE ADMITTED THAT THE TOP OF SAJAMA WAS AN UNCOMFORTABLE place and he mildly regretted that they had not used their usual dome over the drill, to protect themselves from the wind. It was certainly colder than I thought it would be. At night the temperature dropped to the single digits. If I had gone without a sleeping bag, I would have had to run in place or do calisthenics to keep from freezing to death—and God knows what that would have done to the pressure in my head or the blood in my lungs. Having not adjusted to the altitude, I was especially sensitive to the cold. Ten days is a short time for going from sea level to 21,500 feet—and especially for staying there for the night. On a climb, you usually blitz for the top, spend an hour or so enjoying the view, and then race down as far as possible by sunset.

We would retire with the sun and wait for it to warm our tents before stirring in the morning. Since the nights were about twelve hours long, that added up to about fourteen hours without sleep. Well, maybe it was sleep, but whatever it was seemed worse than none at all. My particular brand of acute mountain sickness features an unsettling symptom known as Cheyne-Stokes breathing: when I drift off to sleep, I let out a sort of sigh and hover for a while without inhaling, then startle awake gasping for air. My nights on Sajama consisted of thousands of split-second naps, an endless semi-hallucination, sponsored every once in a while by an electrifying dream sequence. Gusts of frigid air snaked through the tent seams to bite my exposed ears or wrists and sometimes seemed to be lifting my tent off the ground. More than once I honestly believed I was flying.

In the morning, we would appear in our own time at the mouth of the snow cave (in my case, slightly dazed), squeeze down in, and sit hunched against the cold, waiting for the cook to hand pots of hot water or porridge down from his tent above the opening.

My one full day on the summit was fine—sunny, not much wind—and spirits were high. The team had refined the drilling, packing, and logging-in of the cores to the purity of a dance after two weeks of repetition—not to mention decades of previous improvement to the design of both the drill and the process.

In the long silences between words I sensed the kinship common to mountaineers, but without the usual raucousness and ribaldry. These folks had different reasons for visiting the high, white wilderness; and Lonnie, Bruce, and to a lesser extent Vladimir had spent years, literally, drilling together in similar locations. They didn't waste energy on jokes or stories. (I've noticed that they don't tell stories much anyway.) It is true that by this time on an expedition, everyone has usually used up all the stories he knows, but there was more to their silence than that. It had a reverence to it. The feeling was almost monastic. Then again, standing on a spectacular summit on a clear day almost always brings on a sense of clarity.

THE DRILL HANGS BY A STRONG CABLE FROM A PULLEY AT THE TOP OF A FIBERGLASS pole, hollow to save weight, which stands on the very summit—wide views of brown desert, blue mountains, and distant crystalline ice caps all around. On the pole side of the pulley, the cable runs down to a winch, which is mounted on a platform at the pole's base. The cable is used not only to lower and raise the drill from the coring hole, but also to supply it with electrical power.

The drill itself is cylindrical, made mostly of aluminum, and divided in two. Four flexible metal flanges, "anti-torque skates," are mounted longitudinally along the outside of its upper half, which contains the motor. The "anti-torques" keep the upper half from twisting in the hole, while a "screw" in the lower half spins and cuts out a core segment. The lower half has a smooth outer housing, but the hollow screw inside it is threaded on the outside and has diamond-coated teeth along its bottom edge.* As the screw spins, it cuts its way into the ice, fills itself with ice core, and sends shavings into the space between itself and the housing. Meanwhile, the whole unit slides down—with the anti-torques "skating" down the cylindrical wall of the hole.

Bruce, who runs the drill and who designed the first high-altitude ice core drill almost thirty years ago, monitors the tension and movement of

*The screw is basically a large version of an ice-climbing screw.

the cable. When the drill stops moving down he knows the screw is full of core. He then works the cable for a while to break the set of the drill in the ice, switches on the winch, and pulls the drill and its fresh, cold contents to the surface. He and one of the others, usually Vladimir or Victor, swing it onto a long, thin, half-cylindrical stand resembling a horizontal slide, and extract the screw containing the ice core from the lower housing. They never touch the ice.

Lonnie and usually Patrick, wearing rubber gloves to minimize contamination, then extract the cylindrical ice core segment from the screw and place it on a second stand. Lonnie lays a ruler and a special meter-long spiral pad next to the core segment, sketches a picture of it, jots down a few notes, and simultaneously calls out a description to Patrick, who also takes notes. They number the segment, pack it in a plastic bag and then into a cylindrical tube, and place it in an insulated core box, which they will later stow in the deep freeze of the ice pit.

The silence of solar power is absolutely magnificent.

It so happened that the day I was there they were reaching bedrock at the bottom of the first of two deep cores they would drill on Sajama, so their silence and sense of drama may have derived from the fact that they knew they were getting down to the heart of the matter once again. Lonnie had done this a few times before, and he knew from counting the layers and reading other signs in the ice as they went down that this core reached back to the last ice age, more than twenty thousand years ago. This was only the second time anyone had found ice that old in the tropics, and he, Bruce, and Vladimir had all been there the first time.

That was on Huascarán, the highest mountain in the tropics, in Peru's Cordillera Blanca, and had resulted in their first big climatological discovery. Huascarán showed that the tropics have played a bigger role—perhaps even the lead role—in past climate changes than was thought at the time by Lonnie's colleagues, focused as they were on the polar regions.

There, in the clear air with the world spread below us, Lonnie pointed his mittens to opposite horizons and said, "To be relevant, you must work in the tropics. They contain fifty percent of the earth's landmass, which absorbs the sun's heat, and three-quarters of the human population. Tropical heat is the engine that drives climate change. It powers big storms and hurricanes and heightens El Niño. If a subtle increase in carbon dioxide raises tropical temperatures even slightly, moisture will enter the atmosphere from the tropical oceans, and the greenhouse effect of that moisture will raise temperatures worldwide."

He and his friends did get a little loose near the end of the day, joking and chucking each other on the shoulder and raving about the quality of the dark, brittle ice—maybe because it actually warmed up enough to unzip your parka there for a while. But I suspect the main reason was that they had struck a vein that was more valuable to them than gold.

When Lonnie seemed free, I would ask questions and scrawl notes with my cold-stiffened fingers. Despite his disturbing cough and the large number of tasks he was juggling in conditions that sometimes cause mountaineers to make foolish or even fatal errors in judgment, he would fix his striking blue eyes on me and rattle off facts as if he were sitting at his desk in Ohio. He pointed out that not only is every known glacier in the tropics retreating, but they are retreating more rapidly each year; that there used to be six glaciers in Venezuela, now there are two; that the altitude at which the average temperature drops to the freezing point—below which glacial ice will melt—is rising at the rate of fifteen feet per year; that sea level is rising, too, and the small island nations in the South Pacific stand to disappear; that tropical storms are gaining in intensity. Each of the last three El Niños had set a record of one sort or another, and storm-related insurance losses were increasing exponentially. In the eighties they averaged two billion dollars a year; in the nineties, twelve billion, he told me.

But he hid behind these details. He would not come out with a grand pronouncement about global warming. On the one hand, it was clear that his interests ran deeper than that. The quest he has joined, one of the more challenging scientific endeavors of all time, is to solve the riddle of climate change, not simply to answer that one yes/no question. He may have been holding back out of fear that I would distort his words, but I think he was also looking over his shoulder at his academic peers, aiming to duck the potshots that inevitably flew in those days when anyone walked out on a scientific limb and said in public what nearly all of them knew inside. Scientists have an annoying habit of backing off when they are asked to make a plain statement, and climatologists tend to be worse than most—and not only because of the complexity of their field. The economic interests who fear any sensible discussion of global warming have succeeded in politicizing this branch of science more, perhaps, than any other. In the early nineties it was not unheard of for officials at the National Science Foundation or the National Oceanic and Atmospheric Administration, Lonnie's main funding sources, to warn scientists not to publish data about the dangers of the greenhouse buildup, for fear that a conservative Congress would slash all science funding. It would be almost a year and a half before Lonnie would carefully open up.

But I still took a message home. A snow cave is a strange place to begin worrying about global warming, but that's what he had me doing after just one day.

THE LAST PIECE FOR THE BALLOON, A LARGE STEEL CARABINER THAT HAPPENED TO PLAY a central role in holding the whole thing together, arrived on the summit that same sunny day. (Unfortunately, the previous day, on which I had made my spontaneous ascent, would have been a good day for a flight. It had dawned clear and calm, and just this one piece had been missing.) Two balloonists, Seth and Fred, had joined us as well, so we spent part of that day preparing the balloon for a flight.

Bruce Comstock, the designer, had instructed the pilots on the flight plan. They were to stand outside the basket while the balloon was being inflated, prepared to let it fly off alone until the very last second, when they would make a quick decision on whether or not to jump in and join it for the ride. If so, they would gun the burners and, with luck, gain altitude as quickly as possible. Comstock believed the main danger was that a gust of wind might drag them into a downdraft in the lee of the summit and send them bouncing down the mountainside.

Aware that I was a climber, Lonnie asked me to join Victor in finding a way to anchor the balloon as it was being inflated. A solid anchor is critical for safety, and it has to be what climbers call "bomb-proof." A full balloon can drag a pickup truck with its brakes on, even in a light wind.

In my vague, unacclimatized state, I am not sure I appreciated the full dimension of our responsibility as Victor and I dithered away with a snow saw, carabiners, climbing webbing, and a few plywood plates to bury as "deadmen" in the snow. A few years later, he assured me that our anchor would have worked.

That evening, in crimson and violet light, we attached the basket to the anchor and unfurled the balloon on the snow in the direction of the prevailing wind. At dinner, Seth and Fred briefed us on our responsibilities during liftoff, which would take place just before dawn. They described the risks quite vividly. Anxiety about the strength of the anchor made a significant contribution to my third consecutive night of insomnia.

False alarm. None of us needed a wind gauge to know we could sleep late the next morning.

Having a strong hunch the balloon would never fly, I decided that morning that I'd seen what I needed to see on the summit. After another ersatz breakfast, I thanked Lonnie, Bruce, and the rest and climbed from the protection of the snow cave into a gale that was strong enough to knock me down once or twice. As I fumbled with the crampons I had dropped on the wind-scoured snow the day I arrived, Benjamín emerged from the cave and pushed against the wind to give me a hug.

"Be careful!" he yelled in my ear. "Good luck!"

BUT LONNIE WAS NOT ABOUT TO LET THE SUCCESS OF THIS VENTURE HINGE ON A LONG shot like the balloon. Unbeknownst to me, one of the million and one tasks he had been juggling while I was on the summit had been to organize the first epic ice carry. As I descended, thirty or more porters were climbing to the summit to fetch loads of ice, some up to a hundred pounds, and carry them down to three trucks that the balloonists had driven to their illegal road head on the ridge below High Camp. A few residents of Sajama village started walking from the hacienda that morning and returned to sleep in their own beds that night, covering an incredible thirteen thousand vertical feet—all above fourteen thousand feet—in a single day. Young Genaro, of the twenty-three summit trips in eighteen days, was first to deliver his load, shouting "Genaro, primero!" as he arrived.

In order to keep the ice cold, the schedule called for the descending porters to reach snow line just at sunset, when temperatures would quickly drop. Since I descended through High Camp before the porters from there had started up and, in veering to base camp, missed all the activity on the ridge below, I remained completely oblivious to the entire operation. Reaching base camp before dark, I wolfed down a quick meal and crawled into my tent, barely strong enough to remove my jacket and hat before worming into a sleeping bag.

It is not unusual to have a tough time on one's first night down; it's like the insomnia of a runner the night after a marathon. I lay in my tent, shattered but glassy-eyed . . . and soon vaguely surprised to hear the porters ambling randomly in. To make my agony complete, their banter quickly bloomed into a full-on celebration.

Upon delivering their loads to the rendezvous, they had been instructed to walk cross-country to base camp, where camp manager, Ping-Nan Lin, a.k.a. Ahnan, an analytical chemist from Lonnie's group, had prepared a feast. They soon discovered a few large cans of pure alcohol for a backup thermal drill lying in plain view in the cook tent,* and all hell broke loose. (Lonnie even had that base covered. This wasn't the first time porters had broken into the alcohol. He always takes ethanol, rather than methanol, so they won't go blind, and he always packs more than he

*Thermal drills melt their way into the ice. The solvent prevents them from freezing in place.

needs.) Long after sunrise, when I finally gave up and crawled out of bed, a few staggering diehards were still going strong.

TWO DAYS LATER, HAVING FINALLY SLEPT WELL, EATEN WELL, AND SPENT ONE FULL afternoon soaking in a hot spring, gazing across the desert at Sajama and Las Payachatas towering in the sky, I rode in a truck to the rendezvous for the second ice carry.

Even that low on the mountain, it was inspiring to watch the sun go down, and it was that much more inspiring to watch the porters come in. Few had lights. They staggered toward us in the dark, tripping on small boulders and occasionally falling to the ground. I was unable to keep myself from rushing up the ridge to lift the loads from as many backs as I could, producing cries of relief that were indistinguishable from cries of pain. A few of the men collapsed. Amazingly, one of the more spectacular of the early-morning drunks from the first time around stumped in first, dropped his pack to the ground without even a grunt, and strode on purposefully to base camp. He must have been crestfallen when he found Ahnan guarding the alcohol, on strict orders from Lonnie.

In the darkness and confusion, I found "Genaro Primero" sitting stiffly on the ground, his legs sticking straight out before him, a dazed look on his face, and his throat so sore from a bug blazing through High Camp that he couldn't speak or swallow water. I pulled him to his feet and found someone strong enough to help him down to dinner.

THE BALLOONISTS, SETH AND FRED, ALSO TRUDGED IN THAT NIGHT, HAVING PERSEVERED on the summit for five nights and risen in the dark three mornings in a row to check the ceaseless wind. That was the last nail in the coffin for the balloon.

There was a lot of bitterness about that. Bruce Comstock complained that Lonnie had told him it would be calmer—a prediction Lonnie had based on almost a year's worth of readings from the summit weather station—but that was before El Niño stepped in. Three members of the Massachusetts group that operated the station had reached the summit with Lonnie at the beginning of this expedition and discovered the station blown down. This had probably happened in May, when it had stopped sending data to its satellite. (Bernard Francou was evidently unaware of this development when he told me in La Paz that El Niño would soon bring high winds.)

So, yes, Lonnie had not predicted the future accurately. On the other hand, by Bruce's own estimate, in the twenty-two days between Lonnie's arrival on the summit and Seth and Fred's descent, four days were calm enough for a flight. Lonnie guesses more. They had barely missed a fine opportunity on my summit day only because one piece of the balloon had been missing, and by that time the drillers had been running much more sophisticated a rig than the balloon for nearly three weeks. My guess is that if the balloonists had worked *with* rather than *against* the rest of the expedition, they would have flown more than once. Over the decades, Lonnie's group has reached equally or more challenging objectives against higher odds, and the only explanation is their remarkable teamwork.

AFTER I LEFT, THE DRILL CONTINUED TO HUM. THE CREW REMAINED ON THE SUMMIT for only five more days, thereby putting in a total of twenty-eight consecutive days in that frigid place and setting a record for the highest ice cores ever recovered at the time. They and their six tons of equipment reached Columbus, Ohio, five days later, and three tons of Sajama's ice flew in the day after that.

It took about a year for Todd Sowers to analyze the trapped air, and when he did he found that the porters had done a magnificent job. These were the first tropical cores to produce definitive air bubble results. Lonnie had read the ice well, too: Sajama's basal layers turned out to be twenty-five thousand years old.

I LEARNED THAT THIS WAS THE SECOND OF THREE CONSECUTIVE EXPEDITIONS THAT Lonnie undertook that year. In April, Victor and Vladimir had drilled in Franz Josef Land, in the high Russian Arctic. Lonnie was principal investigator, or scientific leader, in that effort and had visited the site on a previous reconnaissance, but sometime after the reconnaissance (and maybe because of it) the Russian bureaucracy realized that there was a military base near the drilling site and barred Americans from going back.

Sajama came next, in June and July; and Lonnie was home less than two weeks between that and his next expedition, to the Dasuopu ice cap on the Tibetan side of Xixapangma, one of only fourteen peaks in the world more than eight thousand meters high.

Vladimir flew in from Moscow for all three expeditions. Bruce Koci, as an American, was excluded from Franz Joseph Land but went to Bolivia, of course, and Tibet, and ended up seeing home even less than Vladimir that year. After Tibet, Bruce returned to Alaska to share a few weeks with his wife, then left for three months at the South Pole, where he was lead engineer on a monstrous project aimed at drilling a series of mile-and-a-half-deep holes in the West Antarctic Ice Sheet. ("When you think about it, South Pole is a good place to get rid of seasonal affective disorder," he wrote: as his neighbors in Alaska sunk to the depths of a dark Arctic winter, he enjoyed perpetual daylight at the bottom of the world.)

I WENT STRAIGHT FROM SAJAMA TO A PREVIOUSLY ARRANGED CLIMBING TRIP IN THE Bugaboo range in British Columbia, then managed to catch Lonnie in Columbus on my way back east, the morning before his flight to Tibet. This also gave me a chance to put Ellen's face together with her voice. She picked me up at the hotel and shuttled Lonnie and me around in their minivan as they attended to last-minute errands. I also met another of the inner circle, graduate student Mary Davis, who had packed for Tibet while Lonnie was in Bolivia.

He looked years younger than he had on the mountaintop. He was clean-shaven; his hair was less gray. The only adverse effects from Sajama were frost-nip and some lost nails on a few of his toes. Calmly, he answered questions in his office. Calmly, he walked me to the large cold room down the hall to see Sajama's ice cores (soon, perhaps, the only place on earth where tropical ice will be found). I offered to duck out in the early afternoon, and I have a feeling he made the transition to hyperspeed the moment I left.

In Tibet, a team of Chinese, Russian, and U.S. scientists, the four Peruvian guides I had met on Sajama, several Sherpas from Nepal, a few Chinese porters, and one or two Tibetan yak herders with their yaks worked together for the better part of three months. Bruce, Vladimir, Mary, a graduate student named Keith Henderson, and Jorge Albino S., one of the Peruvian guides, drilled for more than thirty consecutive days above 23,500 feet—2,000 feet higher than Sajama—and broke the record for high-altitude ice coring they had set only three months earlier. This also appears to be the longest continuous stretch that anyone—including mountaineers—has spent at that high an altitude. But this group did more than simply survive up there: they accomplished world-class science. Two years later, while poring over the Dasuopu data in preparation for a talk, Lonnie would have a flash about the grand workings of climate—something we have been calling his "asynchrony theory"—that may turn out to be as important as his Huascarán discovery. More on that later.

BUT DEATH REMINDED US OF THE DANGER OF THE HIGH MOUNTAINS WITH UNCOMMON frequency during that period. In the Bugaboos I witnessed my first death in the mountains: a lovely young girl crushed by a large rock as she approached the base of a superb granite spire. And the story of the tragedy that unfolded over the following months, on and in the wake of Lonnie's Dasuopu expedition, will be told here in due course.

Meanwhile, the El Niño that had blown down Sajama's weather station went on to become the most destructive in the 130 years that records have been kept. Bolivia had girded for its usual drought, but in a twist that highlights the difference between climate and weather, this El Niño made its debut with a few weeks of violent storms. The rainy season arrived in August rather than December. One day, the town of Sucre was buried in three feet of one-inch hailstones. The next night the town received three inches of rain in three hours—its worst storm in twenty-three years. In the end, however, the climatological forecast held true. Bolivia saw less than its usual amount of snow and rain that season.

Since the country's weather is usually extraordinarily stable during the dry season, August is the peak season for mountaineering. Thus, when a massive storm early that month dumped up to six feet of snow on the mountains, droves of visiting climbers retreated to the bars of La Paz to numb their disappointment and count the days to their flights home. However, a guided group on Sajama made the unfortunate decision to stick it out, and a French woman on that expedition died of edema a few days after the storm, as she was slogging down from the summit in deep snow.

Sadly, Yossi Brain's number would come up, too. Two years later, he would die in an avalanche in the remote Apolobamba range in northern Bolivia. I never did get back to climb with him.

Lonnie has told me that he would gladly leave the mountains if he could get the information he needs any other way. Having seen him in his element, I'm not sure I believe the "gladly," but I do believe that he finds the excitement and sense of exploration in the science, not in the great outdoors. He's not into adventure sports—in fact, he scoffs at risk-takers. He generally chooses the safest and easiest route up a mountain and calls upon Benjamín's Peruvian crew to fix a way up the hard parts. Indeed, his safety record is remarkable, considering the number of very high mountains he has climbed and the fact that his time at high altitude is measured in years.

He has also told me that the only reason he ventures into these dangerous places is that he believes the risks are justified by the social importance of climate research. That dedication was palpable as I watched his team work on Sajama. He had caught me up in the drama of his research. I soon asked if I could tag along on his next expedition.

PART II

EARLY DAYS

ALL TRUE PATHS LEAD THROUGH MOUNTAINS . . .

—GARY SNYDER

· 3 ·

WHERE NEXT?

Lonnie always has a plan. "That's the key to life," he says. "If you don't have a plan to measure yourself against, how will you know if you're making progress?"

His own is to gather a complete collection of high-mountain ice cores while the glaciers last and he is still young enough to get to them. As his group brings them in, they perform the basic analyses to set the timescale and present trends in significant parameters, then they publish the results, usually in *Science* magazine, the premier general-interest journal, in Lonnie's opinion. And as he and Ellen dig deeper into the data, they publish syntheses and more subtle interpretations. They continue to mine information from cores they retrieved as many as twenty years ago.

These are the pieces to a puzzle. In their later years, the Thompsons plan to lay them all out, paying due consideration to the host of pieces gathered by other scientists—seabed sediments, tree ring records, corals, and others—and put the whole puzzle together. Lonnie believes "the real science is still coming."

After almost thirty years and more than forty expeditions, he has most of the pieces he needs. Just a handful remain.

THE OLD-TIMERS CUT THEIR TEETH AT QUELCCAYA IN PERU, THE ONLY TRUE ICE CAP IN the tropical zone. Lonnie can't count the number of times he has been there. Twenty? Thirty? In 1983, he, Bruce Koci, and a few others gave birth to the field of tropical ice core research by retrieving two cores there from above eighteen thousand feet. The massif's heavy snows—a few yards per year—permitted annual layer counting all the way to bedrock, fifteen hundred years ago, and gave its glacial margins, draped as they were over the Andean foothills, the appearance of a topsy-turvy layer cake. Quelccaya's record of ancient El Niño events and other, longer and greater, climatic oscillations has given archeologists insight into dozens of cases of dislocation and collapse among the pre-Incan societies that once lived nearby. Quelccaya has been cited more than any other of Lonnie's studies, not only by climatologists and archeologists but also in a few popular books.

Where Lonnie once took a photograph of the layer-cake margins that has been widely published as an example of annual stratification, only the jumbled rocks of a glacial moraine now remain.

Exactly ten years after Quelccaya, his team drilled farther north, closer to the equator, on Huascarán. The black, glacial-stage ice they pulled up from the bottom there made quite a splash in climatological circles and helped bring the tropics for the first time into serious consideration as an *agent* of climate change. At Sajama the team took a reading on the other side of Quelccaya, to the south. Lonnie's life plan calls for extensions in both directions, for a pole-to-pole transect of the Americas.

As we stood on Sajama, he pointed due south, to the spot where the dotted line of snowcapped Andean peaks met the distant horizon, and said, "The Patagonian ice caps are the next big glaciers in that direction. We're trying to get there in a few years." The southwestern coast of South America is home to some of the worst weather in the world. Lonnie is making arrangements with the Chilean navy to approach the godawful place by sea.

At Cape Horn, the Andes drop into the Drake Passage for six hundred miles, then reemerge as the Antarctic Peninsula, to writhe like a serpent to landfall on the western half of the frozen continent. Ellen has drilled at Siple Station, near the peninsula's base, and a few other spots on Antarctica, including the South Pole and my personal favorite, the Pole of Relative Inaccessibility: the geometric center of the continent and, presumably, the coldest place on earth.

At the end of the eighties, Lonnie succeeded in drilling a few short cores on the peninsula over successive austral summers, but he hopes to return and retrieve a long core to bedrock. He figures a well-timed Antarctic record might answer one or two current questions in his field—as well as any question about events that took place thousands or tens of thousands of years ago can be answered, anyway—and the peninsula seems to be a good place to get such a record. It snows like mad down there; again, layer counting should be accurate for thousands of years.

The most rapid warming of the last fifty years has taken place at high latitudes, both north and south—an effect scientists have been predicting since the late nineteenth century, when the first rigorous theory of the greenhouse effect was proposed. A modern-day glaciologist from the University of Washington points out that the peninsula is "really getting hot, competing with the Yukon for the title of the fastest warming place on the globe." At Britain's Rothera Station, on the peninsula's western shore, for instance, temperatures rose an astounding twenty degrees Fahrenheit in the last quarter of the twentieth century. But climate is subtle; the change on the peninsula may have been carried by the water or air currents that slither up and down the coasts of South America and the peninsula itself and squeeze through the Drake Passage.

But tiny Heard Island, an Australian protectorate at the same high latitude, lies east of the peninsula, in the southernmost Indian Ocean, about as far from anywhere as a place can be. (Hence, it has become a holy grail for ham radio operators, and there is a plan to set up a base there for the study of ocean temperatures. It is one of the few spots on the planet from which sonar signals can be sent through all the world's oceans. Pulses could potentially be bounced off every continent but Europe. Delays in their reflections would provide a measure of the average temperature of the water along their paths.) Heard is so far from any major land mass that its snow should tell a purely maritime story. If the island has experienced a warming similar to the peninsula's, the effect is probably general to that latitude—and a result of the greenhouse effect rather than peninsular currents.

This is another ferocious spot. "It roars around there—nothing but low after low blowing through," says Lonnie's student Keith Henderson, who has prepared a proposal to drill there. ("And besides, it's a miserable place," smiles Lonnie.) They would work in the center of the mile-wide caldera of a dormant but possibly active volcano. ("That could add some excitement," adds Keith.) A boat would drop them off at the start of the austral summer. They would hope for another to return before winter makes the island inaccessible again; but they'll take supplies for a full year, just in case.

Following the transect north from Huascarán, the next site of interest is Ecuador, very close to the equator and with at least three intriguing mountains to choose from. There is a great gap between Ecuador and the southernmost candidate in North America, Bona-Churchill glacier in Alaska's Wrangell range, which should again yield good resolution: pure westerlies blow across the Aleutians from Siberia to sugar the Wrangells with nearly constant snow. Bona-Churchill may also bear the signature of the Pacific

Decadal Oscillation, an El Niño–like seesaw in the low-pressure system that hangs over the northern Pacific.

Alaska has warmed by about five degrees in the last thirty years, and Alaskans have witnessed many obvious changes: nearly all of the state's glaciers have begun to retreat; permafrost that has been frozen for a hundred thousand years is melting to produce sinkholes in the tundra, which have damaged roads and undermined buildings; spruce forests are dying. First the trees are weakened by the warmer climate and more damaging snow, which falls wetter and heavier than it used to; then tree-eating insects, which proliferate in the warmth, move in to finish them off. In southern Alaska, a swath of boreal forest three to four hundred miles wide is turning, piece by piece, from green to red, where the trees were recently killed, then to gray, where they have long since died. Between a third and a half of Alaska's white spruce have died in the last twenty years.

At similar and higher latitudes on the Atlantic side of North America, Ellen and many others have drilled in Greenland, and, of course, Lonnie's group has drilled on Russia's Franz Josef Land.

Back down at the equator, following El Niño's path west past its first landfalls in New Guinea and Indonesia, it next visits India, where it modulates the strength of the monsoon. Sir Gilbert Walker, the visionary director-general of the Indian Meteorological Service at the beginning of the twentieth century, first realized that the strength of the monsoon was related to the extent of snow cover on the Tibetan Plateau, which lies just to the north. Snow cover has a strong effect on the reflectivity of the land and, therefore, on its ability to store the sun's heat; so, since the monsoon is also tied to El Niño, it is possible that the snow in Tibet plays a role in the delicate shifts that cause El Niño to occur.

The Tibetan Plateau is one of the earth's dominant features; it is sometimes called the third pole. Not only do its unique size, height, and variable reflectivity give it a strong role in climate, but it also lies near the center of Eurasia, the earth's largest landmass. The centers of the continents, particularly this one, lying so far from the mitigating effects of the oceans, have felt the earliest effects of global warming.

Lonnie's program in China began in the Qilian Shan, a range that forms the border between the northern edge of the Tibetan Plateau and the Gobi Desert. In 1987, his group retrieved three deep cores on the Qilian Shan's Dunde ice cap that revealed the last fifty years to have been the warmest in the last forty thousand in that part of the world. But the Chinese government refused to let foreign nationals take an intact core out of the country back then. The team had to melt and bottle the top century and a half's worth before taking it home. Lonnie believes he can weave his diplomatic way to getting a complete core now, and he'd like to revisit Dunde with his improved techniques.

His next major effort—and the most logistically difficult of his career—took place in remote western Tibet, in the mysterious Kunlun Shan. In the early nineties, the team retrieved a core on the extensive Guliya ice cap that went back three-quarters of a million years—farther than any that had been recovered at the time, even in the polar regions. Guliya stood alone in that regard until 2003, when, after eight years of effort, the European EPICA collaboration (European Project for Ice Coring in Antarctica) completed the drilling of a core with an equally long memory on the East Antarctic Ice Sheet.

Lonnie is working on a second transect in China, hoping, among other things, to gain insight into the notion of tropical asynchrony. Since Guliya, he has drilled at Dasuopu, on the southern edge of the Tibetan Plateau, and at Purogangri, midway in latitude between Dasuopu and Dunde. And he has a fifth Chinese site in mind.

THESE DRILLING SITES HAVE BEEN SELECTED THROUGH A COMBINATION OF SCIENTIFIC INquiry; geopolitical serendipity; a free-ranging "Have drill, will travel" approach; and the spirit of pure exploration. In spite of the fact that Lonnie's drill design has been a matter of public record for two decades (ever since Quelccaya) and copied, perhaps even improved upon, by others, no one else has succeeded in obtaining a significant climate archive from high mountain ice—and not for lack of trying. It is the seasoned and committed team that Lonnie and Ellen have built over the decades that has set them apart. In his mid-fifties—the decade of life in which the best mountaineers usually shift from technically difficult ascents to feats of endurance—and with a remarkable record for bringing home the bacon, Lonnie is presently at the top of his game. His is the only team in the world capable of preserving these vanishing archives, and there is not much time.

AS WE LOOKED OUT ON THE WORLD FROM THE TOP OF SAJAMA, HE SPOKE OF YET another out-of-the-way place: "There, ahead, all he could see, as wide as all the world, great, high, and unbelievably white in the sun, was the square top of Kilimanjaro," wrote Ernest Hemingway of the delirious visions of a dying man.

This legendary mountain stands three-quarters of the way around the world from Bolivia, at the western end of El Niño's thin equatorial strike zone, just three degrees south of the equator, in the heart of the tropical zone. Lonnie knew that the snows of Kilimanjaro were the only snows in Africa that held any possibility of yielding a good climate record—and also that that record would soon disappear.

WHEN I HAPPENED TO SEND HIM AN E-MAIL A YEAR AND A HALF LATER, HE RESPONDED by inviting me on a quick reconnaissance to Africa—with about twice as much warning as I'd had for Sajama: ten days instead of five. After a night flight across the Atlantic, I rendezvoused in the Amsterdam airport with him and a surprise bearlike accomplice: Keith Mountain, an Australian and a founding member of the inner circle. This was Keith's first trip with Lonnie in eight years.

It was an odd flight to Tanzania; there was no hint of the cultures we were about to encounter—just two Africans among the safari-bound passengers. But when we landed after dark at Kilimanjaro International Airport and I stepped through the jet's door to the top of the gangway, a warm dry wind brought the sweet smell of Africa to my nostrils: dust, dung, woodsmoke, low-grade diesel exhaust. A big yellow moon hung low and full in the east. The windows of the small terminal building cast dim rectangles on the tarmac below. The land beyond was completely dark. I felt a thrill as I crossed the threshold into that free and chaotic world.

It was immediately clear that Lonnie felt quite at home in such a world. Not only were we the only foreigners in the thronged terminal who didn't have a safari guide, we also had no hotel reservation and weren't even sure which town we should use as a base for our climb. Our fate was decided in the darkness and sweaty disarray of the parking lot, by the porters who had happened to pick up our fifteen large duffel bags: they dropped them on the ground by two vehicles labeled "Key's Hotel, Moshi."

We were soon bouncing in a 4×4 in the direction of the moon, the unmistakable shape of Kilimanjaro hovering behind a haze of dust and smoke to our left—its square, mythic summit outlined by a shaving of silver, moonlit snow.

When we had found our rooms, the three of us repaired to the dining room, where photographs of big-game animals and views of the mountain hung on the wood-paneled walls. Shebani, the courtly maître d', recommended Kilimanjaro lager beer, not for its taste—being a devout Muslim, he wouldn't have known it—but for its label, on which the head of a giraffe peers out from below an image of the mountain. "This is the best beer," Shebani pronounced gravely. "For the giraffe is a polite animal."

Taking him at his word, Keith proceeded to down Kilimanjaros at a prodigious rate and with no apparent effect, while he and Lonnie told war stories long into the night. Keith did most of the talking. Lonnie laughed and played the straight man. They reminisced about the early days: Peru and Quelccaya, where they both had their formative experiences, and some miserable horse trip in the mud and rain in China. Lonnie made occasional reference to his youth in West Virginia, which is still very much with him and in which he takes great pride.

As I lay awake in the middle of my first African night, sweating under mosquito netting that made the still, close air only more claustrophobic, Keith lay sprawled on his stomach on a bed in the corner, snoring enthusiastically, nine-tenths exposed to the tsetse flies and anopheles mosquitoes, one arm hanging to the floor. In the vast silence beyond our walls, soft voices murmured close at hand and roving bands of dogs howled in the distance. It was good to be headed for the hills.

· 4 ·

BEGINNINGS

Lonnie grew up in the 1950s in a small town named Gassaway, in Braxton County, West Virginia. The only major employer in the area was the coal mine in Widen, a few towns away. He remembers that trains carrying coal from Widen used to come through Gassaway and that there was a circular turntable for switching locomotives in the center of town. Braxton County was poor then and hasn't prospered since. The coal mine in Widen has closed. The population of Gassaway has shrunk since Lonnie was a boy.

His father died when Lonnie was in twelfth grade. Out of pride, he had never allowed Lonnie's mother to work, so she had no marketable skills—though after he died, she did return to high school and earn a diploma. The family grew so cash-poor that they sometimes resorted to foraging in the woods for food. They lived in the poor part of a poor town, "the wrong side of the tracks," in Lonnie's own words. "You go without eating; pretty soon you decide, 'I don't want to be here,'" he confides.

He credits much of his success to his mother's influence, and she remains his greatest fan. "She always told me, 'You've got to get an education; the only hope is an education, otherwise you'll end up like the people here.' She was always pushing me. . . . I think part of my optimism comes from that. Because I've been there, I can function on all the levels in between. Sometimes I don't think you really come to grips with who you are unless you've been forced to the wall in certain situations. There's a tremendous amount of confidence that comes from overcoming that."

Neither he nor his wife, Ellen, is a great believer in material success, and they have no interest whatsoever in luxury. The money from the many awards and prizes they have begun to win in recent years goes straight to research. Ellen claims that they work so that their dog, Precious, "can live in the style to which she has become accustomed."

LONNIE WAS FASCINATED BY THE WEATHER EVEN AS A BOY. THE FAMILY BOUGHT GRAIN for their farm animals in large cloth feed bags, and his mother once collected all the bags bearing weather-related images—clouds, weather vanes, maps of high and low pressure zones, lightning bolts—and sewed the pieces into a shirt for her son. He still keeps it in a drawer in his office.

When he was in high school, he built an amateur weather station in his barn and made lunch money by providing forecasts for his schoolmates—crucial information in farm country, where a cut of hay might be lost if it wasn't baled by the next rain. He was paid only if he got it right. He beat out the radio forecasters and was known locally as "The Weatherman."

For all his gallivanting about, Lonnie remains a West Virginia boy. His office at Ohio State is only 140 miles from Gassaway, as the crow flies. About fifteen years ago, he and Ellen bought property on the Elk River, bordering his mother's farm. They talk of retiring there.

He attended Marshall University, in Huntington, the town where he was born. For the first three years he majored in physics, believing it would provide a good foundation for graduate work in meteorology, but he ultimately found the subject "too theoretical," so he let go of his cloud-filled dreams for a while and switched to the more practical field of coal geology. At a party given by the geology department, he met his future wife, Ellen Mosley—a physics major, ironically, and a West Virginia native as well. In time, her theoretical background and his practicality would prove a winning combination.

After he graduated and did a short hitch in the military he and Ellen were married. In 1972, he entered a graduate program at Ohio State, still focused on coal, and Ellen took a job as a secretary in downtown Columbus to support them both while he went to school.

"We moved here with everything we owned in the back of a borrowed pickup, and it wasn't full," she recalls. "All we owned was books and a desk. We rented a furnished apartment, and I worked for two years, so I've got both my Ph.D. and my Ph.T. That stands for 'Putting Hubby Through.' "

At the beginning of his second graduate quarter, Lonnie found a flyer in his mailbox for a research assistantship at the university's Institute of Polar Studies, to perform lab analyses on an ice core from Antarctica. He wasn't much interested in ice or the poles at the time (he probably didn't know quite what he was interested in), but an R.A. pays tuition and a stipend and generally gives rise to a thesis. With these lofty goals in mind, he applied—and thus met his first scientific mentor, Colin Bull.

Bull, a British former director of the institute, was chairman of the geology department and well known both as a glaciologist and as a vital force in polar research. And his background was somewhat typical for this interdisciplinary field: he had earned a doctorate in physics on a topic that had nothing to do with weather or climatology but as a young man had met a few members of Robert Falcon Scott's Antarctic expeditions, and this had given rise to a lifelong fascination with the poles.

One of Bull's research efforts at the time had to do with the so-called microparticles in ice cores—a fancy word for dust, basically. He and his students were improving upon a method, first developed in Florida, for counting dust particles individually, ice layer by ice layer, using an instrument that had been invented to count human blood cells. Bull recalls that "Lonnie, initially, was just tagging along behind in the usual way."

They were analyzing the longest ice core that had been recovered at that time, the 2,164-meter, or one-and-a-third-mile, Byrd core, which had been retrieved in January 1968 at a field station in West Antarctica named for Admiral Richard E. Byrd, the most famous polar explorer of the mid-twentieth century. Byrd had attained glory in 1929 by flying an airplane over the South Pole for the first time and had also written the bestseller *Alone,* chronicling his near-fatal attempt to spend the winter of 1934 alone on the frozen continent.*

Lonnie's very first project happened to result in an important discovery. Bull and Wayne Hamilton, the senior student in the "dust lab," with Lonnie "tagging along," found that the dustiness of Byrd's ice went through a yearly cycle—it rose and fell with the seasons—so that it could be used to count back the years. It would have been a Herculean task to count the layers all the way to the bottom of that mile-long core, but by testing selected samples at different depths the team found ways to estimate the thinning of the layers with depth and made the first solid calculation of the age of the ice at the base of the West Antarctic Ice Sheet: between twenty and thirty thousand years—much younger than previously thought.

Hamilton moved on to Cleveland State University; Lonnie took over

*During Lonnie's first year in graduate school, Colin Bull wrote a proposal to the Byrd family that succeeded in bringing the admiral's papers to the institute and resulted in changing its name to the Byrd Polar Research Center. The Center also holds the papers of the Frederick A. Cook Society, which was formed to uphold the reputation—though it has probably long since been lost—of the brilliant and controversial physician, explorer, and mountaineer who visited both polar regions in the decades on either side of 1900, the last golden age of geographic exploration. Some—and certainly the members of this society—believe Cook was first to reach the North Pole.

the dust lab; and somewhere in there—mid-1973 to early 1974—Lonnie caught fire. Several important events took place more or less simultaneously around then: in the fall of 1973, Ellen entered graduate school with an instant R.A. in the dust lab. ("When she saw what I was doing, she wanted to get involved, too," says Lonnie.) Though still a graduate student, Lonnie was named principal investigator on a grant from the National Science Foundation to upgrade the dust lab; and two months after his wife joined up, Lonnie went off to Antarctica on his first expedition.

He lived "on the ice" at Byrd Station for two months, from early December 1973 to early February 1974, helping a fellow student, Ian Whillans, measure the strains and movements in the ice cap and digging pits in the snow and taking samples from their walls so that he could take them back to Ohio and measure them for dust.

Lonnie's second study, published later in 1974, compared the dust levels at Byrd to those in the first core to bedrock that had ever been drilled in the polar regions, at Camp Century, Greenland, a U.S. base so named because it lies a hundred miles from the coast. That core, retrieved in 1966, was 1,387 meters, or about nine-tenths of a mile, long.

THOSE FIRST GREAT EFFORTS AT THE POLES—CAMP CENTURY TOOK FIVE YEARS—HAD been initiated in the paranoid atmosphere of the Cold War. The United States established Camp Century mainly for military purposes, both to develop cold-weather technologies and capabilities and simply to establish a presence in the Arctic; while in Antarctica, the U.S. Navy handled the logistical aspects of all science conducted by Americans and Western Europeans. ("It was almost a parable," writes the historian Spenser Weart, "the coldest of Cold War science.")

The drills used at both Camp Century and Byrd were developed by the Cold Regions Research and Engineering Laboratory (CRREL), an arm of the U.S. Army Corps of Engineers, located in Hanover, New Hampshire. Besides establishing the bases, the United States developed unique aircraft to fly into them and carried out the actual drilling. Since the latter task was something of a fiasco, particularly at Camp Century, the U.S. members of the Camp Century collaboration were preoccupied with what most scientists would consider mundane details, while their European counterparts, primarily Danish, Swiss, and French, were free to focus on the important things. Indeed, the Europeans made off with all the scientific coups—perhaps the most fundamental being the 1964 discovery by the Danish scientist Willi Dansgaard of a way to estimate the temperature of the overlying atmosphere at the time a particular layer of ice fell as snow.

Dansgaard's so-called temperature proxy is based on the mix of oxygen isotopes in the ice itself (the frozen water, the H_2O). In its predominant form, ^{16}O, an oxygen nucleus comprises eight protons and eight neutrons, while the next most prevalent form, ^{18}O (about one-one thousandth of all naturally occurring oxygen), has two extra neutrons. It takes more energy to shift a heavier molecule from the liquid state to the gaseous state; so when water evaporates from the ocean, for instance, the resulting vapor is lighter isotopically than the water.

As this same bank of moist air rises, it is further depleted in the heavier molecules, because whenever the air drops in temperature a portion of its vapor will condense into clouds, which are basically large collections of water droplets. Both isotopes condense during cloud formation, but the heavier isotope is preferred: it has a stronger desire to return to the liquid state. The colder this bank of air gets, then, the lower the ratio of ^{18}O to ^{16}O in its remaining vapor. Thus, the isotope ratio in a dusting of snow that ultimately lands on a mountaintop or a continental ice sheet is a measure of the temperature of the overlying air at the time the snow fell.

Since temperature generally varies on a seasonal basis, Dansgaard's isotope ratio can also be used to count annual layers; in fact, when Lonnie first started working with Colin Bull, the isotope method was the principal way of determining the age/depth profile of an ice core. The method worked only in the upper, younger layers, however, because ice molecules tend to diffuse ever so slowly between adjacent layers, thus smoothing out seasonal isotope variations over time.

Lonnie's earliest work showed that dust particles don't diffuse; hence they can be used to count the layers all the way to bedrock. Surprisingly, however, this came as a jolt to the big guns in climatology.

Sometime during the fruitful period between mid-1973 and mid-1974, Lonnie was invited to a meeting at the Maryland headquarters of the National Science Foundation to present a plan for upgrading the dust lab. Evidently, NSF also wanted to understand how the Ohio State plan might fit into the big picture, for they also invited Chester Langway, the former leader of drilling at CRREL; Claude Lorius, the Frenchman who is credited with the realization that the air bubbles in an ice core sample ancient atmospheres; and none other than Willi Dansgaard. These three old friends, who had been working together in Greenland since the early sixties, could be seen as . . . well, *core* members of what some have called the "polar mafia." All three were internationally prominent by then, and it would be fair to say that Dansgaard is legendary now.

The night before the meeting, Jay Zwally, program manager for NSF's Office of Polar Programs, invited Lonnie and the three senior scientists to his home for dinner. Now, even though Lonnie was only a graduate student, with fewer than two years' experience in climatology, he had seen enough data to convince himself that he could use dust to date the Camp Century and Byrd cores. Over dinner, he announced his intention to present the idea at the meeting the following day.

As he remembers it, "Langway took me aside and said, 'You know, if you say you can date ice cores using dust layers, your career is over.' I mean, those were his exact words: 'You're finished.' And it really worried me. I couldn't sleep. But, eventually, early in the morning, I said to myself, 'Well, young man, if this is all it takes to end this career, then it probably didn't have much going for it in the first place.'"

He stuck to his guns.

"It's a standard technique now. Dust is used to date Greenland ice all the time. It's the best dating parameter they have up there. . . ."

From the very beginning, then, Lonnie found himself an outsider and a bit of an upstart in the field of ice core climatology. His first run-in with the top dogs of the polar mafia may very well have excited a pressing need for him to get out from under their thumbs. It may even have set the tone for his entire career.

HIS EARLIEST WORK PRODUCED NOT ONLY THAT RATHER FUNDAMENTAL METHODOLOGICAL discovery but a climatological discovery as well—one that did not make much of a splash at the time but that resounded for the next few decades, both in Lonnie's work and in the wider field of paleoclimatology: he, Bull, and Hamilton found in the Byrd Station and Camp Century cores—the only cores that had been retrieved from the top and bottom of the planet at the time—that peaks in dust concentration, extending for centuries and even millennia, coincided with dips in Dansgaard's temperature proxy. In the bottom two hundred meters of the Camp Century core, for example, which represented ice from the most recent ice age, there is a huge peak in dustiness and a dramatic dip in the isotope ratio. This correlation would also seem to make sense, for most forms of dust reflect incoming sunlight and tend, therefore, to cool the atmosphere—in a process known as the "parasol effect."

As it happens, airborne dust was a hot (or perhaps cold) topic just then. For a few years in the 1970s, a small, though in retrospect not particularly perspicacious, group of scientists was predicting that the massive amount of particulate matter humans were spewing into the air—either directly from tailpipes and smokestacks, or indirectly, through slash-and-burn agriculture, overgrazing, the destruction of the earth's forest cover, and so on, which were exposing vast tracts of open, usually dried-out and eroded soil to the air—would soon trigger a global cold spell. On the basis of a primitive computer model written by his student Stephen Schneider, a NASA scientist named S. Ichtiaque Rasool even went so far as to state in *Science* magazine that the increased dust in the air "could be sufficient to trigger an ice age." This unleashed a spate of doomsday books and the first television special ever dedicated to climate change, produced by the respected science writer Nigel Calder, who also issued a book, *The Weather Machine,* in which he suggested that a full-fledged ice age might "start next summer, or at any rate during the next hundred years."

While the Ohio State crowd did not hop onto that bandwagon, Lonnie and Colin Bull did mention Rasool and Schneider's work in connection with the *paleo*climatic changes they had observed at Camp Century and Byrd.

BULL RECALLS "SITTING DOWN IN THE INSTITUTE, OH, IN ABOUT 1970 OR THEREABOUT, and contemplating how we could extend this work to the tropics. It was at that stage that somebody, perhaps me, asked John Mercer, who'd already done quite a lot of work in Argentina, what he knew about the high stuff on the Patagonian ice cap. John, I have frequently said, was the one person who was touched by genius among the folk that I knew at that stage. John was an incredibly wise old bird, able to take data from one field and look at its implications in another, and so on. He was a very, very bright individual."

Mercer, also British, was unique in many ways. He preferred to do his fieldwork in the nude, for example, and was periodically hauled into the dean's office for including explicit self-portraits in his lectures to his students. He didn't always wait until he reached the hinterlands, either: he was once arrested at five in the morning, jogging naked through a Columbus city park.

But despite his eccentricities, Mercer was a restless, prolific, and visionary scientist, and an unsung explorer with an encyclopedic knowledge of the world's glaciers. He had studied many of the most remote, sometimes as the first European to visit them, and the questions on his mind in the mid-1970s went straight to the heart of what was then still the fledgling field of climatology.

In the 1940s, he had begun a pioneering series of expeditions to Patagonia—which was nearly unexplored then and remains infamous for its weather—seeking dates for the successive advances and retreats of the region's two enormous ice caps, north and south. He would gather organic debris, such as dead wood, from the moraines below the glacial tongues and estimate its age using carbon dating.

TO GIVE SOME IDEA OF MERCER'S CREDENTIALS AS A SIMPLE GEOGRAPHIC EXPLORER: IN 1958, when he joined Eric Shipton—widely considered one of the greatest such explorers of the twentieth century—on an expedition to Patagonia, it was Mercer's third visit to the region and Shipton's first. And it is an interesting coincidence that Shipton provides Lonnie's connection to the splendid lineage of British exploration (both through Mercer and through Cedomir Maranguníc, whom we'll meet in a moment), because Shipton and Lonnie introduced similarly revolutionary, lightweight styles to the fields of mountaineering and ice core drilling, respectively.

Shipton and his frequent traveling companion, H. W. Tilman, were the first to take the light, so-called alpine style of climbing into the world's greatest ranges. Rather than resort to the massive militaristic siege—requiring dozens if not hundreds of porters, tons of equipment, and a series of numbered camps filled with a pyramid of supplies, all designed to place two or three presumably lucky individuals at a high camp, fit enough and well enough supplied to risk a dash for the summit—Shipton and Tilman would simply gather a group of friends with not much more than the equipment they could carry on their backs and make an attempt on a major Himalayan peak, often with breathtaking success. Tilman's 1936 ascent of 25,645-foot Nanda Devi, the highest mountain ever climbed at the time, is seen as the outstanding mountaineering achievement of the period between the wars. He and Shipton bragged that they could plan an expedition on the back of an envelope. This lightweight approach, not the sort that led to the gruesome and much-publicized fiasco on Everest in 1996, defines the cutting edge of Himalayan climbing today. It is the most elegant and dangerous game in the sport.

Shipton and Tilman opposed the siege on both tactical and philosophical grounds. Speed often equates to safety in the mountains, and a light team tends to be more flexible, free, and egalitarian. The reason 1958 found both men yearning for Patagonia—Tilman would die in his eighties at the helm of a small boat in the Southern Ocean—was that they had given up Himalayan climbing years before, finding the range overcrowded even then and the new vogue of huge expeditions too stifling and bureaucratic.

In the early fifties, Shipton was the most experienced Everest climber by far, having made four attempts on the mountain in the decades between the wars; and he was the original choice to lead the expedition on which Edmund Hillary and Tensing Norgay eventually made the first ascent. In 1951, Shipton invited Hillary, an upstart colonial from New Zealand, to join the expedition on which Shipton discovered the South Col route by which Hillary would later climb the mountain. The British Himalayan Committee then unanimously chose Shipton to lead an attempt in 1953.

But Shipton proclaimed quite openly "that fanaticism and nationalism should not form the basis of a mountaineering enterprise," and when he followed through by sharing photographs of the proposed route with a Swiss team that would ultimately fail on the mountain in 1952, a certain competitive faction on the committee succeeded in replacing him with John Hunt, a military man and a devout proponent of the siege. Hunt's successful expedition benefited greatly from a lucky stretch of stable weather. Most everyone agrees that he did a good job, but nearly all the climbers, Sir Edmund included, believe emphatically that they would also have succeeded had they gone alpine style with Shipton. Indeed, Shipton, in all likelihood, would have made it to the top.

Gracious as always, but deeply wounded, Shipton retired from exploration for a few years before striking out for the free land of Patagonia; just before he left, he wrote one of the more resonant phrases in twentieth-century mountaineering: "On great portions of the maps of the southern Andes of Patagonia there are blank spaces with the word 'Inesplorado' written across them."

Lonnie's greatest contributions to his field would also spring from a nimble and somewhat risky lightweight style, quite out of step with a bureaucratic and nationalistic prevailing culture. And he, too, would take his approach to every blank on the map he could find.

BUT BY 1958, PATAGONIA WAS NOT AS BLANK TO JOHN MERCER AS IT WAS TO ERIC SHIPton. The two men had very different temperaments as well. Shipton's biographer describes a close friend as believing Shipton "neither understood science nor trusted scientists, although he regarded them as an evil necessity both for funding and for high climbing." Shipton was as charming and free-wheeling as Mercer was dry and serious of intent. They did not get along. True to form, Mercer walked out on the expedition right in the middle of the south Patagonian ice sheet. He had a different sort of exploring to do. . . .

Fifteen years later, as a young Lonnie Thompson was just catching fire, Mercer was rooting around in the archives of the American Geographical Society in New York City when he discovered an aerial photograph of a twenty-seven-square-mile ice cap in southern Peru, new even to him, blanketing the 18,600-foot Quelccaya massif in a high cordillera near Cuzco. (This gives some idea just how unexplored South America could be even that recently.) He may have been spurred by Bull's interest in extending ice core work to the tropics; he may have been following his own nose; it was probably a little of both. In any event, he decided to visit the place in order to extend his Patagonian studies to lower latitudes.

AS LONNIE REMEMBERS IT (EVIDENTLY UNAWARE OF HIS ADVISOR, BULL'S, PREVIOUS conversations with Mercer), Mercer mentioned his idea to David Elliot, the director of the institute, and Elliot suggested that someone join Mercer to do some preliminary ice core work. Then Mercer just happened to strike up a conversation with Lonnie, and Lonnie agreed to go.

As Lonnie left for his first expedition to Antarctica, Mercer flew to Washington to meet with Jay Zwally, the NSF program manager in whose home Lonnie either had or would soon endure a vexing dinner conversation. Grantsmanship was looser in those days; the managers spent more energy on science and less serving the bureaucracy, and they even took the lead in launching a project now and again. Zwally remembers being intrigued by the Quelccaya idea, but that Mercer had no operational plan in mind.

"I asked, 'So, John, how would you go about doing this?' We got on the phone and called some climber . . . John Ricker? . . . in Canada, and he said he'd go. It was nearing the end of the year, and I had fourteen thousand dollars left in my budget. I think I gave half to John [Mercer] and half to Lonnie."

That wasn't much money for an enterprise of that magnitude, even then.

The grant came through at the end of 1973. On the ice at Byrd, Lonnie received a telex from Mercer.

"Now, Byrd sits on a flat white plain," Lonnie says. "It's minus forty nearly all the time—and that's in summer, the daytime. [There's only one "day" a year so close to the pole.] The place seems exotic for the first two weeks, but after a month or two the thought of a tropical glacier with mountain views begins to sound appealing."

Lonnie's only experience in the mountains had come from Boy Scouts and hikes in the hills around Gassaway with his brother as a kid; but another grad student on the ice that year, David Rugh, had done some mountaineering. On the way home from Byrd, participants in the U.S. Antarctic Program always stop in New Zealand. Thus, on the boredom of the ice, Rugh and Lonnie hatched a plan to get Lonnie some practice in the New Zealand Alps, a range famous for its snow and ice climbs. They chose Malte Brun, a ten-thousand-footer near Mount Cook.

It required the usual alpine exertions. They left the hut at three A.M. Lonnie remembers humping along the knife-edged summit ridge with his feet dangling on either side as he watched the morning sun crest the horizon. They enjoyed the view for a while and then started down. Below the sharp ridge the slope eased off, and Lonnie let down his guard. This is when accidents happen.

He doesn't remember how—most likely, he never knew—but one moment he was slogging lazily down the hill and the next he was careening down the ice on his stomach, headed straight for a cliff. He flew over the edge . . . and moments later found himself twisting in the wind at the end of a rope, with a sickening perspective on a crevasse-torn glacier a thousand feet below.

"At that moment," he says, "I made a pact with God that if I ever got off this mountain alive I would never climb another for the fun of it."

He seems to have kept his promise.

The accident happened in February, and Lonnie went to Quelccaya, which is nearly twice as high, only four months later. Asked why, he responds, "Well, a person can have two reactions to an experience like that. He can retreat, or he can learn and go on. I just had the insight to go to Peru and see how it might develop. My philosophy is if you want to find something new, go where no one has ever been. You're guaranteed to find something new if you look around once you get there."

SO, IN JUNE 1974, LONNIE THOMPSON, JOHN MERCER, JOHN RICKER, AND CEDOMIR Maranguníc* landed in Lima, Peru.

Lonnie's memories of his first trip to the developing world confirm Jay Zwally's hunch regarding Mercer's lack of a plan:

"We didn't know where we were going," Lonnie says, "although we did know what general part of Peru we needed to get to. We took a *collectivo* from Lima down to Arequipa. I distinctly remember the fact that the car had bald tires, and the driver—every time we would go over one of these passes, there'd be a church—he would stop, go in, do his genuflecting, and come back into the car. It was like the Indianapolis Five Hundred sliding around those curves. . . .

"From Arequipa, we took the train up to Puno, and John Mercer got robbed on that train. That's where our limited finances suddenly became very limited.

"We went on up to Cuzco and bought food and supplies, and then back down to Sicuani. From Sicuani we caught a truck up to the end of the road, which at that time was Succa Pulca, a little Quechua Indian village before you go across the pass. Crossing that pass you get the first view of the Quelccaya ice cap, sixty kilometers [forty miles] away—really an impressive place. It's the biggest ice cap in the tropics. In fact, I've never

*Cedo Maranguníc, Yugoslav by birth, had emigrated with his parents to join a large community of their fellow nationals in the unlikely location of Punta Arenas, Chile, the southernmost city in the world. (It lies on the northern shores of the Strait of Magellan, facing Tierra del Fuego.) In contrast to Mercer, Maranguníc got on well with Shipton, whom he met the year after the other two had their row, and went on to become his regular partner on Shipton's subsequent attempts to fill in the blanks on the map of southernmost South America. Those adventures took place in Maranguníc's college years. He then went to graduate school at the Institute of Polar Studies. After earning his doctorate, he returned to the adopted country he loved and continued in research. Lonnie is consulting with him as he plans to drill in Patagonia over the next few years.

seen a true ice cap in the tropics outside of this one. It's remarkable in that sense."

When a week of negotiations on hiring horses to carry their gear into the back country came to naught, Nickolas "Big Nick" Caritas stepped into the void—"a remarkable guy in many ways," according to Lonnie, "very much of a survivor, always coming up with new ways to extract the maximum amount of money from his guests." Caritas once asked them to pay for a stillborn foal that a mare produced on the trail, when Lonnie and his colleagues had no idea that the mare had been pregnant.

Keith Mountain, who would meet Caritas on his first trip to Quelccaya, a few years later, says, "Yeah, we would work with Big Nick because he could orchestrate fifteen to twenty horses at a time. They weren't all his, but he could wield enough influence. And Nick had it all worked out: 'I'll use your horse, and I'll give you two dollars a day for it, but I'm going to charge Lonnie Thompson ten dollars a day.' . . . So ol' Nick had that entrepreneurial spirit . . . but he was a very nice man. As long as you made a deal with him, he would stay with the deal—he was honest. A good trick for these folks is to get halfway up there and say, 'Well, we're going home now; we don't want to do this; we need more money.' Nick only did that a couple of times. All in all, he was a good guy to work with, and I don't see how we could've gotten all that stuff up there without him."

Reaching agreement on the horses, the 1974 team walked in for two or three days and set up base camp at the edge of Qori Kalis, the main, west-facing tongue of the ice cap, by a large boulder at the edge of the snow, which appeared to be moving downhill under the force of the advancing glacier. Lonnie would return to this site for decades.

While Mercer worked in the valleys below, mapping the moraines, collecting samples for carbon dating, and treating himself to full-body tans in the intense high-altitude sunlight, Lonnie, Marangunic, and Ricker climbed to the 18,600-foot summit to sample ice the old-fashioned way: by digging a snow pit and scratching at its walls and by drilling a few shallow cores with a hand auger.

Once he'd gotten the scientists to the top and assured himself that the path was safe and well marked, Ricker asked Mercer if he could take a walk around the ice cap while they finished their work. Mercer said no, but Ricker went anyway; as it happened, their work was accomplished and Big Nick's horses had arrived to take them out before Ricker returned. They left him in the backcountry.

And when the scientists got to Cuzco, Lonnie recalls, "Mercer came to me and said, 'Lonnie, I've run out of money; I don't have any money for you.'

I'm only a graduate student; this is my first time in a third world country; and I'm thinking, 'Well, aah . . .' "

Then, in Lima, Mercer simply caught a plane home, leaving Lonnie behind with the equipment. He had to pay an excess luggage charge out of his own pocket to ship it back, which left him with only fifty cents. During his last few days in Lima, he lived off the kindness of Paul See and his wife, the owners of the pension where he stayed. They even packed him a lunch for his flight to Miami—where he discovered, to add insult to injury, that Mercer had given him the wrong phone number for his Miami in-laws. Lonnie spent the night on a chair in the airport.

Ellen remembers being "livid" upon learning of Lonnie's plight by phone. However, having watched her husband survive many more hair-raising episodes over the decades, she quickly waxes philosophic: "He always survives, though. Something would have happened. He's always going to struggle through."

Lonnie also admits to being upset, but "not as mad as John Ricker was when he showed up six months later."*

With the perspective of decades, Lonnie now says, "When you first interact with a person you don't realize that there may be a long history there. It took me a while to learn that those incidents on Quelccaya were not isolated. Things like that were happening to John Mercer all the time. If anything could go wrong it would always happen to him. I remember once in Chile—we had to get into these rafts to cross a channel and get to the shore, and out of thirteen rafts John's was the only one that sunk. Once you realized that, you realized there was no intent on his part; it was just the way it was. . . . Once I really got to know him and his humor—he had a very dry sense of humor—we became great friends; but after a very stormy beginning."

"John redeemed himself over the years," adds Ellen. "We got to know him and love him and realized that there was absolutely nothing malicious about this man. This was John: totally oblivious, always being robbed, always losing things, and just barely squeaking by himself. Yet, he is one of the most eminent glacial geologists in terms of South America. We're still working on the Mercer Problem."

When Mercer died of a brain tumor, thirteen years after the fiasco on Quelccaya, his widow asked Lonnie to scatter his ashes at the informal gathering she held in his memory.

"It was quite an honor," says Ellen. "We had a basic service—it wasn't

*A few years later, Ricker published what is still the definitive reference to climbing in the most spectacular range in Peru, the Cordillera Blanca, about a thousand miles north of Quelccaya.

really a service. It was a group of us, John's closest friends. Keith Mountain was there. Ten, twelve people who were very close to John went down to his favorite place at Lake Hope, in southeastern Ohio, down in Hocking Hills, where he and his family would spend weekends—to this big rock where he and his wife, Judy, used to sit and drink beer and watch the sun go down.

"You know, Lonnie went from thinking this John Mercer was the worst person, to scattering John's ashes ten years later."

He may have been less practical and more eccentric than Lonnie, but in his iconoclastic and fiercely independent way Mercer was close in type to the sort of scientist Lonnie would become. Ellen believes "he was a man ahead of his time. John Mercer planted the seed, but a seed only grows in fertile soil. Lonnie was the soil."

IT'S HARD TO KEEP UP WITH LONNIE WHEN HE SPEAKS ABOUT THE SCIENCE BEHIND THE first Quelccaya expedition. For in many ways he is still working on the same problem. He'll range back and forth across the decades, touching on expeditions he's taken all over the world, realizations he's had on mountaintops in the Andes, Tibet, Antarctica, Central Asia. . . .

Little did Lonnie know when he first went to Quelccaya in 1974 that the main question in John Mercer's mind that long ago would remain at the center of his own research—and indeed climatology as a whole—for the next thirty years. Mercer's discoveries in Patagonia and with Lonnie in southern Peru still happen to raise serious questions, for instance, about the explanation for the Younger Dryas that had risen to the level of climatological gospel by the time Todd Sowers related it to me on our walk together at Sajama.

THE EARTH IS MUCH WARMER NOW THAN IT HAS BEEN FOR MOST OF THE PAST TWO MILLION years: it has been in the grip of the Great Ice Age, or Pleistocene geological epoch, for all but the most recent tiny slice of that period. The Pleistocene was characterized by periodic advances and retreats of the northern continental ice sheets, called glacial cycles, the last of which is known in North America as the Wisconsinan, after the state on which it left its greatest "impression."

The Wisconsinan ice sheet grew, halted, and shrank many times in its one-hundred-thousand-year life. After reaching a final maximum about eighteen thousand years ago, it began to withdraw for good, but reluctantly. It shrank for a thousand years or so, then turned abruptly and grew for a time, then shrank, then grew, and so on. One of the first of those relatively short "cold reversals" is named the Older Dryas, after *Dryas octopetala,* the mountain dryad, a small alpine relation of the rose, with eight cream-colored leaves and a yellow center, whose range in the northern hemisphere expands dramatically with drops in temperature. About twelve thousand years ago, when the Wisconsinan seemed to be giving up the ghost for good, the longest, strongest, and last of these fitful coolings set in, the Younger Dryas. It lasted for about a thousand years, and when it ended, with astounding rapidity, the present, relatively congenial, Holocene epoch began.

"Congenial" because a few hunter-gatherer tribes in the Fertile Crescent (present-day Iraq and Syria) and China's Yangtze River Valley coped with the renewed harshness of the Younger Dryas by resorting to agriculture and animal husbandry for the very first time—and because the subsequent twelve millennia of the Holocene have provided those first seeds of civilization with unusually warm and unusually (though not entirely) stable climatic conditions in which to take root, spread, and grow.

As the scientific terminology would indicate, all the early evidence for the Wisconsinan and its final death throes was discovered in the northern hemisphere. In the 1940s, however, when John Mercer first hunched his shoulder against the wet winds of Patagonia, he was searching for signs of the Wisconsinan—and the Younger Dryas in particular—in the southern hemisphere. He was still on that trail three decades later, when Lonnie Thompson joined him on his first expedition to high altitude a bit farther north.

After finding carbon dates for the advances and retreats of the glaciers in Patagonia and at Quelccaya, Mercer concluded that either there was no cold spell at all in South America at the time of the Younger Dryas or that if there was one, it was small, and it began about five hundred years *before* the cooling began in the northern hemisphere—quite a significant lead. It seemed to Mercer that the Younger Dryas, so-called, was specific to the northern hemisphere and to the North Atlantic region in particular, since it left its strongest marks in North America, Greenland, and northern Europe.

Now, the possibility that this dip in temperature may have been worldwide and that it may have *started* in the southern hemisphere or even in the tropics would seem to pull the rug out from under the story Todd Sowers told me on Sajama about a great pulse of cold water from Lake Agassiz pouring into the North Atlantic, shutting off a great ocean conveyor belt, and plunging the entire earth into a cold spell. Indeed, certain members of what I will call the North Atlantic School have referred to this fly in the ointment as the "Mercer Problem."

It would take Lonnie twenty-three years to find a solution to that "problem," and he would find it in the ice he would mine on Sajama.

·5·

A DECADE ON QUELCCAYA

It says something about the social history of ice core climatology that the results from the first ice coring expedition to the tropics were published in the *Antarctic Journal of the United States*.

All four members of the expedition, even John Ricker, are credited as authors on the first of two consecutive papers, in which, besides pointing out what Quelccaya had to say about the problem that would eventually bear his name, John Mercer demonstrated remarkable prescience with regard to the ice cap's deep drilling potential. Illustrating its "polar" quality with a photograph (taken by Lonnie) of the vertical, clifflike walls along its outer margin, he predicted that Quelccaya's ice was "likely to contain the best stratigraphic record of any glacier in the tropics" and that Lonnie's new method for using dust to count annual layers would work very well there. Indeed, the photograph makes it difficult to understand how anyone, especially an expert in the field, could still question the new method: the vivid annual snow layers look as though they could easily be separated with a hand shovel; one layer occasionally protrudes like the eave of a house over the one just below it; and the top layers are ten feet thick.

In the second paper, Lonnie and Willi Dansgaard, the sole authors, revealed an interesting conundrum about Andean ice. Lonnie had looked at the dust, Dansgaard at the oxygen isotope, and they had found, as expected, that both varied seasonally and could, therefore, be used as "calendars." Oddly, however, the isotope ratio rose in summer and fell in winter, when, if it bore the expected relation to temperature, it should have done the opposite. It would take Lonnie almost as long to figure that one out as it would for him to solve the Mercer Problem.

BY 1976, LONNIE HAD EARNED HIS PH.D., HE HAD THE CLOUT TO SEEK HIS OWN GRANTS and separate his finances from Mercer's, and a slight but portentous cultural change had transpired in climatology. A new entity had come into being at NSF, the Office of Climate Dynamics, which was not tied to a specific region, as the Office of Polar Programs was. Moreover, its director, Hassan Virgi, whose doctoral work had centered on air circulation over South America, realized both how little was known about tropical climate and how important it might be. As a native of India, Virgi was also intensely aware of the immense human impact of the Asian monsoon—and its relation, through the transpacific El Niño/Southern Oscillation, to the climate of Peru. From 1976 on, Virgi's office would fund all of Lonnie's Quelccaya work.

But Lonnie still had to convince the broader scientific community, upon which NSF relies for expert proposal reviews, that a deep ice core from the tropics might have something interesting to say about global climate. Not only did the huge polar drilling projects compete for the same funds, but most scientists, assuming the tropics were as stable over the long term as they are on a yearly basis, sincerely believed that they were a dull place climatologically. Tropical temperatures don't change much over the course of a year; seasons are characterized by more or less predictable rainy seasons. The thinking was—and, to a large extent, still is—that the action took place at high latitudes, where seasonal variations are more significant.

NEVERTHELESS, LONNIE SET HIS SIGHTS ON RETRIEVING AN ICE CORE TO BEDROCK ON Quelccaya's summit. Between 1976 and 1984, he would visit the mountain ten times. One might say that he was obsessed with the idea.

Over the first three years, he laid the ground by studying virtually every aspect of the ice cap. As Mercer collected his radiocarbon samples in the valleys, Lonnie and various colleagues hand-drilled short cores on each of the three summits; they placed stakes in the snow to measure annual accumulation; they surveyed the glacial surface in the usual way and the rocky surface beneath with a bootlegged radar device, then put the two together to produce a full three-dimensional map of the shape and thickness of the ice cap. Each year before they left they placed a few simple meteorological stations near the summit to record temperature, wind speed and direction, precipitation, and the number of daylight hours over the ten or more months they'd be gone. These "met stations" stood on fiberglass poles ten to fifteen feet high, guyed in with supplemental stakes and cables and fortified with piles of snow—in the hope, observes Keith Mountain, "that they would survive the accumulation [snow] season with some dignity and trustworthiness. Typically they did not."

On one of these early forays, a meteorologist from the University of Wisconsin named Stefan Hastenrath (who was nearly as widely traveled and equally as eccentric as John Mercer) conducted an experiment that would take on added meaning later: he filled bowls with snow and placed them on the surface of the glacier, then measured their weight for a few days in order to ascertain whether or not snow could survive in the intense high-altitude sunlight. He concluded that it could.

Lonnie went greedy on the place: he wanted to know everything he could about it. In 1977 he decided to go down at a different time of year:

"We had no observations from the wet season in that part of the world, so I had this idea: I'll just go down at Christmas, and we'll climb up and check it out.

"That was one of the worst expeditions I've ever had, because it was just me and a Peruvian cook, and I got sick on the trip in, and the cook ensured that I stayed sick all the way through. I lost forty pounds in two weeks—most of it puking between the ears of a horse."

He found the ice cap covered in chest-deep snow, which he could not have slogged through even if he had been healthy—and returned to Ohio with his tail between his legs.

This series of studies helped him form a picture of Quelccaya's "mass balance"—the annual tally of its growth due to snowfall and refrozen rain and its contraction due to melting and sublimation—which then allowed him to predict whether the ice cap should advance or retreat. In 1978 he joined Henry Brecher, a mapping specialist from Ohio State, in taking metric photographs of the Qori Kalis outlet glacier, just above base camp, to get a baseline measure of the glacier's shape and position.* Since the glacier seemed, and probably was, healthy at that point, and the global cooling craze happened to be in vogue, the two men expected Qori Kalis to advance over the next few years.

The 1978 expedition was Mercer's last to Quelccaya and the last he would take with Lonnie, although he seems to have passed the baton, for Lonnie began to gain some stature of his own around then. In March 1979 he published his first paper in *Science*, with Hastenrath and the eminent

*Brecher, who had quite a bit of field experience from Antarctica, found Peru "a lot less comfortable. It was third world to the nth degree! When we stayed in the hotel at the trailhead, there were drunks all over the place. They walked right through our rooms to get to their own."

Peruvian glaciologist Benjamín Morales Arnao. The paper summarized the work of those early years and reads primarily as a justification for future deep drilling on the summit. Perhaps the imprimatur of *Science* helped, because NSF granted approval for that very thing just as the paper was submitted.

Nineteen seventy-eight was an eventful year. Lonnie also found time to join Hastenrath on a trip to Africa to assess Mount Kenya's Lewis glacier. (They found that its top layers had been destroyed by recent melting and decided it wasn't worth drilling.) NSF, in its bureaucratic wisdom, established PICO, the Polar Ice Coring Office (named by the ubiquitous Jay Zwally); and Bruce Koci wandered onto the scene.

Having participated in virtually every ice drilling project and helped to develop virtually every ice drilling technology that the United States has come to grips with since that time, Bruce stands as a walking, talking encyclopedia and historical guide to this unusual branch of human endeavor.

I'm sitting with Bruce in what he calls "the PICO main heating plant," a little less than a mile from the small pole sticking out of a trackless white plain that marks the geographic South Pole. It is a few weeks before the millennium, and we are working as drillers for AMANDA: the "Antarctic Muon and Neutrino Detector Array," a telescope embedded in the ice about a mile below us, designed to detect the strange, ghostlike particle known as the neutrino. *Scientific American* has dubbed AMANDA the weirdest of the Seven Wonders of Modern Astronomy. Bruce was the main designer of the enormous drill used to construct AMANDA and is the unlikely Zen master of AMANDA drilling. Since this is much more of an engineering challenge than ice core drilling—an art that was perfected in the early nineties, in Bruce's opinion—AMANDA is his main occupation right now.

Our present job is to drill holes more than a mile deep in the two-mile-thick West Antarctic Ice Sheet with a drill that, according to Bruce, "represents the whole history of Antarctic drilling." It is somewhat bigger than the drills Lonnie uses: the heating plant in which we are sitting is one of two buildings filled with boilers burning jet fuel, which together generate the two million watts of power necessary to heat the water that is sprayed into the ice to make the holes.

"The contrast between the two projects is pretty interesting," says Bruce. "With Lonnie we go in with light equipment and bring out heavy ice core. With AMANDA we have a huge drill—two hundred thousand pounds—and the data doesn't weigh anything. We're trying to learn something from nothing, if you want to think about it that way.* Hot water drilling is not for the faint of heart, either. You never forget you're on the wrong side of the phase diagram—and a long ways on the wrong side of the phase diagram. That always comes home to haunt you."

By this he means that we are spraying water into ice about one hundred degrees below freezing, in air temperatures between fifty and a hundred degrees below freezing. We spend most of our time lurching around like the Keystone Kops, dragging water-filled hoses upwards of a mile long across a flat, white, oceanic surface, which stretches to the four horizons. If the water in a hose happens to stop flowing for even a few minutes, it will freeze solid, and the hose will become a large and extremely unwieldy piece of detritus. Thankfully, it is possible to take refuge in here with the boilers every once in a while.

IN THE LATE SEVENTIES, AS LONNIE WAS LAYING THE GROUND FOR DEEP DRILLING AT Quelccaya, Bruce was pursuing a graduate degree at the University of Minnesota, his home state. Something of a ringer, he had already had careers in aerospace engineering and analytical instrument design. He tells me how he got involved in *this* unlikely profession:

"That was a rather strange happening, because I was getting somewhere on a degree in glaciology—and rapidly running out of time—when all of a sudden I got this call that the University of Nebraska was looking for someone with a degree in engineering and some understanding of glaciology. So I called them up, got hired over the phone in, like, mid-October, and was on a plane two weeks later for the Ross Ice Shelf."

The U.S. ice drilling program was at a low ebb at that time. Owing to both the drilling fiasco at Camp Century and the resentment of U.S. granting agencies (not scientists) at the disproportionate scientific success of the Europeans, the United States had pulled out of polar ice core drilling after the Byrd effort. Not until 1981, after a fifteen-year hiatus, would the next deep polar core be retrieved: Dye 3, at Greenland's south dome. The United States would provide air support, while the drill would be designed by the Danes.

For the time being at least—global warming being a critical factor there—Antarctica's Ross Ice Shelf is the world's largest body of floating ice, about the size of France. The plan on Bruce's first visit was to drill a hole through the two-hundred-foot shelf in order to enable a group of scientists

*When the Austrian physicist Wolfgang Pauli first postulated the existence of the neutrino in 1930, he believed it was weightless. This belief held for about sixty years, until shortly after my conversation with Bruce, when the neutrino was shown to have an infinitesimal mass. The leader of the collaboration that made that discovery soon received a Nobel Prize.

to study the ocean below. The first order of business was to extract the previous year's drill from the hole. (It had frozen in and spent the winter there.) These are not delicate operations. They used a bulldozer, and it didn't work.

Their backup was a "flame jet," conventionally used by the mining industry to cut crystalline rock. Theirs consisted of two ten-thousand-pound compressors, which fed air at a thousand pounds per square inch to a modified jet engine, basically a huge Bunsen burner, which was lowered, spitting out flame and partially combusted diesel fuel, into a roiling, water-filled hole.

"Well," says Bruce, "it makes an awful racket . . . lotta noise, lotta smoke; it's real dirty . . . but it did drill through the ice shelf, and relatively quickly. Provided a hole about eighteen inches in diameter, so the scientists were then able to lower their things down and do their experiments."

As with any engine that burns fossil fuel, however, the flame jet produced prodigious quantities of carbon dioxide, which dissolved easily in the freshwater in the hole as well as the seawater beneath, turning both into something like Perrier. At one point during the drilling, a kink formed in the hose that delivered air to the drill. When it un-kinked, a large air bubble entered the freshwater column; the pressure dropped; and the dissolved carbon dioxide began to erupt like the fizz from a champagne bottle.

"We had four thousand pounds of hose, a fifteen-hundred-pound drill on the end of it, and we were lowering it down the hole with a bulldozer. All of a sudden it quit going down the hole . . . and then it started coming back *out* of the hole . . . and everybody took off! The drill, the hose, and everything came out the top, and then we had a geyser about forty feet high and four feet in diameter. And we got salt water! The fact that it was salt water really blew us away, because it meant that we'd bailed the whole hole."

NSF canceled the project soon thereafter—and then, somewhat surprisingly, awarded Nebraska the PICO contract when it was put out for bids the following winter. When Lonnie applied for PICO's help on the deep drilling at Quelccaya, Bruce was assigned to the job.

He was a good fit for two reasons: for one, his diverse engineering experience (not to mention the outrageous nature of his first drilling experience in Antarctica) had prepared him well for the sort of inventiveness that the development of a new technology for high altitudes would require; for another, he had spent plenty of time in the remote wilderness. Bruce had taken six canoe trips of four hundred miles or more in the Canadian Arctic, which he claims to have dreamed of visiting since the age of five.

"I was never in the drilling business to be a driller," he says. "I hate machines—maybe one of the few engineers in the world you'll ever find that feels that way about them. I hate them. If they don't serve me, they die awful deaths. I take things to them and I deep-six them and I burn them up and I do all kinds of terrible things. That's one of the few times I'll ever fly into a rage is over a machine that does something that it shouldn't do.

"I'm here for the experience. I came into this thing as a canoeist. I walked out of a good aerospace job and decided to go into ecology and then got back into engineering through glaciology, starting at Minnesota. I've always come for the place. I haven't come to do the drilling. I'll do my damnedest to make sure the drilling goes well, because that means I can go to another good place.

"I have only a love relationship with rivers or mountains and always ask their forgiveness for our trespass and delving into their inner secrets."

HE AND LONNIE DECIDED TO USE A HELICOPTER TO FLY A STANDARD POLAR ICE CORING drill to the top of Quelccaya, along with a gas-burning generator and the gas required to power it. But to give some idea of the sorry state of the U.S. drilling program at the time, PICO had no drills of its own, only a few ancient relics that had been built by the Swiss. Lonnie went down to Antarctica to pick one up over the 1978/1979 austral summer, then packed it for the deep drilling expedition in Peru, which took place over the subsequent austral winter.

Just getting the equipment to Sicuani was an adventure. The Peruvian customs authorities confiscated a snowmobile, believing it was a "troop-moving vehicle, because it had a track on it," according to Lonnie. "It took us a week to get that back." Then it took the helicopter thirteen hours to fly through the mountain passes from Lima to Cuzco, which was just partway to Sicuani. The vast quantity of fuel required by both the chopper and the drill was transported in drums by train. And a helicopter was such a novelty in remote Sicuani that Lonnie found himself managing about three thousand onlookers as he cleared a spot for the craft to land in the field behind their hotel, which served as the staging area for the flights to the ice cap.

And that was the easy part.

Lonnie remembers that they "flew two missions, one early in the morning and one late in the evening, when convective activity would be low. We'd be flying along at nineteen thousand feet, and the chopper would start dropping like a stone. These lights on the control panel lit up, and the pilots were terrified. We couldn't convince them to land. We flew all the way back and landed in the field behind the hotel; never put down on the mountain or anywhere else at high elevation—I just couldn't believe they couldn't do it; the weather was beautiful—and then there was no option as far as getting the drill up there. You couldn't put it on a horse. The generators were too heavy. . . ."

The expedition was a total failure; however, its seeds produced a brilliant success:

"From that disappointment we came up with the idea that we had to have a lightweight, portable system where you wouldn't be dependent on aircraft. That's when we had the idea for solar power, and it all fell into place."

About two months after our conversation in the PICO main heating plant and twenty-one years after the failure on Quelccaya, I'm chatting about that failure with Bruce. The millennium has passed, the Y2K demon has not swallowed the earth, and we have moved from the pole to the equator. We're sitting on the sand at about nineteen thousand feet in the summit crater of Kilimanjaro with our backs against our packs, facing west to soak in the slanting rays of the sun as it sinks swiftly toward the crater rim. We've just come down from a day of drilling with Lonnie on the mountain's sprawling northern ice field, which gleams white and aquamarine on the sand to our right.

"Found out helicopters couldn't go up to that level," Bruce recalls, "but it was definitely beneficial to see this type of ice cap just to experience the altitude . . . which doesn't sound very high now. It was eighteen thousand seven hundred feet. There were too many machines involved. That's a problem in a project like this. We had a standard generator which we couldn't get up there—we couldn't take it apart into enough little pieces, anyway, to get it up the mountain.

"But the important thing there was that we learned enough, and—maybe it's the Midwestern upbringing or something like that, or maybe I hadn't been involved in glaciology quite long enough to know that it should have been impossible, but when we came back from Quelccaya in '79, I saw absolutely no technical reason why we couldn't do that. And, more important, we didn't need ultralight stuff to do it. We needed stuff that was robust and that would work in the field, because the drill doesn't weigh a lot. It's a thousand pounds, and we come out with eight thousand pounds of ice core. So we went back, licked our wounds. NSF was kind enough to give us another chance. . . ."

They didn't miss a beat. Lonnie kept visiting Quelccaya every year, and successive hand-drilling expeditions provided an opportunity to bootleg a solar-powered drill that could be carried to the top of the mountain. This was the start of what Bruce calls "one of the longest-standing friendly relationships between science and engineering ever."

"The two disciplines require different approaches," he writes. "Science needs proof; engineering tends to go further out on the tree limb in predicting results. Drilling also requires work in the present time, which means that if the driller screws up on either the design or the operation, the result is no science. Another thing that seems to make this work is the simple fact that we are both from the Midwest, which tends to make people of even temperament. And there's also what I would call the passion factor: I find the science interesting and will go to great lengths to get the best science possible. Lonnie and I have certainly had disagreements on things but neither one of us loses sight of the 'mission.' We are both concerned about the state of the planet and what we humans are doing to it. We both feel that good science derived from these small tropical ice caps is an excellent way to get the word out."

SO LONNIE PRODUCED SCIENTIFIC RESULTS THAT JUSTIFIED GOING BACK, AND BRUCE made engineering improvements that benefited the ice coring community as a whole. One recognized need—particularly acute for Lonnie at his eighteen-thousand-foot field site—was to lighten the hand auger. The current state of the art, designed almost twenty years earlier by a forerunner to CRREL named SIPRE (Snow Ice and Permafrost Research Establishment), was fabricated of solid metal. Bruce designed a new auger out of composite, which allowed Lonnie to drill deeper holes at Quelccaya and was quickly adopted by other groups.

And when NSF "tasked" PICO to build its own electrical drills, Bruce incorporated lighter components—without necessarily telling his superiors what he was up to:

"On the first drills we built, instead of going with steel cable, which is what we had in '79, we decided to go with Kevlar. That cut the weight of the cable by a factor of three, which meant we could actually carry it up Quelccaya. It also meant the winch could be a little lighter—we went with gear reducers and things like that. But I was told specifically not to design an ultralight system—and it wasn't *ultra*light, it was *reasonably* light. The trick was we could make a lot of little pieces out of it.

"It was the sunburn through sun block nineteen, or whatever was available at the time, that turned on the light suggesting solar power. On one expedition Lonnie sunburned the roof of his mouth. Yuck! So we learned through getting fried that maybe solar power wouldn't be a bad idea. Instead of one generator plus spare parts, you have fifty generators with all those panels, so we could take one out and it wouldn't hurt us. They also have no moving parts; they don't make any noise; and as we found out, they were wonderful. I calculated we would need about nine hundred pounds of fuel to do the job. The solar panels weighed six hundred pounds at the time."

Bruce selected the panels; Lonnie took them to Quelccaya to see how they worked: "Lo and behold, the performance was twenty percent above what it was at sea level," Bruce recalls. "There was lots of light up there. They looked real good."

KEITH MOUNTAIN BARRELED INTO COLUMBUS ABOUT SEVEN MONTHS AFTER THE HELIcopter disaster, with a fresh bachelor's degree from the University of Oregon, where he had first heard the sirens' call of the great, white spaces on a research trip that had him living on the Collier glacier in the Cascades, more or less alone, for three and a half months. In Columbus, he grabbed his first chance at fieldwork when someone pulled out of the 1980 Quelccaya expedition just a week before it left. "I was like a rat up the drainpipe," he says. "Not a chance. Just right at it."

As a specialist in "radiation balance"—the tally of both visible and invisible light flowing in and out of the ice cap—Keith can vividly describe how bright it was up there:

"By day, you had to have warm clothes, but at the same time you're being blasted by eleven hundred watts a square meter in this unbelievable sun and reflection. The sunburns that would occur there! You get pretty creative after a while. Guys got duct tape over their nose, shirt over their ears, balaclava. . . . Even though you're dying in there, it's better than getting really burnt. You could not function without good-quality UV glacier glasses. It was just impossible. It was a very difficult environment to work in."

With the exception of Bruce, who seems to love any place as long as it's far enough from a city, every member of the inner circle has a special place, and for Keith and Lonnie it is Peru—Quelccaya in particular.

Keith believes "it's where you have your first really valuable experience. It sets a tone that's hard to overlook. Quelccaya was a hell of place; and despite the enormous physical difficulties of working there, when time or your physical state permitted, you could just sit and look at it with a sense of awe and outright amazement. It was stunning. You were always left with the impression that this was certainly one of God's finer creations. . . .

"When we arrived at base camp, typically early in the season, early to mid-June, there was usually snow on the ground. It was cold as hell, and you were pretty much exhausted from the three-day walk in. The porters were almost all Quechua Indians and, by the time we made it to the camp, would include a few of the locals and some children. These were incredibly tough people. The children always had shit in their eyes, runny noses, unwashed faces, but seemed not to notice. . . . You could not help but marvel at their human resilience in the face of their harsh everyday lives: living in peat or stone huts; no wood, just alpaca shit for fuel; certainly no medicine. The next thing that even resembled a town was three days away. You could not have any feeling other than deep respect for these people. I always tried to be very cautious in their presence.

"One of the guys Big Nick would hire was this smaller Peruvian who we called Superman—actually *Superhombre*. This guy was incredible. He would do things like carry the heaviest, most awkward load that we couldn't put on a horse, like the gas bottles or the engines for the drills or—stuff you couldn't believe. And he would be miles and miles behind the rest. He'd come into camp at four in the morning; he was just tougher than hell. He blew a sandal out there, at one point. He was incredible!

"These guys sit around huddled together in blankets, eat, chew coca leaves, drink booze if they had it; next morning, they'd be up and gone again—just incredible people. But old *Superhombre* was definitely a piece of work."

ALL THROUGH THIS SECOND PHASE OF QUELCCAYA EXPEDITIONS, LONNIE AND ELLEN put out a steady stream of papers to various journals, mostly co-authored and related to dust in polar cores, with a few reports on Quelccaya sprinkled in. In 1981, they presented a dust analysis, in *Science,* of a 905-meter (half-mile) core that had been recovered by the French two years previously at Antarctica's Dome C.* Although this core did not reach bedrock, it did penetrate the Wisconsinan glacial stage; and the Thompsons' analysis showed that the air over Dome C had been extremely dusty during the glacial, thus supporting the conclusions Lonnie and Colin Bull had previously reached about the Camp Century and Byrd cores. Lonnie and Ellen were the sole authors of this paper, her first in this highly regarded journal.

Few of their colleagues remember that the Thompsons also produced some of the earliest evidence for abrupt climate change around then. Lonnie recalls being "pooh-poohed" at a talk in Grenoble in 1981 when he suggested that the Wisconsinan may have receded in fewer than a hundred years. By the mid-1990s, "abrupt change" would evolve into the biggest buzzword in climate science.

Lonnie kept exploring as well. On Keith's first visit to Peru, they took a thousand-mile side trip to the Cordillera Blanca and drilled a short core in

*This was the one major polar core produced during the fifteen-year withdrawal of the United States from polar drilling. Since the French scientists could resupply Dome C with snow cats from Dumont d'Urville, their base on the coast, they accomplished their task with a minimum of U.S. support. American aircraft were used only to transport the core segments quickly to the coastal base once they had been brought to the surface.

the 19,800-foot Garganta col on Huascarán, the highest mountain in the country. The next year, Lonnie drilled another short core on Chimborazo, the highest mountain in Ecuador, and spent a month in Greenland with Bruce, testing their new solar-powered drill. With a few more years' legwork under his belt and his reputation partially rectified, Lonnie then asked NSF for a second chance at deep drilling on Quelccaya.

None of the pundits—members of the polar mafia all—backed his proposal. Even Lonnie's former collaborator Willi Dansgaard asserted in his review that "Quelccaya is too high for human beings, and the technology does not exist to drill it."

Dansgaard and his fellow polar—or, more specifically, North Atlantic—enthusiasts might be forgiven for this oversight, for they were especially self-absorbed at that moment. They had recently achieved their first big success in fifteen years, the Dye 3 core to bedrock in south Greenland, but this had led, unfortunately, to a second round of infighting between the U.S. and European funding bureaucracies.

Not only was Quelccaya very far from the North Atlantic radar screen, but the North Atlantic scientists may also have borne a certain resentment at the competition it signified. They had accepted Lonnie's dust calendar by then; he and Ellen had described the method in the literature; they had advised other groups on how to set up their own dust labs; and Dansgaard had built one in Denmark. In spite of the fact that the Thompsons' lab was certainly the preeminent center for dust analysis in the world, Chet Langway, who was still in charge of the American side of things in Greenland, chose Dansgaard's lab for the analysis of Dye 3's dust.

Ellen believes she and Lonnie "were cut out." It was the only kind of analysis they could do in their labs at the time.

NEVERTHELESS, AND DESPITE THE PUNDITS, HASSAN VIRGI AT THE OFFICE OF CLIMATE Dynamics took a risk and allocated funds for a second try on Quelccaya. These monies covered only the science aspect, however. Bruce's employer, PICO, the Polar Ice Coring Office, was part of Polar Programs, which refused to allocate money for a solar-powered drill. ("This was a pretty big white elephant project," Bruce admits. "When you think about it, a solar-powered ice drill at eighteen thousand six hundred feet is pretty nuts.")

But the University of Nebraska's office of solar power research was also willing to take a risk. Bruce convinced them to front the money for the panels, arguing that the university could always keep them if NSF didn't come through. With the drill built, he and Lonnie then managed to persuade the bureaucracies at NSF and PICO to let Bruce join a deep drilling expedition in the summer of 1983.

They both knew they were risking their careers.

As the sun sets behind Kilimanjaro's crater rim, Bruce squints and bows his head. "That was a real solo journey for both Lonnie and me. We looked at each other and said, 'If this doesn't work, we'll be doing something else in a year . . . in less than a year.' "

Figuring he would be drummed out of climatology if he blew it a second time, Lonnie applied and was accepted to a few business schools. Corporate America missed out on a superb leader.

THE 1983 TEAM WAS A SMALL ONE: SIX SCIENTISTS AND DRILLERS, A "DRIVER" NAMED Eugenio, and Benjamín Vicencio, the Peruvian mountaineer who would wish me luck fourteen years later as I left the top of Sajama. Dave Chadwell and Mike Strobel, two undergrads from Ohio State, were newcomers.

Strobel had some experience in the Antarctic and Greenland, where there are "field stations" and life is comparatively civilized, but twenty-three-year-old Chadwell, a civil engineering major, had never left the forty-eight states. He had answered Lonnie's ad for a surveyor after spending the fall semester on a work-study project surveying dams for the Army Corps of Engineers. Dave had never even flown in an airplane, much less seen anything like the chaos of Lima, so Lonnie, who loved Peru, decided to reward him for all the effort he had put in packing and so forth back in Ohio by giving him the full educational experience. He let him ride in one of the trucks that carried their gear on the five-day trip "down the coast of Peru," as Dave describes it, "then over the hump there, through Arequipa and up, eventually, to Sicuani. I was the only one in our group that made that trip. You get to see a lot of the countryside that way."

They might have been more wary had either realized that the violent Maoist guerrilla group Sendero Luminoso, a.k.a. Shining Path, had risen to power in southern Peru over the previous few years. That very season, in fact, while the team was working on Quelccaya, Sendero blew up a station that kept records of local alpaca farming, not far from the mountain. The scientists learned about it on their way out.

Bruce, on the other hand, holds the iconoclastic view that Sendero's presence actually improved things: "I think in '83 day-to-day life was probably safer than it was in '79, because everybody was afraid to go out on the streets, so there wasn't all this craziness and traffic. In Cuzco, in particular, the traffic in '79 was just nuts. There were bank riots and all kinds of things. There was sort of a Shining Path presence in the air in '83, but we never really knew it. We were pretty remote back in there, and they would have had to deal with Big Nick and the boys—the guys that took us in by horse. They wouldn't have done very well against those boys."

Dave laughs to remember the negotiations with Big Nick and says Nick reminded him of a savvy old horse trader in a John Wayne movie.

LEANING AGAINST HIS PACK IN THE CRATER ON KILIMANJARO, BRUCE GETS A FARAWAY look in his eyes. "That was one of the most beautiful walks in the world. You could see it almost immediately upon leaving the camp where the trucks dropped us off. You just kind of walked over the hill, and there was the ice cap. It looked close, but it wasn't.

"Big Nick and the boys took us in. The walk took three days. We got there with all the solar panels and everything intact. Believe me, I sweat bullets over getting the panels there. We had special packaging made up for them, to make them horse-proof. Horses did roll on them, and buck on them, and kick at them—and we didn't lose any. Also, I should indicate that we conducted a test of the solar drill on top of the parking garage at Ohio State, which is not a very auspicious beginning, I guess. . . ."

"The nice thing about going back to a place like that, time and time again," adds Keith, "is that you get anxious to get into the mountains. You want to get to the ice cap. When you first go it's all new—the towns you go through, the people you meet, the things you see—but then finally, after a while, it's the science that's really important."

WHEN THE HORSE CARAVAN LEFT THEM AT THEIR BASE CAMP BY THE BIG BOULDER, THEY found themselves quite alone. They had brought a mountain of food, with the intention of sticking it out for as long as it would take to get the job done, but they had little idea how long that would be. They had no porters; they did their own cooking and ferrying of loads up and down the glacier. Aside from a shortwave radio that Lonnie tuned in to the BBC once a day, they had no way of communicating with the outside world. Isolation of that sort creates a unique mind-set. In the best of circumstances, as it seems this was, a sense of mutual trust begins to build.

They fell into a routine in which one person remained in base camp each day while the rest worked up top. That person would melt and bottle the ice that had been extracted the previous day, wash clothes, and have "sort of a day off," according to Dave, "but your main duty was to fix the evening meal, and the highlight of the meal was the soup we made every night. Basically, it was whatever you could find in terms of fresh vegetables and some hodgepodge of soup stock or any kind of seasoning that may have been left over from some other meal—some freeze-dried meal, that kind of thing. You'd also fix a couple of freeze-dried entrées—some godawful spaghetti and something else I don't remember.

"I'd stuffed a bag of popcorn in the bottom of my duffel and taken it all the way to Quelccaya. So every Saturday night we threw in a cup of that popcorn and popped it up and opened two bottles of beer and passed it all around—a couple of beers and a few handfuls of popcorn. . . ."

KEITH WAS MAKING HEADWAY ON HIS STUDY OF THE ICE CAP'S RADIATION BALANCE BY then, based on the information from the summit "met" stations from previous years and measurements from a radiometer, which he carried to the summit on this trip. There was no such thing as an automated data-acquisition system in those days, so he spent a few nights alone up there, rising every hour to record the radiometry readings in his lab notebook. So as not to miss any, he kept himself awake by sticking his foot outside the tent—in temperatures that dropped, according to his own measurements, to as low as five below zero.

Thus he learned just how delicate the health of a high mountain glacier can be—and that this one, in particular, robust as it seemed, was very close to a tipping point. The dry air at eighteen thousand feet allowed the snow (as well as his foot lying on top of it) to radiate away significant amounts of heat and to cool dramatically at night, while the intense light of day had the potential to warm it far above the melting point. Keith's measurements showed him that each day was a race between the rising temperature of the snow and the clouds that moved in from the Amazon cloud forest, more than ten thousand feet below and many miles away, to shelter the ice from the sun at some variable time in the afternoon.

"What was saving this ice sheet at that time was its height and its enormous reflectivity," he says. "It was the sort of pure snow that you would expect to find in a place like Antarctica—reflectivities of eighty-two, eighty-three percent.

"At that altitude, ninety to ninety-five percent of the sun's energy would make it through the atmosphere to the surface—enough to fry anything. But even in the middle of the day so much of this energy was being reflected that you would have only seventeen percent—sometimes even less than that, ten percent—for melting. You go from this staggeringly large amount of incoming energy to a minuscule amount.

"But all the sun has to do is raise that surface temperature enough to get a little melting; then the structure of the snow changes and so does its reflectivity. That's catastrophic, because now you've got twenty-seven percent of the energy coming in—three times what you had before. It's a slight change with unbelievable consequences.

"We were looking at this incredibly delicate balance. It was right on the edge. Some afternoons we would see physical melting of the layers. Then the clouds would come in and cut the sunlight way back, and the air temperature would drop below freezing. There was just this little window of opportunity. Another very important consideration for this glacier was the snowfall. Sometimes the clouds would drop just a dusting and give it a fresh, bright surface—a new start for the next day."

Basically then, in the innocent early eighties, the only true ice cap in the tropics was hanging on by its fingernails.

THE CENTRAL TASK WAS THE DRILLING, OF COURSE, AND WHILE THE POLAR MAFIA MAY not have been smiling upon the idea of tropical ice coring at the time, it seems that the object of study, climate itself, had a different point of view: an El Niño set in that year, the strongest ever recorded, in fact. Since Quelccaya, like Sajama, sits at the southern end of El Niño's north–south seesaw, it experiences drought during El Niño events. The weather was amazingly stable that summer.

"Every year, every season is different, and it may work in your favor, and it may not," says Lonnie. "That's a variable you can't control. . . . There we were drilling on Quelccaya with solar power during the biggest El Niño of the twentieth century, and it could not have worked out better, because it was one huge drought in southern Peru—day after day, clear sky. And we weren't capable of predicting El Niños in those days. It was pure serendipity. You couldn't have planned it better if you *had* the information. Sometimes you get to feeling like these things are meant to happen."

The bright sun enabled Bruce's solar array to put out two kilowatts of power. "People found out this was serious solar energy," he points out, "because several got bit by the two-forty [volts] we were generating—it hurt!" Lonnie remembers a few hummingbirds flying all the way up from the Amazon River Basin to check for nectar in the orange electrical connectors.

At first they used a cranky old Swiss mechanical drill. ("Mechanical drills are like racehorses," says Bruce. "They're fast, but they're temperamental. The core wasn't great at that time.") When they hit water at about thirty-five meters they were stopped dead, because water would short out the engine inside the drill.

"The story gets a little weird here," says Bruce, "because NSF tasked us to do this with an electromechanical drill. Now, I had recommended as a safety—because the ice was so close to the melting point—that we really should take a thermal drill along. I had developed a thermal drill back in 1974 or something like that for Minnesota, when we worked on the Barnes ice cap. So they threw this thermal drill into the overhead account, and we kind of chucked it in a tube and took it down and hoped we wouldn't have to use it.

"Well, anyway, we slipped the old thermal drill on there and sent it down the hole and started drilling with it, and Lonnie at that point wasn't very optimistic about what we were going to get. . . . So I drilled for about forty minutes. Then we pulled it back out again and undid the core dogs [which hold the core in the drill] and started to slip this thing out. . . .

"I wish I could have gotten the sequence of Lonnie's eyes as this core came out, because his eyes just kept getting bigger as the core got longer. It was a two-meter core section, 'round about perfect. And it's interesting, because in the review of the proposal someone had said thermal drills don't take good core. We took more than one hundred and sixty two-meter lengths of core that were perfect. And the stratigraphy was so good that I could see it from about thirty feet away. The thermal drill just polishes the core perfectly."

It was all completely new to Dave Chadwell, of course, but he got caught up in the excitement anyway:

"We were counting the dust layers as we went back—every time we would bring up a core—and we were trying to keep a running count of the years. I remember when we found the Huaynaputina eruption—in 1600, I think it was; I've forgotten the details now—and thinking, 'Okay, this is a five-hundred-year-old core. What was happening in the world at that time?' "

Actually, he remembers the details pretty well. According to the diaries of Spanish priests living in Peru at the time, this volcanic eruption, the largest in Peru's recorded history, took place from February 19 to March 6, 1600. The town of Arequipa, fifty miles to the southwest, was completely dark for a week. Houses collapsed under the weight of the accumulated ash.

THE HIGH ALTITUDE, THE DAILY EXERCISE, THE BRILLIANT SUNSHINE, THE VISIONARY surroundings, and the headiness of success all seem to have had an intoxicating effect on what Keith calls "the collective mind." Spirits ran high. Everyone but the day's cook would race to be first to the drill site each morning, three miles and 1,600 vertical feet away. "We got along strikingly well," says Dave. "Everyone pulled their weight and knew everyone else would, too. There was a lot of trust involved."

They began to develop a sense that they were making history as well.

"We even told stories. . . . We knew it was working pretty well, and every day we'd put the samples in a backpack and take them down. Everybody had a pretty heavy load. This was when the Indiana Jones movies were just coming out, and we'd joke about Lonnie's archrival,

Willi Dansgaard, coming over the horizon and grabbing up all the samples, because he'd heard we were pulling it off."

Actually, if Lonnie ever did bear a grudge against Dansgaard, it wasn't for long. He later pointed out in a speech that "upon returning to [Ohio State], I contacted Willi and asked if he would like to join us in analyzing the Quelccaya ice cores. He graciously accepted and he also admitted that he had been wrong in his assessment of our proposal. This led to a long and productive collaboration that has lasted for decades." Indeed, Dansgaard appears as an author on one of the papers that resulted from the 1983 expedition.

One hundred fifty-four meters down, they came across a sudden tilting of the dust layers and the emergence of elongated air bubbles in the ice: signs of dramatic horizontal flow at that depth sometime after those layers had been laid down. Radar measurements from previous years indicated that they were still thirty feet from bedrock, but they stopped drilling. For the layer count had been disrupted, rendering the chronology below that depth ambiguous.

That first core, taken from the very summit of Quelccaya, turned out to be 155 meters long and reach 1,350 years into the past, more than twice the age Lonnie and Ellen had estimated in the grant proposal and *Science* paper they had prepared prior to the helicopter disaster in 1979. Performance was finally beginning to exceed expectation.

They'd been in the field for about two months now, and they were getting low on food.

"Now, I was tasked to do just one core," says Bruce. "Well, we got done with the one core, and Lonnie and I just looked at each other and said, 'Two cores is really what we want.' He had enough bottles for that. . . . This is one of those things, because if NSF tasks you to do one core and you try and do a second and stick the drill, maybe you've got a problem. Anyway, I said, 'Yes, we'll go for it.' "

They moved to a col about six hundred yards downhill, began drilling again, and gave Eugenio some money to make a run out to Sicuani and buy more food. But when he reached civilization, he gave some of the money to his very poor family and spent most of the rest celebrating. He returned to base camp with only a small load of vegetables, when what they really needed was meat, protein. (That he had the temerity to come back at all says something about the good nature of the scientists.)

"I think we were all so surprised that he came back without anything that it took us a while to realize," says Dave. "I mean, you know, you'd been on the mountain with the guy for sixty days, and you could certainly understand his need to cut loose a little bit."

So Keith volunteered to walk out and get what they really needed. In town, he called Ohio with the news that they had gotten one core to bedrock and were going for a second.

"The interesting thing was," says Bruce, "that the funding had been pending and pending and pending, and finally, when Keith went out—I don't know who he called, but anyway the word got out, especially the fact that we were going for a second core—and the funding somehow miraculously appeared, within hours.

"That put PICO back on the map as far as a drilling entity. The U.S. was once again the drilling power. We had been a nonissue ever since the Ross Ice Shelf project, which I worked on a little bit at the end. They had stuck the drill and done a whole bunch of things that basically killed off the U.S. ice drilling program."

But Keith also returned with the news that Bruce's mother had died almost a month earlier. Lonnie gave him the option of going home, but he replied that it was too late anyway, it had already been a month; he wanted to finish the drilling.

THE TASK NOW TOOK ON THE CHARACTER OF A BATTLE, AS THEY RACED AGAINST BOTH the coming of fall and a second dwindling of supplies. This may have been a factor in one very close call:

"Mike and I were pulling out the drill at one point," says Keith. "We got it out, and you have to clean the back end of it to get all the shavings out before you send it back down. The drill kicked back at us—and it slipped. It came down, and the bit caught me right on top of the goddamned nose and just sliced it right open. Jesus! And blood just went pfft!

"So I'm standing there holding the goddamned nose, and there's blood just going everywhere in the snow, and I'm thinking, 'Damn! *Damn!*' I wasn't sure whether the nose was broken. I put my hand out, and there's blood coming out, and I looked at Mike and said, 'Is it bad?' And he goes, 'Yeah, yeah!' And I'm like, 'That's not what I wanted to hear, Strobel! That's not what I wanted to hear!' Anyway, he sort of patched it up, and it wasn't a problem."

When they thought they could see a light at the end of the tunnel, Lonnie set a date for Nick Caritas and his horses to come in and carry everything out. They were now going on three months in the backcountry, and as they worked feverishly to prepare for departure, they ran completely out of food.

Dave remembers having nothing but coca, "so we threw all the coca leaves we had in a pot, and I think we had some old half-beaten-up orange we threw in the pot, and we just boiled all this stuff together to make this very powerful tea."

And Keith went off on another quest for protein.

"I went off with Eugenio over the moraine, about ten miles down the next valley. We're going to buy a sheep. God, we need food. We're going to need meat; we need stuff with a bone in it—sheep or a llama or something. Well, we couldn't find anybody that would sell us plain meat, but it looked pretty promising that we could buy the entire goddamn sheep. Well, fine. But the question is how are we going to kill this thing? We've got squat; we've got a couple of Swiss Army knives. And Eugenio says, 'Oh, I'll take care of it. Not a problem.' So we're going to cart this poor goddamn helpless animal back and gut it right there, just festoon the camp in intestines. My image was that this was going to look like a slaughter yard—intestines hanging around like a Renaissance festival. It turned out that never happened, but we did send some of the folks out early because we had no food. At the end we'd just scour around in the rocks, picking up old tea bags and throwing them back in, so we could get something in our stomachs. Finally, a guy came up and gave us a guinea pig [a staple in Peru]. Throw a guinea pig in the pot! God, it was just desperate, just desperate!

"But these are beautiful places, and they're—it has been just great. For me it has been a great privilege to be part of all of this, and I find it extremely interesting that I probably—I mean at work—that I wouldn't know what else to do. There are a lot of options. I could teach high school. I could fly airplanes.* I could put gas in cars. I could fix things . . . but being up there watching my skin fall off and bleeding is what I'm designed to do. That's why God put me together.

"I ate it up. I never saw anything wrong with it. There were times that I'd sit there and just watch the shit come down. . . . You'd be cold and wet and miserable, and there was just nothing you could do but sit there and watch. And it was just—there was something unbelievably humorous about it. There was no point viewing it any other way than in a real positive light."

At the end they began to sprint. Dave remembers Keith "off to the north dome putting up accumulation stakes; Phil piecing together the met pack; and on one of the last days I have notes that Keith and I were making two or maybe three trips up to the summit and back with the sled. We would pile the solar panels on the sled and drag them down to the edge of

*He owns one.

the glacier, then start back up for another trip. Somebody else was coming from camp up to the edge of the glacier with different boots on. We had a kind of an assembly line going.

"The sleds were pretty heavy, so once in a while we would break through a crevasse. I can remember one time having my feet dangle below me and my arms spread out sort of plugging up the hole so I wouldn't go all the way through."

The second core made bedrock without any breaks in the strata. It was longer than the first, 164 meters, and reached back farther in time, 1,500 years.

"We finished within minutes of when we had to leave," says Bruce, "but that was the best thing we ever did was get two cores. . . . We even made it into the high-tech section of *USA Today*, of all the learned journals on earth. I should have just quit right then, while we were ahead and I had a few more brain cells left, I guess. . . .

"My main thought at the end of that thing was, 'Boy, am I glad I don't have to analyze those six thousand million bottles of water. I'd kill myself.' I mean, I looked at those boxes and boxes and boxes of water and said, 'You've got to be kidding me. I couldn't do that.' . . . And Lonnie was so happy about having all that stuff. . . ."

ON THE VERY LAST DAY, DAVE REMEMBERS, "WE GOT UP EARLY AND GOT OUR STUFF rigged up. I can remember walking up to the moraine that's sort of equal height with Quelccaya, looking back at the field camp and the glacier. We just started walking, and we walked all day. I don't know how long it is, but by that time we were all very fit, so we made it to a construction outpost by late that night.

"Mike Strobel and I [the two young Turks] got up the next morning right at sunlight, and we didn't have any food or anything so we just started walking down the road. We thought a vehicle was coming to meet us that day. We walked several hours—it was a Sunday morning, I remember, beautiful morning and lots of neat geology to look at—and we got to this construction camp, and a guy gave us canned chicken to eat and boiled us some tea or something like that; got us set up. He must have had a phone or something, I don't remember. Eventually, he helped us get to the Land Cruiser that had been parked nearby, our Land Cruiser, so we pulled all the stuff out that we didn't need and got it started and went back and got Bruce, and the three of us drove into town, to Sicuani—took us two or three hours to get down there—and we just basically went wild on the food. We went into one restaurant and we'd eat roast chicken and then we went to another and had a steak, and by that evening we were all almost sick from eating too much. It was a 'kids in the candy shop' thing: everything we'd look at we thought, 'Oh, man, we *have* to have one of those.' "

"SO WE WENT BACK," SAYS BRUCE, SITTING IN THE CRATER ON KILIMANJARO, "AND Lonnie started publishing; and people came out of the woodwork wanting to do this and that and the other thing, correlating it with anthropology and all kinds of things. It was a unique record. It wasn't a long record, but, boy, it sure was detailed, and it was perfect to the year.

"There was a pause after that while Lonnie went to look at China."

PART III

THE WARMING SETS IN

THIS IS A DIFFICULT SUBJECT: BY LONG TRADITION THE HAPPY HUNTING GROUND FOR ROBUST SPECULATION, IT SUFFERS MUCH BECAUSE SO FEW CAN SEPARATE FACT FROM FANCY.

—GUY STEWART CALLENDAR

· 6 ·

FIRST SIGNS

Lonnie remembers the 1983 Quelccaya expedition as "a life-changing experience." It established his international reputation in a single stroke.

He and Ellen published papers about the summit cores in *Science* for three years in a row. The first dealt with the signatures and dates of major El Niño events. (Colin Bull recalls the "great fun" he had arguing with Lonnie about how to distinguish those signatures, adding, "He did a good job of that.") The second was the paper of record, presenting the full climate archive going back 1,500 years. The third reviewed Quelccaya's record of the Little Ice Age, a cold spell that gripped most of the earth from the sixteenth century to the late 1800s—ending, oddly enough, just as mankind first began to send invisible clouds of heat-absorbing carbon dioxide into the atmosphere.

As it happens, Quelccaya's ice yielded the very first evidence that South America had participated in the Little Ice Age; and with the perspective of twenty years, it is now clear that it is also the best (that is, most distinct) evidence for that widespread event to be found in any climate record that has been retrieved anywhere—with the possible exception of Huascarán, where Lonnie would drill ten years later. This was an early and still largely disregarded sign that the tropics aren't quite as insensitive to global climate changes as scientists have long believed.

The most amazing thing about Quelccaya, though, was its quartz crystal timing accuracy. The published layer count, for example, puts the ash layer from Huaynaputina at A.D. 1600: precisely the year in which the priests saw it.

Lonnie, Keith, and Dave took their last trip to Quelccaya (for the time being) in 1984—sort of a cleanup operation. Though they spent six weeks on the ice cap, this season did not have the high drama of the previous year's. In fact, they picked up their most important piece of information as they crossed the final pass and got their first view of base camp: Qori Kalis, the outlet glacier, had pulled back a few hundred yards since the previous year—quite the opposite of what Lonnie and Henry Brecher had expected when they had taken the first mapping photographs six years earlier. The glacier's margin now stood a good distance above the large boulder it had been pushing downhill when Lonnie first saw it in 1974. He says this was the moment when "it really started to click" in his mind that a major climate change was taking place. When they returned to civilization, Lonnie began asking around.

"We talked with people working up in central Peru," he says. "This very small ice cap completely disappeared in that period of time. Gone. Then we started to look around at Electroperú's records of the retreat of these glaciers. You could see this acceleration in the retreat rates kicking in in the late seventies. On Quelccaya we seemed to be seeing the same thing, even though at that time we didn't have a very long observational record."

Electroperú, the country's main electric utility, relies on glacial runoff to generate hydroelectric power. Its owners evidently took heed of the ice caps' advice: a few years later they sold their dams to a group of naïve investors from the United States.

And when Lonnie returned to the States, he found that the retreat was not restricted to South America:

"I had interactions with people like Stefan Hastenrath and his student Phil Kruss [both of whom had joined previous expeditions to Quelccaya]. While we were working on Quelccaya, they were measuring the retreat of the Lewis glacier on Mount Kenya. You look at that data, and you say, 'Gee, the same thing's going on over there.' And then you get reports out of New Guinea—Carstensz—and that ice cap is going. There are photographs from there. The reports coming back since the '73 and '74 Australian expeditions say these ice masses are disappearing."

Thus, ironically, just as Lonnie taught himself how to drill on them, the earth's high mountain glaciers began to disappear. As he and Keith Mountain knew, they stand in a precarious balance, which makes them unusually sensitive to climate change. And by leaving the fold of polar ice core drilling to study what were thought to be the boring and unchanging tropics, Lonnie found himself in exactly the right place at exactly the right time to watch a great climate change begin.

With the perspective of time it would become clear that the sudden retreat of the earth's mountain glaciers was one of the first signs that global warming, caused by humans, had begun. In the first year of the new millennium, Mark Dyurgerov, a Russian scientist who had joined Lonnie on an expedition to Soviet Central Asia in the early nineties, would publish a retrospective study demonstrating that mountain glaciers all over the world had undergone a sudden transition to melting in the mid- to late seventies; and in 2001—twenty years after things first "clicked" in Lonnie's mind—the first United Nations report to conclude unequivocally that human activity had caused the warming would place the transition, similarly, in the mid-seventies.

Furthermore, quite a few leading climatologists see that transition as one of many indications that our planet has now passed from the congenial Holocene epoch—the only epoch human civilization has known—into a new and unpredictable one, which some call the "Anthropocene," for they believe humans have caused it.

Some background:

The sun sends its energy to the Earth mainly in the form of visible and ultraviolet radiation, at short wavelengths. The planet's surface absorbs this radiation, heats up, and sends energy back into space in the form of "blackbody radiation," primarily at longer, infrared wavelengths, which are invisible. The hotter the surface, the more blackbody radiation it will send out, seeking the temperature that will balance the incoming shortwave and outgoing longwave energy flows. This is the principle of radiation balance.

A straightforward calculation shows that an object with our planet's reflectivity, at our average distance from the sun, will seek a surface temperature of about one degree below zero Fahrenheit. The average temperature at our surface, however, is almost sixty degrees higher than that, owing to the so-called greenhouse effect of our atmosphere.

Certain atmospheric gases, existing in trace amounts, let the visible and ultraviolet sunlight in, but absorb the infrared going out. They reemit that absorbed energy in all directions: roughly half continues out, and half is sent back toward the Earth's surface. This process effectively adds to the inflow and subtracts from the outflow. Each layer of the atmosphere, therefore, acts as a sort of blanket over the layer just below it. Temperatures are warmer lower down and warmest at the surface.

From space, the Earth would appear to be radiating at roughly the correct blackbody temperature of –1°F, but a close look would reveal that that radiation originates between six and eight miles from the surface, at the outer edge of the lowest and densest layer of the atmosphere, which is known as the troposphere.

Garden greenhouses, incidentally, don't work this way. They take in the sun's energy through their glass walls, of course, but the main way they hold it in is by preventing convection, the flow of wind that would mix the cool air outside with the heated air inside. In spite of its imprecise name, however, the greenhouse effect is well understood. It is neither controversial nor new. Indeed the temperature at the balance point and the precise number of degrees by which the measured levels of greenhouse gases in our atmosphere should raise the temperature at the surface were both estimated quite accurately more than a century ago. Even now, the vanishingly few climate scientists who remain skeptical about the dangers of global warming agree with the rest of their colleagues on basic greenhouse physics.

It is also worth noting that the effect is quite strong and that, generally speaking, this is a good thing. If there were no greenhouse gases in our atmosphere, the surface of our planet would be completely covered in ice, and life as we know it could not exist.

HISTORIES OF GREENHOUSE SCIENCE GENERALLY BEGIN WITH BARON JEAN-BAPTISTE Joseph Fourier, an eccentric French polymath who is best known for developing a mathematical method called Fourier analysis. As early as 1824, he suggested that "the [Earth's] temperature can be augmented by the interposition of the atmosphere, because heat in the state of light finds less resistance in penetrating the air, than in repassing into the air when converted into non-luminous heat." Given the primitive understanding of heat and light at the time, however, Fourier admitted that "in this examination we are no longer guided by a regular mathematical theory."

He likened the Earth and its atmosphere to an old scientific instrument known as a heliothermometer. This was a box lined with black cork whose cover consisted of three layers of glass separated by air spaces, with a thermometer inside. This quaint and somewhat misleading instrument had been invented by Horace-Benedict de Saussure, the great Swiss mountaineer and natural scientist of the Alps, who had used it to measure the power of sunlight at different elevations in the mountains—most famously in 1788, when he conducted experiments for nearly two weeks on the Col de Geant, near the summit of Mont Blanc. In his own investigations of the greenhouse effect, Fourier carried out experiments with modified heliothermometers and demonstrated that if air, rather than vacuum, filled the spaces between the glass plates it was more effective at retaining the "non-luminous heat" given off by the sun-warmed inner walls of the instrument. Fourier's earliest reference to previous work on the subject was to Edme Mariotte, who had noticed in 1681 that sunlight passes easily through glass and other transparent materials, while the infrared radiation given off by a fire ("chaleur de feu") does not.

THE UNDERSTANDING OF HEAT AND LIGHT BEING MORE POETICAL THAN MATHEMATICAL in Fourier's day, it would be another fifty years before the Irish physicist John Tyndall made the first steps toward putting greenhouse science on solid footing. Tyndall was not only a fine scientist but also one of the leading public intellectuals of the Victorian era. He was a close friend of Thomas Carlyle, Charles Darwin, Darwin's champion Thomas Henry Huxley, the botanist Joseph Hooker, Lord Tennyson, and other leading lights of the day. In the early 1870s, Tyndall's lecture tour of the United States, replete with the dramatic laboratory demonstrations for which he was famous, was successful enough to permit him to endow fellowships at Harvard, Columbia, and the University of Pennsylvania on the profits. (He took no money for himself, keeping only enough to pay for his expenses.)

And as with de Saussure and a surprising number of key contributors to the understanding of atmospheric phenomena over the centuries, Tyndall's interest in meteorology was kindled by a great love of the mountains. In 1862, after a successful climb in Switzerland, he wrote in a letter to Michael Faraday, a towering figure in nineteenth-century physics,

> We did not hope to see any thing from the summit, nevertheless we attained it and were rewarded: not only by the clearness of the prospect, but by the changes of the atmosphere, which were quite marvelous, sometimes shrouding all, sometimes melting as if by magic and revealing the mountains. The atmosphere is a wondrous factory: the grand origin of all its power being overhead, lifting the snows and driving the clouds by its individual might.

Tyndall first visited the Alps as a very poor student in 1849, walking all the way from the university he attended in Marburg, Germany, "trusting to my legs and my stick, repudiating guides, eating bread and milk, and sleeping when possible in country villages where nobody could detect my accent." Seven years later, having risen to the position of professor at London's Royal Institution and been recognized as a Fellow of the Royal Society, he visited Switzerland with T. H. Huxley (also an avid mountaineer) to study alpine glaciers, an interest that had arisen from a line of research Tyndall had begun into the effects of pressure on magnetic crystals.

He seems to have caught the bug on that trip, for he reached the summit of Mont Blanc in each of the following three years. On his second and third ascents, moreover, he mixed business with pleasure by making meteorological measurements on the mountain's 15,800-foot summit, the highest in the Alps. And the third time, he even prefigured Lonnie by spending the night up there. Enduring a "raging headache," he, with the help of his guides and porters, placed twenty minimum thermometers in various spots around the summit. His biographers state, however, that "it was a labour in vain; not one of the twenty survived to tell its tale."

AN UNDAUNTED ABILITY TO FACE THE UNKNOWN SERVES BOTH THE EXPLORATION OF large mountains and the pursuit of science; and, as it turns out, John Tyndall was one of the finest and most successful mountaineers in Europe during the first golden age of Alpine exploration. According to a certain Lord Schuster, the author of a chapter on Tyndall's mountaineering accomplishments that appears as an appendix to his biography, "Before 1854, very few of the great summits in the central Alps had felt the tread of human foot. Pacard and Balmat's ascent of Mont Blanc in 1786 [for which they claimed a prize offered by de Saussure] had not given the impulse to mountaineering which might have been expected, because the French Revolution, and the wars which followed it, made the Alps difficult of access and turned men's minds to other things."

In 1858, four years after the golden age began, Tyndall made a daring and controversial solo ascent of Monte Rosa, the second highest peak in the Alps and the highest in the Pennine group, which is perhaps the most spectacular range in western Europe. (Ten of the twelve highest peaks in the Alps, including the Matterhorn, are found in the Pennines.) Afterward, Tyndall wrote words that not only reveal his eloquence, but also ring very true to anyone who has climbed alone: "[I]t is an entirely new experience to be alone amid these scenes of majesty and desolation. The peaks wear a more solemn aspect, the sun shines with a purer light into the soul, the blue of heaven is more awful . . . the feeling of self-reliance is very sweet, and you contract a closer friendship with the universe than when you trust to the eye and arm of your guide."

His supreme mountaineering achievement was his first ascent, in 1861, of the Weisshorn, in a single nineteen-hour push. According to Schuster, this was "perhaps the most attractive and at the time of its conquest the most coveted, next to the Matterhorn, of all the summits in the Alps." The Weisshorn is one of the more remote peaks in the Pennines, higher than its famous neighbor and more symmetrical, with three sharp ridges leading to its pyramidal summit. Modern climbers often prefer it to the Matterhorn for its solitude and its more challenging and aesthetic climbing.

But a man of Tyndall's skill and inclination could hardly have ignored the Matterhorn. Although today the Matterhorn might seem "easier" than the Weisshorn, it is very different to climb a mountain for the first time or by a new route than it is with a guidebook, the accumulated knowledge of thousands of previous ascents, and—as is usually the case on the Matterhorn today—with a dozen or so parties climbing at the same time as you are. Tyndall played a major role in the Matterhorn's exploration and very nearly made its first ascent.

His first reconnaissance took place in 1859, with his trusted guide, Johann Josef Bennen of Zermatt, the town on the Swiss side of the mountain. The following year, they made a very early attempt at the summit and reached the highest point yet.

In 1862, a young landscape illustrator named Edward Whymper made three or four attempts on the mountain and was in the midst of failing on the last when Tyndall arrived in the town of Breuil, on the Italian side, with a view to having another go himself. According to Schuster, "Tyndall was forty-one, a Fellow of the Royal Society, and a man of high standing in general society and particularly in Alpine circles. He was accompanied by a guide who was universally respected, and in whom he had illimitable trust," whereas "Whymper was twenty-two years of age. He had no particular Alpine achievement to his credit . . . [and] his reputation was coloured by [a previous] accident which must have seemed . . . the result of pure foolhardiness."

Nevertheless, and in spite of Bennen's reservations about Whymper's "rashness," Tyndall invited the younger man to join his team. According to Tyndall, Whymper replied, "If I go up the Matterhorn I must lead the way," whereupon the offer was withdrawn. Whymper later claimed that he had merely asked leave not to have to follow Bennen if one of the guide's decisions was "evidently wrong." In any event, Whymper did not go along.

Tyndall's 1862 party got to within a few hundred feet of the summit, another high point, and was the first to reach the second-highest summit, later named Pic Tyndall. They then proceeded along a ridge, now named the Tyndall Grat, which connects Pic Tyndall to the final summit cone. Where the ridge meets the cone, they encountered a wide gap separating the ridge from a daunting "final precipice."

"So savage a spot I had never seen," Tyndall wrote. He claims that he and Bennen wanted to continue, but that their companions, among whom was a French guide named Jean Antoine Carrel, refused. "It took [Bennen] half an hour to make up his mind, but he was finally forced to accept defeat. What could he do? The other men had yielded utterly and our occupation was clearly gone."

After this frustrating defeat, Tyndall lost interest in the Matterhorn for a few years; but the decision to turn back took on added weight later.

Whymper finally climbed the Matterhorn in 1865. On the descent, in one of the most infamous episodes in mountaineering, four members of his party fell four thousand feet and died. Three days later, Jean Antoine Carrel made the second ascent by nearly the same route he had attempted with Tyndall in 1862, making use of a rope they had left high on the mountain on the earlier attempt.

Tyndall, who happened to be vacationing in the Swiss Oberland at the time, received news of Whymper's accident in the form of a strange visitation:

"I was accosted by a guide, who asked me whether I knew a Professor Tyndall. 'He is killed, sir,' said the man; 'killed upon the Matterhorn.' I then listened to a somewhat detailed account of my own destruction, and soon gathered that though the details were erroneous, something serious had occurred. [Soon] the rumour became more consistent, and immediately afterwards the Matterhorn catastrophe was on every mouth and in all the newspapers." Tyndall went to Zermatt and attempted to recover the one body that had been lost in the fall, out of sympathy for the man's grieving mother, but eventually gave up, owing to both the danger of the undertaking and the onset of bad weather.

Three years after Whymper's "success," Tyndall finally ended "the long contest between me and the Matterhorn" by reaching its summit from the Italian side, more or less by the route he had attempted in 1862. He then descended to Zermatt, thereby accomplishing the first complete traverse of the mountain. It was the fifth ascent overall and only the second or third from the Italian side. Describing the feature that had turned his party back in 1862, he wrote, "I think there must have been something in the light falling upon this precipice that gave it an aspect of greater verticality when I first saw it than it seemed to possess on the present occasion. . . . I cannot otherwise account for three of my party declining flatly to make any attempt upon the precipice. It looks very bad, but no real climber, with his strength unimpaired, would pronounce it without trial insuperable."

Their ascents of the Matterhorn made heroes of Whymper and Carrel, and Whymper's book *Scrambles Amongst the Alps* went on to become a classic of mountaineering literature. Although it purports to tell the story of the mountain's exploration, it obscures Tyndall's contributions and tells only Carrel's account of the fateful decision to turn back in 1862. (Carrel claimed, not surprisingly, that Bennen, and not he, had lost heart.) When Tyndall happened to publish his own climbing memoirs, *Hours of Exercise in the Alps,* in the same year as *Scrambles,* Whymper saw the older man's conflicting remembrances as a threat to his own reputation and raised a row in mountaineering circles. (Some things never change.) Tyndall did not stoop to respond. Climbing was just a hobby to him.

JOHN TYNDALL'S UNDAUNTED SCIENTIFIC IMAGINATION BEGAN TO FOCUS ON THE "wondrous factory" that is the atmosphere just as his alpine explorations were reaching their crescendo. In 1859, the year of his first reconnaissance of the Matterhorn, he applied his renowned experimental acumen to an investigation of the absorption of infrared light (known as "obscure" or "radiant heat" in his day) by various atmospheric gases. After designing an elegant differential spectrometer specifically for the task, it took him less than six months to come out with definitive results, which he promptly presented in a demonstration to the Royal Society, "with the Prince Consort in the chair."

His most intriguing discovery was that "perfectly colourless and invisible gases and vapours" had vastly different abilities to block the obscure heat. Oxygen, nitrogen, argon, and hydrogen, which constitute 99.9 percent of the atmosphere, were nearly transparent to it, while the trace constituents water vapor, carbon dioxide, and ozone were nearly opaque. These fine experiments enabled Tyndall to conclude—with far more precision than Fourier—that "[t]he solar heat possesses, in a far higher degree than that of lime light, the power of crossing the atmosphere; but, when the heat is absorbed by the planet, it is so changed in quality that the rays emanating from the planet cannot get with the same freedom back into space. Thus the atmosphere admits of the entrance of the solar heat, but checks its exit; and the result is a tendency to accumulate heat at the surface of the planet."

Tyndall found that water vapor was the most potent greenhouse gas by far, and carbon dioxide the second most. From that he deduced that water vapor "must form one of the chief foundation-stones of the science of meteorology." He understood, for instance, that temperatures would drop precipitously on clear, dry nights, as the ground underfoot radiated its heat unhindered to the stars, and that such a drop in temperature would enhance the likelihood of fog or dew in the morning, as the tiny bit of vapor that remained in the air would condense into droplets as the air cooled. "Remove for a single summer-night the aqueous vapor from the air which overspreads this country," he wrote of England, "and you would assuredly destroy every plant capable of being destroyed by freezing temperature. The warmth of our fields and gardens would pour itself unrequited into space, and the sun would rise upon an island held fast in the grip of frost."

If you've ever watched a sunset in the desert, you may have noticed that it cools off dramatically the moment the sun sets. This is the greenhouse effect at work—or, actually, taking the night off. In clear, dry desert air there is nothing to prevent the thin layer of sand that stores the sun's heat during the day from radiating it away at night, when there is no inflow. On the other hand, as you may also have noticed, humid or cloudy nights tend to be warmer.

NEAR THE TOP OF THE SCIENTIFIC AGENDA IN TYNDALL'S TIME WAS THE QUEST FOR AN explanation of the ice ages. The notion that large portions of Europe and North America were once covered in ice had been conceived in the 1830s by Louis Agassiz, a Swiss zoologist widely considered to be the father of glaciology, who eventually moved across the Atlantic to Harvard University. (The great lake that once sat on the Wisconsinan ice sheet was given his name.)

Tyndall sought an explanation in the greenhouse effect: "If . . . the chief influence be exercised by the aqueous vapor," he wrote, "every variation of this constituent must produce a change of climate. Similar remarks would apply to the carbonic acid [his name for carbon dioxide] diffused through the air. . . . [C]hanges [in the atmospheric levels of these gases] in fact may have produced all the mutations of climate which the researches of geologists reveal."

He wasn't far off. The current belief is that while something else provided the trigger, subsequent changes in carbon dioxide were the proximate cause of the changes in temperature that ultimately led to the advances and retreats of the polar ice caps of the Pleistocene era.

TYNDALL WAS SICKLY MORE OR LESS ALL HIS LIFE, AND A BACHELOR FOR MOST OF IT. HE was finally married at the age of fifty-five to a woman fifteen years his junior, and they were utterly devoted to each other. Once he retired, in fact, they planned to write what they called "our autobiography" together. They never got the chance.

As his wife testified at the necessary inquest into his death, Tyndall had been lying in bed and she had been administering his daily medicines when,

> "I measured a teaspoonful of magnesia, as I thought, and added water. He took this at a gulp, then, according to custom, a gulp of ginger. All he said was, 'There is a curious sweet taste'. I tasted the drop left and saw there were two bottles on the table and that what I had taken came from the full bottle of chloral standing near which had recently come from the chemist."

> The coroner asked whether she took the bottle supposing it was magnesia and Mrs. Tyndall assented, adding: "I said, 'John, I have given you chloral', and he said, 'Yes, my poor darling, you have killed your John'."
>
> Mrs. Tyndall here broke down under intense emotion and the coroner and jury were visibly affected.

Despite the best efforts of the local doctors, Tyndall was gone in less than twelve hours.

BY THE TIME OF THAT TRAGEDY, THE SEARCH FOR AN EXPLANATION TO AGASSIZ'S DIScovery of the ice ages had become something of a Holy Grail to the gentlemen (and it was almost exclusively men) who pursued science in the genteel Victorian manner of the day. Meanwhile, fundamental developments in thermodynamics—to which Tyndall had made contributions—led to a rigorous understanding of blackbody radiation, the process by which any object that sits above absolute zero gives off energy simply by virtue of its temperature.*

Thus, by 1896, the great Swedish chemist Svante Arrhenius had both the motivation and the necessary tools to take a first swipe at a comprehensive theory of the greenhouse, which he based on the principle of radiation balance. In April of that year, he presented a dense twenty-nine-page treatise, summarizing an estimated ten thousand to one hundred thousand hand calculations, in *The London, Edinburgh and Dublin Philosophical Magazine*. "I should certainly not have undertaken these tedious calculations," he wrote, "if an extraordinary interest had not been connected with them. In the Physical Society of Stockholm there have been occasionally very lively discussions on the probable causes of the Ice Age."

Arrhenius took his cues from Tyndall, who had placed central importance on the greenhouse effect of water vapor but, given the huge reservoir of liquid water available in the oceans, could not imagine a way for humidity to change without some other driver. Tyndall's notion was that a change in carbon dioxide would cause an initial change in temperature, humidity would then change accordingly, and the change in humidity would lead to a second change in temperature. This is the first step in a multistep process known as "water vapor feedback," which was more fully explored later. It is based on the fact that warm air has a greater capacity to hold water vapor than cold air, for the same reason that it is more humid in summer than in winter.

*Even a vacuum radiates, the most famous example being the Cosmic Microwave Background Radiation, which permeates all space and is believed to be a remnant of the Big Bang. The spectrum of the CMBR indicates that the temperature of empty space is about 3° Kelvin.

Arrhenius was no experimentalist, but he made extensive use of previous experiments by others, particularly the American Samuel Pierpont Langley, who had invented an instrument with the precious name "bolometer" to measure the intensity of radiation in the infrared absorption bands of carbon dioxide and water vapor that were known at the time.

And Langley takes us back to the mountains again. In 1883, he conducted the very first field experiments with his bolometer on the highest point in the continental United States, the summit of Mount Whitney in California. His purpose was to answer a question Tyndall had posed years before: the great man had wondered why sunlight contained so little infrared light at sea level and had speculated that it was being absorbed by water vapor in the overlying atmosphere. Langley sent Tyndall a letter from the top of the mountain to let him know that he was right: Langley had measured more infrared in the sunlight up there.

In his 1896 treatise, Arrhenius was concerned not with the absorption of light coming *in* from the sun but with the absorption of what he called "ultra red rays" flowing *out* from the Earth. He made ingenious use of another series of observations by Langley, on the absorption of infrared *moonlight* by the gases of the atmosphere.

After observing—quite accurately by today's thinking—that the earth's average temperature was about 15°C (60°F), Arrhenius wrote,

> In order to get an idea of how strongly the radiation of the earth (or any other body of the temperature +15°C.) is absorbed by quantities of water-vapour or carbonic acid in the proportions in which these gases are present in our atmosphere, one should, strictly speaking, arrange experiments on the absorption of heat from a body at 15° by means of appropriate quantities of both gases. But such experiments have not been made as yet, and, as they would require very expensive apparatus beyond that at my disposal, I have not been in a position to execute them. Fortunately there are other researches by Langley in his work "The Temperature of the Moon," with the aid of which it seems not impossible to determine the absorption of heat by aqueous vapour and by carbonic acid in precisely the conditions which occur in our atmosphere. He has measured the radiation of the full moon, (if the moon was not full, the necessary correction relative to this point was applied) at different heights and seasons of the year. This radiation was moreover dispersed in a spectrum. . . . Now the temperature of the moon is nearly the same as that of the earth, and the moon-rays have, as they arrive at the measuring-instruments, passed through layers of carbonic acid and of aqueous vapour of different thickness according to the height of the moon and the humidity of the air.

In other words, Langley's measurements of "obscure" radiation from the moon—undertaken for an entirely different purpose—were all Arrhenius needed to make sound estimates of how much the same radiation emanating from Earth would be absorbed by the water vapor and carbon dioxide in the Earth's atmosphere. It was an elegant finesse.

Of the five sections of Arrhenius's treatise, the one that has left the most enduring impression is found under the not particularly pithy heading "Calculation of the Variation of Temperature that would ensue in consequence of a given Variation of the Carbonic Acid in the Air." It was there that he made the first estimate of what has become the central parameter in global warming research, so-called climate sensitivity: the average change in temperature that would accompany a doubling of the carbon dioxide level in his day, which is now called the "preindustrial level." He put the change at between five and six degrees Celsius, or nine and eleven degrees Fahrenheit.

More than a century and tens of billions of dollars worth of research later, scientists are coming up with slightly lower numbers, but not by much.

A few modern researchers, especially certain modelers who have added an incredible amount of detail to their own computer simulations, argue that Arrhenius was simply lucky. As a physicist myself, I don't think so. Reducing a system to its basic elements is a standard approach in physics. It usually reveals the underlying dynamics and leads to surprisingly accurate predictions of that system's behavior. Although I haven't checked Arrhenius's ten thousand calculations, I suspect his work, which is about as close as one can get to a "back of the envelope calculation" in this complicated field, nails down the fundamentals of the greenhouse pretty well.

Many of his asides about secondary effects have also stood the test of time. He points out, for instance, that temperature changes should be greater over land than over the ocean, for land has a higher albedo (or reflectivity), and also that the melting of snow will tend to amplify changes at high latitudes, because melting decreases albedo, allowing more sunlight to be absorbed by the land and transformed into heat. This effect has become extremely obvious in the Arctic especially, over the last few decades.

Unfortunately, however, when it came to his main purpose, which was to explain the shifts between what he called "glacial" and "genial" epochs, Arrhenius's arrow landed very wide of the mark. Having established that variations in carbon dioxide would certainly have had the power to cause such shifts, he grasped at the nearest geological straw for a basic mechanism that would change carbon dioxide: random variations in volcanic activity. This speculative and utterly untestable idea would eventually prove wrong.

He based this flimsy argument primarily on the work of one of his Swedish colleagues, the geologist Arvid Högbom; while Högbom's work is simplistic in retrospect, it does demonstrate that a fundamental understanding of the so-called carbon cycle, that is, the various pathways by which carbon is added to the atmosphere and removed from it—into the oceans, the biosphere, and the solid earth—was coming into focus at the time.

Arrhenius's treatise includes a four-page excerpt from an 1894 memoir by Högbom in which the latter explains that on geological timescales carbon dioxide is added to the atmosphere mainly by volcanism and taken out mainly by the weathering of exposed rocks and minerals through rainfall. Water droplets draw the gas out of the air and carry it down to the rock surfaces, where it reacts to form carbonates, which are washed in streams and rivers to the oceans, where they sink to the bottom, then sink further still, first to form sedimentary rocks and finally to join the magma in the deep Earth and rejoin the volcanic cycle. These are still believed to be the primary natural processes in the carbon cycle, although research remains active in this field and many details have been added in the intervening century.

Högbom notes, by the way, that at that time coal was being burned at a rate of "in round numbers 500 millions of tons per annum" and that "[t]ransformed into carbonic acid, this quantity would correspond to about a thousandth part of the carbonic acid in the atmosphere"—a negligible percentage, both he and Arrhenius evidently believed.

IN 1903, ARRHENIUS WON A NOBEL PRIZE FOR UNRELATED WORK IN THE FIELD OF ELECtrochemistry. His reputation established, he then wrote a popular book called *Worlds in the Making*. His view of coal burning had changed in the seven years since he'd published his treatise, partially because the rate had almost doubled, to nearly a billion tons per year. He now believed that "the slight percentage of carbonic acid in the atmosphere may by the advances of industry be changed to a noticeable degree in the course of a few centuries." From this he anticipated a gradual warming, which he saw as a good thing: it might prevent the onset of "a new ice period that will drive us from our temperate countries into the hotter climates of Africa." He also guessed, on a less dramatic note, that "[b]y the influence of the increasing percentage of carbonic acid in the atmosphere we may hope to enjoy ages with more equable and better climates, especially as regards the colder regions of the earth, ages when the earth will bring forth much more abundant crops than at present, for the benefit of rapidly propagating mankind."

This cheeriness typified the prejudice common among nineteenth-century Europeans that their form of civilization had a beneficial influence extending even to the cosmos. Colonialism was at its height. There was a long history to the notion that the marks of civilization on the land actually improved the "air" and climate. The eighteenth-century French philosopher Charles-Louis de Montesquieu and the Scotsman David Hume believed that the clearing of the primeval forests had improved the climate in Europe since Roman times and that it held the promise of doing the same in North America—which, they realized, suffered much colder weather than similar latitudes in Europe. (They even believed the improved climate would make America more fit for Europeans and less so for the "savage" races they had found there.) In his *Notes on Virginia,* in which he frequently defended the flora and fauna of his homeland to his European contemporaries, Thomas Jefferson also bought into the idea that agriculture improved climate. His heartfelt patriotism even led him to believe that such an improvement was already happening: "A change in our climate . . . is taking place very sensibly. Both heats and colds are become much more moderate within the memory even of the middle-aged."

Thus one can understand Arrhenius's blind optimism. Besides, even this great man could not foresee the explosion in population and industrialization that would take place in the twentieth century, nor the astounding appetite for fossil fuel that would come with it. In a lecture in 1896, he estimated that it would take three thousand years for fossil fuel burning to double the preindustrial carbon dioxide level. It is now virtually certain that he was off by an order of magnitude. The level will probably double sometime this century, less than two hundred years after Arrhenius gave his lecture.

THE PREEMINENT AMERICAN GEOLOGIST OF THAT TIME, THOMAS CHROWDER CHAMBERLIN, was working on his own explanation for the ice ages when Arrhenius's treatise came across his desk. Chamberlin had developed a sophisticated model of the carbon cycle, which took considerably more geological evidence into account than Högbom's, so he found it difficult to accept Arrhenius's simplistic explanation involving volcanoes, but the treatise did convince him that variations in carbon dioxide might have the power to change temperatures by the required amount. Bringing the comprehensive scientific approach that was his trademark and a broad understanding of geological dynamics to the problem, Chamberlin soon came up with the idea of water vapor feedback.

In a 1905 letter, he wrote, "[W]ater vapor, confessedly the greatest thermal absorbent in the atmosphere, is dependent on temperature for its amount, and if another agent, as CO_2, not so dependent, raises the temperature of the surface, it calls into function a certain amount of water vapor which further absorbs heat, raises the temperature and calls forth more vapor, thus building a dynamic pyramid on its apex with consequent instability and demonstrative effects that attend the aqueous phenomena of the atmosphere."

This eloquent description of a positive feedback mechanism marks the first of what has now grown into nearly an endless list of feedbacks in the climate system, both positive and negative. The present uncertainty in estimating climate sensitivity, in fact, arises mainly from the difficulty in taking the feedbacks into account, not in estimating the direct effect of greenhouse gases. (Moreover, the uncertainty in estimating future temperature changes, so frequently underscored by greenhouse contrarians, stems as much from the difficulty in predicting the economic developments that will affect fossil fuel use as from climate science.)

So, as carbon dioxide increases, temperature rises, which adds more water vapor to the atmosphere, which raises temperatures again, which brings more vapor into the atmosphere—and more carbon dioxide, because the gas will tend to bubble out of the ocean as its surface warms. But if increased moisture were to make more clouds, the dynamics would change: white clouds reflect the sun's rays back into space before they can reach the earth's surface and heat it up—a counterbalancing negative feedback—whereas dark clouds might absorb sunlight and promote warming.

Increased carbon dioxide, humidity, and the attendant increase in precipitation will tend to enhance plant growth, building more extensive forests, for instance. This creates a "carbon sink," which is a negative feedback, because all plants "inhale" carbon dioxide during photosynthesis. As they grow, they incorporate it into their structures and tissues, making it unavailable to the air—unless the plants are slashed and burned, of course. On the other hand, forests and grasslands tend to add greenhouse water vapor to the air around and above them, and their dark leaves absorb sunlight—in contrast with bare earth, which generally reflects light. These are smaller positive feedbacks.

Arrhenius mentioned in passing snow's positive feedback effect, but he overlooked the extremely strong positive effect of sea ice, which plays a major role in one of the more intriguing ideas to have emerged in climate science lately: the so-called Snowball Earth hypothesis.

IN 1998, IN A SINGLE BOLD STROKE, A GROUP OF GEOLOGISTS AND CLIMATOLOGISTS, primarily from Harvard, unified a number of puzzling observations in fields as diverse as geology and evolutionary biology by suggesting that the earth may have experienced a series of "freeze-fry" cycles, lasting ten or so million years each, during the late Neoproterozoic era, six or seven hundred million years ago.

First, they suggested, tectonic and chemical processes conspired to scrub carbon dioxide from the atmosphere. This diminished the power of the greenhouse, cooled the air, and caused the great ice caps to begin creeping from the poles toward the equator. A positive feedback then set in: since the ice would reflect more and more sunlight as it grew, its growth would have sped the cooling. Simulations show that when the ice cover on the oceans reaches about thirty degrees latitude, north and south, the feedback runs away: temperatures plummet, the ocean surface freezes all the way to the equator, and the planet's average temperature drops to about –60°F (–50°C).*

During this "freeze" phase of the cycle, however, the volcanoes that still poked their smoky heads through the ice would have continued to add carbon dioxide to the air. With the global blanket of ice snuffing out the usual processes for removing the gas from the atmosphere (covering the rocks that react to form carbonates, mainly), the volcanoes, according to this argument, would slowly have boosted the level to about 350 times that of today and produced a scorching greenhouse effect. And as the air warmed, the sea ice would have reasserted its positive feedback by retreating: the dark ocean it revealed would have absorbed more sunlight and sped the warming. On the "fry" side, temperatures are thought to have risen as high as 120°F (50°C).

Ancient rock layers in places as far-flung as Namibia and the Svalbard Archipelago in the Scandinavian Arctic suggest that this freeze-fry cycle repeated itself at least four times.

There has been furious opposition to this theory—as there is to most bold ideas—but its authors have answers to every objection, and it is gaining ground faster than most such outlandish notions. It took ten years, for example, for the majority to accept the idea that a meteor impact may have wiped out the dinosaurs about fifty million years ago; whereas, Snowball Earth's central proposition, that tropical seas were probably covered with ice during the late Neoproterozoic, was verified to the satisfaction of most geologists in 1999, only one year after the hypothesis was proposed.

*The reason this temperature is fifty-nine degrees lower than the radiation balance temperature mentioned earlier is that an ice-covered surface would reflect much more sunlight than today's surface does. This shows just how strong the feedbacks induced by an initial change in carbon dioxide might be.

• • •

THE RETREAT OF SEA ICE IN THE ARCTIC SEEMS TO HAVE BEEN A MAJOR FACTOR IN Alaska's recent warming, and that warming, in turn, has led to another positive feedback: tundra that has been frozen for more than one hundred thousand years is melting to release untold tons of sequestered carbon dioxide into the air.

Feedbacks are everywhere: as the surface of the ocean is exposed, its positive feedback is offset to some extent by two negative feedbacks: the exposed water absorbs carbon dioxide directly; it also provides an environment for the proliferation of plankton, which absorb carbon dioxide through photosynthesis.

A change in temperature begins with a change in radiative forcing—an increase in the carbon dioxide load, for instance, or a change in "insolation," the amount of sunlight coming in. Some feedbacks begin immediately, while others may take decades or even centuries to kick in, because a feedback is really a secondary response to a change in temperature. The oceans also respond slowly, for they are capable of absorbing a tremendous amount of heat before changing much in temperature. Only in the year 2000 did measurements first show that the temperature of the oceans is now rising. That means there is quite a bit of heat "in the pipeline" right now. Temperatures are just beginning to adjust to the forcing that the greenhouse buildup of the twentieth century has already put in place. Not only will they continue to rise even in the unlikely event that greenhouse levels stay constant, but new feedbacks, similar to the release of carbon dioxide from the melting tundra, will continue to unfold. We have started a very large ball rolling.

BUT AT ABOUT THE TIME ARRHENIUS MOVED ON TO OTHER THINGS—AND JUST AS THE automobile began to proliferate and add the burning of oil to the skyrocketing burning of coal—his and Chamberlin's carbon dioxide theory fell out of favor. It is difficult to understand why; in retrospect, the arguments against it were based on flimsy evidence and faulty reasoning, but the excuse came from spectroscopy, which was still in a primitive stage at the time. For more than fifty years, the average geologist or meteorologist would hold the mistaken belief, based on the poor resolution of early spectrometers and a few poorly designed experiments, that the infrared absorption bands of water vapor overlapped those of carbon dioxide, so that the vapor would absorb all the earth's emissions before carbon dioxide got the chance. Other faulty experiments led scientists to believe that there was enough carbon dioxide in the air to absorb all the infrared in its own absorption bands anyway, so even carbon dioxide's own greenhouse effect was saturated: an increase would make no difference. Chamberlin made a few attempts to inject some sense into this discussion, but even he eventually came to believe he had overestimated the effect.

And so industry marched on. By the 1930s tens of millions of automobiles were tooling around on the roads of America, and the northern hemisphere found itself in the grip of a decade-long heat spell commonly known as the Dust Bowl.

CARBON DIOXIDE THEORY WAS REVIVED NEAR THE END OF THAT HARSH DECADE BY AN obscure "steam technologist to the British Electrical and Allied Industries Research Association" named Guy Stewart Callendar, who pursued meteorology as a hobby. In 1938, Callendar presented a paper at a London dinner meeting of the Royal Meteorological Society that began,

> Few of those familiar with the natural heat exchanges of the atmosphere, which go into the making of our climates and weather, would be prepared to admit that the activities of man could have any influence upon phenomena of so vast a scale.
>
> In the following paper I hope to show that such influence is not only possible, but is actually occurring at the present time.

Reviewing historical measurements of atmospheric carbon dioxide from the eastern United States and Kew Gardens, southwest of London, Callendar showed that the average global level appeared to have risen about 6 percent since the turn of the century. This squared with his estimate of the 150 billion tons that fossil fuel burning would have added to the atmosphere in that time, assuming the oceans absorbed the gas with a characteristic time of about two thousand years, so that three-quarters of the emitted carbon dioxide would have remained in the air. (Arrhenius had it the other way around. He believed "the sea . . . acts as a regulator of huge capacity, which takes up about five-sixths of the produced carbonic acid.")

Reviewing temperature records from two hundred weather stations around the globe, Callendar also found that the average temperature at the earth's surface had risen about half a degree in the previous fifty years. With the aid of his own "sky radiation" model, which represented a slight improvement over Arrhenius's, he concluded that fossil fuel burning had probably caused the warming. In fact, the weather stations showed a greater increase in temperature than he calculated from the carbon dioxide increase.

When it came to the general implications of all this, however, Callendar carried on in the cheery tradition of Arrhenius, and for more or less the same reasons: because the residents of northern latitudes would probably welcome warmer air and have an easier time growing crops. He also shared Arrhenius's long-term view that a warming might delay a "return of the deadly glaciers."

Working alone, completely unsupported by outside funds and in his spare time, Callendar studied all aspects of the problem until his death in 1964. His theoretical work throughout was pretty much spot-on, and even at the beginning of his career his insights into the greenhouse effect were roughly half a century ahead of his time.

In a 1939 essay tracking carbon dioxide levels through geological time, he pointed out that a "marked increase [in temperature] commenced from about the time that carbon dioxide production became rapid" (roughly 1890, in his estimate) and that the "five years 1934–1938 are easily the warmest such period at several stations whose records commenced up to 180 years ago."

The centerpiece of the United Nations report issued in 2001 is a graph of the average temperature of the northern hemisphere over the past thousand years. There are many wobbles, but the temperature drops steadily from A.D. 1000 until about 1900, when it suddenly begins to spike. The curve is shaped something like a hockey stick balanced on its edge: a long straight line with a short upward bend at the end, which spans the twentieth century, precisely. And in a contentious debate about global warming that took place on the floor of the U.S. Senate in July 2003, that's what this curve came to be called: the "Hockey Stick."

Scientists are a cautious bunch; I doubt there will ever be a consensus that the human effect began in the late 1800s, but many respected scientists voice that opinion, and it has been confirmed by sophisticated computer models as well.

There was a break in Callendar's output during World War II, when he was assigned to the research staff of a government armaments laboratory, where he would work until his retirement. In 1949, he produced a brief essay for the general reader entitled "Can Carbon Dioxide Influence Climate?" Among a number of near prophesies in this paper is one that Lonnie Thompson would confirm simply by walking to Quelccaya's base camp in 1984: greenhouse warming would be most pronounced at altitudes in the mid-troposphere, so that "a recession of glaciers on high tropical mountains might be one of the first indications." A computer simulation would support this prediction forty years after Callendar offhandedly tossed it out.

Pointing out that "the pitfalls which await the uncritical who venture into this field are numerous," he suggested that there may be some unexpected effects as well: on the Antarctic continent, "where temperatures are below the freezing point at all seasons . . . an increase in the amount of ice . . . might be the ultimate result of higher world temperatures." Recent studies by Ellen Thompson, among others, bear this out.

But those few scientists who paid any attention to Callendar's work—or indeed to the bastard son of meteorology that was climatology at the time—tended to regard it as a curiosity at best. (One suspects it had the aspect of a parlor game even to Callendar.) Most believed, as Arrhenius had in the previous century, that the sea stood at the ready, a sort of benign geological housekeeper, to vacuum away all the excess carbon dioxide we were dumping into the sky. As one contemporary textbook declared, "We can say with confidence [that climate] is not influenced by the activities of man except locally and transiently."

As scientists often do, unknowingly, these early skeptics of the carbon dioxide effect were again dabbling in cosmology, silently invoking what can only be called a divine force for balance. Despite ample evidence that climate had swung dramatically in the past—during the ice ages most obviously—they had sought and found comfort in the faith that Nature works to keep things the way they are. Half a century later, virtually all climate scientists have long since abandoned that faith, while the modern folks who like to call themselves skeptics, but would more aptly be termed contrarians, frequently invoke it at congressional hearings, for example, to appeal to the "common sense" of elected officials. I suspect this faith also lies behind the tired refrain that if there *is* a change taking place, it must be a "natural variation."

Yes, there have been natural variations in the past. Do we have to outpace the fry cycle of Snowball Earth—average temperature 120°—before we'll accept the idea that *we* might be the agents of change?

IT WAS CLEAR TO SOME THAT THE ARCTIC WAS WARMING EVEN IN CALLENDAR'S TIME. AT Callendar's dinner talk in 1938, for instance, a meteorologist in the audience pointed out that temperatures had risen ten times more in the Arctic since the turn of the century than they had at lower latitudes, and that Arctic sea ice had retreated dramatically as well.

Over the next decade or so, other observant individuals would notice the changes in the Arctic, and so would countless mute creatures of other species. In his book *Arctic Dreams*, Barry Lopez relates that the Inuit on Baffin Island first noticed an influx of the American robin sometime around 1942. (They believed the birds had come north to escape "a lot of fighting in the south," i.e., World War II.)

In her 1951 book *The Sea Around Us,* marine biologist Rachel Carson pointed out that many bird species, among them the Baltimore oriole, the greater yellowlegs, and the American avocet, had spread from the midlatitudes into Arctic habitats during the first half of the twentieth century. The common cod first appeared in the waters near Greenland in 1912, and constituted a major North Atlantic fishery by the 1930s. Meanwhile, according to Carson, the populations of certain high Arctic birds, such as the northern horned lark and the gray plover, "showed their distaste for the warmer temperatures by visiting Greenland in sharply decreasing numbers."

"It is now beyond question," Carson wrote, "that a definite change in arctic climate set in about 1900, that it became astonishingly marked about 1930, and that it is now spreading into sub-arctic and temperate regions. The frigid top of the world is very clearly warming up." But she did not connect this change to the greenhouse; she subscribed to the obscure notion of the Swedish oceanographer Otto Pettersson that Arctic climate was controlled by the strength of the tides, which were modulated in turn by periodic conjunctions of the sun and moon.

There were many equally obscure theories to choose from at that point, and the science of that era was not capable of deciding between them. As one meteorologist quipped, borrowing from Kipling, "There are at least nine and sixty ways of constructing a theory of climate change, and there is probably some truth in quite a number of them."

Thus, three years after the end of World War II, the only person who appears to have taken the carbon dioxide theory seriously was an obscure hobbyist who worked for the defense establishment during the day and pored over old documents at his home desk at night, with no resources to do the experiments that might prove the theory right or wrong. Callendar had all the pieces to the puzzle, but they weren't quite the right size and shape and didn't quite fit together. The obvious thing would have been to get better measurements of atmospheric carbon dioxide—an expensive proposition, admittedly. This would have answered the basic question about ocean uptake and rid Callendar of the only truly questionable aspect of his research.

· 7 ·

THE GREAT EXPERIMENT

In his personal calendar, Buckminster Fuller noted the August day in 1945 on which the bomb was dropped on Hiroshima as "the day that humanity started taking its final exam." Oddly enough, atomic bomb research would lead quite directly to the subsequent realization that global warming was a second and equally important question on the same exam.

Physicist Gilbert Plass worked during the war years at the University of Chicago, in a Manhattan Project laboratory devoted to the production of strategic radioactive metals, such as enriched uranium and plutonium. (It was in a squash court under the stands of the university's football stadium that Enrico Fermi had engineered the first man-made nuclear chain reaction on December 2, 1942.) When the war ended, Plass took a position at Johns Hopkins University to pursue a second interest: infrared light.

Upon learning during the war that all warm objects—including warm-blooded animals such as humans—give off invisible infrared radiation, certain military officials became excited about the implications for advanced weapons development. Infrared light can be used to "see in the dark." During the war it was used to aid snipers, for instance. Think also of "heat-seeking" missiles. Industry had its applications as well, so there were big advances in infrared science and technology in the years during and just after the war.

Plass was studying the transmission of infrared light through the atmosphere—a process that was obviously crucial to weapons development and that happened to lie at the heart of greenhouse science as well. Every paper he published on this work, incidentally, not to mention nearly every other paper related to the greenhouse effect published between 1945 and 1960 (90 percent of which came from the United States, since most other industrialized nations were still rebuilding from the war), bears a revealing footnote: "This work was sponsored by the Office of Naval Research." As the Cold War heated up, the navy was first to step into the funding vacuum that had settled in briefly at World War II's end. The other branches of the military soon followed suit, but for the next forty years, until the fall of the Berlin Wall, ONR would lead the way in funding an astounding variety of research projects, many of which could be connected to any military use at all only with the aid of a vivid imagination.

With the infusion of Cold War money for infrared research an error was discovered in the spectroscopy studies that everyone but Guy Callendar had used to dismiss carbon dioxide theory for nearly half a century: the studies had been carried out at much higher pressures than those of the atmosphere. At high pressures, the molecules of a gas collide with one another more frequently, causing the hundreds of infrared absorption *lines* of carbon dioxide and water vapor to "smear out" into *bands*. In the range of wavelengths in which the earth gives off most of its blackbody energy both of the molecules' absorption spectra look something like roller coasters. Not only do the absorption bands for water vapor—that is, the hills on the roller coaster—overlap one another, but if you were to lay the spectrum for water vapor over carbon dioxide's, the water bands would also cover most of those in the carbon dioxide spectrum.

At the lower, more relevant pressures of the atmosphere, the two roller coasters resolve into rather poorly made "picket fences," series of absorption lines in which the former hills and valleys determine the heights of the pickets (that is, the absorbing power of the lines), and the pickets are, for our purposes, more or less randomly placed. So, if the two fences were lined up side by side and you were to try to look sideways through them, together their "pickets" would block most of the view; however, you would still be able to see most of the pickets on both fences. In other words, the two absorption spectra don't interfere much. This means carbon dioxide and water vapor will absorb the earth's blackbody radiation almost independently; absorption by one will not significantly affect absorption by the other—just as Arrhenius and Chamberlin had naïvely supposed.

New research also showed that the percentage of water vapor in the atmosphere falls off very rapidly with altitude, while the percentage of carbon dioxide remains more or less constant very high into the stratosphere. Thus, the two main absorbers of the earth's radiant heat act as blankets lying on top of one another. What passes through the low-lying vapor—or is radiated from it by virtue of its own heat—is free to be absorbed by the blanket of carbon dioxide lying above.

A SEEMINGLY UNRELATED DEVELOPMENT, ALSO FOSTERED BY THE COLD WAR, WAS THE invention of the computer. The U.S. Army had tried its very best to develop the first electronic digital computer during the war, but had missed by exactly six months. The ENIAC (Electronic Numerical Integrator And Computer) was formally dedicated at the University of Pennsylvania in February 1946. Its first task was to calculate trajectories for ballistic projectiles—from large ship- or land-based guns, for instance. Soon after that it was recruited to perform theoretical calculations for the first hydrogen, or thermonuclear, bomb. Its third task was to simulate weather. The last two ideas sprang from the mind of John von Neumann, the brilliant Hungarian mathematician who had been in on the development of the ENIAC from the start. That he was the most highly regarded hydrodynamics expert of the time is demonstrated by the fact that he had been hired as a consultant to the Manhattan Project, specifically to design the so-called implosion lenses for the bombs that exploded over Hiroshima and Nagasaki. These were symmetrical arrays of high explosives, which, when detonated, produced an imploding "shock wave" that triggered the nuclear chain reaction. It was obvious to von Neumann, if not to many others, that the mushroom cloud from a bomb and the swirling clouds of a weather system were similar problems in "non-linear fluid dynamics."

Von Neumann's brilliance also extended to fund-raising, and he understood the greenhouse effect well enough to realize that it could be manipulated to control weather. Arguing that "climatological warfare" might eventually prove more powerful than the nuclear variety, he managed to secure funding from the army, the air force, and the navy simultaneously—a hat trick that any research scientist would envy—to build his own ENIAC at the Institute for Advanced Study in Princeton, and to recruit a formidable scientific team to work on his now-legendary Meteorological Project. The first thing the team did was to simulate, rather successfully, a memorable cyclone that had swamped most of the eastern United States on Thanksgiving Day 1950.

Von Neumann expected computers eventually to allow us to predict and possibly control long-term weather, but this expectation bumped hard against reality in the early sixties, when Edward Lorenz of MIT discovered that a simple weather forecasting program he had written for a primitive computer produced wildly different forecasts when very slight changes were made to the initial weather conditions—the so-called Butterfly Effect.

Thus von Neumann also happens to have planted the seed for the new science of chaos, or complexity.

GILBERT PLASS PUT TWO AND TWO TOGETHER. DURING A SABBATICAL AT THE UNIVERsity of Michigan over the 1954/1955 academic year, he used a second-generation computer to apply the new knowledge in infrared physics and the atmospheric layering of water vapor and carbon dioxide to the study of the greenhouse.

His main purpose, as usual, was to explain the ice ages (and in odd parallel with Callendar, he wrote up his ideas during a long break from academia, while working for a defense contractor in Southern California). Obtaining an estimate for climate sensitivity that lay midway between those of Arrhenius and Callendar, Plass concluded, as had Tyndall and Arrhenius before him, that variations in carbon dioxide could account for all the climate changes in the geological record. The carbon cycle was better understood by then—Plass's picture didn't rely solely on spurts of volcanism. Still, he didn't add much, conceptually, to the other scientists' earlier thinking.

Plass's work was nevertheless important for two reasons. For one, mainstream scientists found his radiation model believable. It was seen as the first rigorous proof that a rise in carbon dioxide would lead inevitably to a rise in temperature. For another, he forsook divine stability. Perhaps the bomb had awakened him, as it had many people, to the possibility that mankind might be as strong as a force of nature, and not necessarily a benign one.

By then, he pointed out, fossil fuel emissions had grown to six billion tons per year, and agricultural activity and the clearing of forests were adding an unknown amount more. The combined contributions of mankind were "several orders of magnitude larger than any factor that contributes to the carbon dioxide balance from the [natural] world at the present time," he wrote. His own model of ocean uptake suggested that it would take about ten thousand years for a quick pulse of carbon dioxide to be reabsorbed; for all intents and purposes, our emissions were staying in the air. If so, "there will be 30% more carbon dioxide in the atmosphere at the end of the twentieth century than at the beginning. If no other factors change, man's activities are increasing the average temperature by 1.1°C per century." He wasn't far off on either count.

And our behavior since Plass made these points seems to confirm his belief that we would ultimately burn all the coal and oil we could lay our hands on. Assuming all known reserves—about thirteen trillion tons—would be burned in less than a thousand years, he found that carbon dioxide would eventually rise to seventeen times its preindustrial level. Even if his radiation model was off by a factor of two, that was a scary thought: temperatures would rise by more than thirteen degrees (seven degrees Celsius), significantly more than the difference between now and the depths of the Wisconsinan glacial stage.

"Thus," Plass concluded about fifty years ago, "the accumulation of carbon dioxide in the atmosphere is seen to be a very serious problem over periods of the order of several centuries. It is interesting that two of the most important methods available at the present time for generating large amounts of power have serious disadvantages when used over long time intervals. The burning of fossil fuels increases the temperature of the earth from the carbon dioxide effect; the use of nuclear reactors increases the radioactivity of the earth. It is difficult to say which of these effects would be the less objectionable after several centuries of operation."

So Gilbert Plass stands as a postwar mirror image of Guy Callendar, reflected across the Atlantic, both working for the defense industry during the day and moonlighting on the greenhouse at night, except that Plass's vision of industrial progress—foul contamination, gross interference in the cycles of nature—was significantly less rosy than Callendar's quaint old Victorian still life.

Plass also had the good sense to realize that it was all still theory at that point: two experimental questions begged clear answers. One would be answered inevitably, he believed: "The temperature trend during the remainder of this century should provide a definitive test of the relative importance of . . . carbon dioxide . . . in determining climate at the present time." But the other required action: "Unfortunately, we can not even say with certainty whether or not the carbon dioxide content of the air has increased since the year 1900."

The glimmerings of an answer to the second question would come from the unlikely direction of radiocarbon dating.

WILLARD LIBBY, ANOTHER ALUMNUS OF THE MANHATTAN PROJECT, WORKED AT COLUMBIA University during the war, also devising ways of producing strategic nuclear metals.* Just before the war's end, he moved to the Enrico Fermi Institute for Nuclear Studies at the University of Chicago, where, in 1947, he conceived of a way to find the age of things that were once alive by measuring their radioactive carbon-14 content. It goes something like this:

*Producing enough weapons-grade uranium and plutonium was the most difficult aspect of making a bomb, and the lion's share of the material resources in the Manhattan Project was devoted to that task.

The Earth is constantly bombarded with cosmic rays from outer space. Some collide with the nuclei of atoms in the upper atmosphere and knock out neutrons. A fraction of those neutrons will be absorbed by nitrogen, the most abundant gas in the atmosphere, and those nitrogen nuclei will subsequently decay into carbon-14. Single carbon-14 atoms are highly reactive chemically and oxidize to carbon dioxide pretty much the moment they are born. As we have seen, carbon dioxide is produced by other processes as well, so that, on average, the radioactive carbon dioxide produced by cosmic rays will make up only about one one-billionth of all the carbon dioxide in the atmosphere.

Libby hypothesized that living plants and animals incorporate carbon into their bodies with exactly the ratio of radioactive, or "live," carbon to its stable, or "dead," counterpart that exists in the air while they are alive. Plants take up carbon through photosynthesis; animals do so less elegantly, by eating plants and one another.

Once an organism dies, the live isotope will disappear from the bones, stems, or shells it leaves behind at a rate determined by the isotope's radioactive half-life: 5,730 years. (Half remains after one half-life, a quarter after two, an eighth after three, and so on.) Thus, the percentage of live carbon found in an old tree limb is a measure of how long ago the tree died. The technique works well for samples between five hundred and forty-five thousand years old. (Now, keep in mind that coal and oil, being the remnants of forests and other large stands of vegetation that have lain in the earth for hundreds of millions of years, contain only dead carbon, the live fraction having long since decayed away. Maybe you can see where this is headed. . . .)

The first fields to make wide use of Libby's technique were archeology and anthropology. Among the first samples he tested were pieces of charcoal and wood from Egyptian graves. Geologists quickly applied the technique to the organic debris that had been left by the retreating ice caps in North America and Europe and discovered that the Wisconsinan glacial stage had ended at the same time, about eleven thousand years ago, as the glacial stage known for historical reasons as the Würm ended in northern Europe; in other words, they were the same event—now sometimes called the Wisconsinan/Würm. Of course, Lonnie's perspicacious mentor, John Mercer, used the technique right away on his explorations in Patagonia.

CARBON DATING MADE ITS MARK ON GREENHOUSE SCIENCE THROUGH THE WORK OF Hans Suess, perhaps the most brilliant individual to veer into climatology in the mid-fifties, and a man who led an exemplary and highly productive life despite severe personal and ethical trials.

Suess was born in Vienna in 1909, the son and grandson of geologists. He earned a Ph.D. at the University of Vienna in 1935 and became a professor of physical chemistry at the University of Hamburg in 1938. One of his research interests was deuterium, or heavy hydrogen, a substance that, unfortunately for him, plays a peripheral role in nuclear fission reactors. Suess had little choice but to be recruited into the fitful Nazi atomic energy program, led by Werner Heisenberg, a towering and, by virtue of his Nazi sympathies, controversial figure in twentieth-century physics.

In the delicate position of scientific advisor to the heavy water* production plant in Rjukan, Norway—which became part of Heisenberg's program when that country fell under Nazi occupation—Suess made a close friend in the Norwegian resistance, through whom he provided intelligence to the British. Despite Suess's assurances that the Rjukan plant could never have led to the creation of a German bomb, the Allies bombed the plant in November 1943. In his book *Heisenberg's War,* Thomas Powers writes, "Of the many German scientists who were the source of information reaching Allied intelligence authorities at one time or another during World War II, Suess is the only one to have confessed the act directly in public." (After the war, of course.)

Even through war's chaos and the desolation that remained of German science once the war was done, Suess's mind was at work on big ideas. Less than two years after the German surrender, he proposed a theory of the relative abundance of chemical elements in the solar system, based on a fundamental understanding of nuclear reactions in stars. The theory's prediction of the fairly regular way in which the abundance of a particular element depends upon its mass has proven to be essentially correct. He then collaborated with a German colleague, Hans Jensen, on a seminal paper concerning the shell model of the atomic nucleus, for the further development of which Jensen later shared a Nobel Prize. Such were the circles in which Hans Suess traveled.

In 1949, Harrison Brown (who figures later) invited Suess to join Chicago's Institute for Nuclear Studies as a research fellow. Libby was there at the time, so Suess began to apply his knowledge of nuclear cosmogenesis to the job of perfecting the new dating technique.

There are subtleties to the interpretation of radiocarbon data. What, for example, if carbon-14 has not always made up the same fraction of atmospheric carbon as it does now? Suess was looking into that sort of question.

In 1953, having moved to a position with the U.S. Geological Survey, he stumbled across a surprising contradiction when he happened to compare

*Heavy water is water that contains deuterium in place of normal hydrogen.

the radioactivity of tree rings from the previous ten to fifteen years with others from the 1800s: the old rings contained about 2 percent *more* carbon-14, when, if they had formed in an atmosphere with the same fraction as the contemporary rings, they should have contained about 1 percent less: that much of the original material would have decayed. From this simple result Suess deduced that dead carbon dioxide produced by the burning of coal and oil must have been diluting out the natural, live portion in the atmosphere since at least the mid-1800s—in other words, some anthropogenic carbon dioxide must be staying in the air. This dilution process has since been named the Suess Effect.

From the modern drop in radioactivity, Suess made what he confessed to be a rough estimate of the characteristic time it would take a molecule of carbon dioxide emitted into the air to be reabsorbed by the sea: between twenty and fifty years—much shorter a time than the estimates of Callendar and Plass, which Suess believed to be "for various reasons . . . somewhat improbable."

Although he admitted that "other factors neglected here will have to be considered for a more quantitative explanation," he eventually realized that this was an understatement. The sea would turn out to be much more complicated a beast than either he or anyone else thought at the time, and in the end it would turn out that the guesses of Callendar and Plass were more accurate than his own. A decade later, after progress had been made by other means, he would run a second, more careful, radiocarbon study and come up with an estimate closer to theirs.

Still, his basic discovery cast the first serious doubt upon the belief that the sea was doing our dirty work for us, and the fact that it came from the left field of radiocarbon dating shows just how little faith the scientific community had in the art of direct carbon dioxide measurement at the time. The resolution of that issue would progress by fits and starts over a period of about six years, involve scientists from around the world, and result in the type of paradigm shift that they live for.

A PROMINENT ROLE WOULD BE PLAYED BY ROGER REVELLE, THE COMPLEX AND NEARLY omniscient director of the Scripps Institution of Oceanography, in La Jolla, California, who held rather an expansive view of his field, having once defined it as "whatever anyone at Scripps does." Such openness would certainly have appealed to a man of Suess's diverse interests, so in 1955 Revelle succeeded in luring him to a permanent position at Scripps. His first task was to set up a radiocarbon laboratory.

Revelle had been interested in the ocean's carbon dioxide uptake long before it would become the central issue in greenhouse science. He had studied carbonate levels in Pacific seabed cores (some of which he had dredged from the bottom himself on research cruises aboard the Research Vessel *Scripps*) in the 1930s for his doctoral thesis. That early work made him aware of the complex chemical buffering mechanism that governs seawater's ability to absorb carbon dioxide. Even that long ago he had begun to suspect that as much as half the carbon dioxide we were sending into the sky might be staying there. His subsequent work wove the next unlikely thread into the greenhouse fabric: nuclear fallout.

Having fallen in love with the sea in his student days, Revelle joined the navy in World War II and eventually rose to a senior position in the management of military research. A few months after the Allied victory over Japan, he was assigned to Joint Task Force One, the coordinated military command in charge of Operation Crossroads, the first postwar atomic bomb test, on the Bikini Atoll (which also led to a social explosion: a certain type of bathing suit was invented and named right then). Revelle directed a team of eighty scientists studying the oceanographic effects of the explosion: everything from its potential to cause tsunamis and destroy fisheries to the diffusion of fallout under the sea—the last of which rekindled his interest in carbon dioxide uptake. At about that time, his private papers began to include calculations on carbonate formation in the shallow pools of the atolls. He would work on this perplexing problem for nearly twenty years without solving it to his satisfaction.

As an example of Revelle's omniscience—and the "revolving door" aspect of government work even then—he was one of the small group of officers who helped found the Office of Naval Research just after the war. Later, as head of its geophysics branch, he allocated many resources to his former institution, including the long-term loan of two naval vessels; then, in 1948, he left the navy to become Scripps's assistant director, thus benefiting greatly from his own largesse.

In 1955 the military set off a "depth bomb" a few thousand feet under the ocean's surface; when ONR awarded a contract to Revelle to study the bomb's fallout, he found, to his surprise, that the radioactive material did not mix in the water very well. It diffused in a layer only three feet thick and spread over an area of forty square miles. "[R]adioactive wastes introduced into the upper layer might remain there for many years," he wrote, "and would be diluted by a volume of water only a fiftieth to a hundredth the volume of the oceans." He realized that this principle would apply to the carbon dioxide diffusing in from the atmosphere as well.

AT ABOUT THIS TIME, THE NEED AROSE TO DISPOSE OF THE PRODIGIOUS QUANTITY OF nuclear waste that would soon pour forth from civilian nuclear power plants. (The first went online in 1957.) In retrospect, it is horrifying to learn that one early idea was to dump it on the ocean floor. Thus Revelle began to pull down funding not only from ONR but also from the newly formed Atomic Energy Commission and from the weapons labs, which produced radioactive waste in the manufacture of strategic nuclear metals.

He was a compulsive collaborator and an ambitious administrator. In the same year that he hired Suess, he hired Harmon Craig, yet another alumnus of Chicago's Institute for Nuclear Studies, also to work on carbon dating. Craig once described the excitement of those days:

"Revelle used to walk around the halls at night; he'd drop in on people's labs and sit down and start talking science. Or he'd bang on your door at midnight and say, 'Hey, I've got an idea!' There was a feeling that the place had a special mission to keep research ships at sea. There were institutional expeditions, in which we'd sail to six or seven different places and measure everything. There were summer expeditions for students. The place had a vision, a feeling that we were pushing at the frontiers of science."

Nowadays, this sort of grand "fishing expedition" is frowned upon by the funding agencies. The influx of management practices from the business world has led them to demand extreme focus in grant proposals, clear explanations of exactly what the research will accomplish and why it will be important. Craig, who in 1998 received the Balzan Prize—the equivalent of a Nobel in the natural sciences—complained, "I'm being damned by reviewers who write that I don't use the scientific method—that is, going in with some nice, carefully outlined hypothesis in which you pretty much know in advance what you're going to find out and what you're going to do with the information after you find it. Well, the scientific method is what you learned in *My Weekly Reader* when you were in grammar school, but no first-rate scientist uses it. I say, if I knew what I was going to find out, I wouldn't do that study; I would do something else." (A bit crankier than Lonnie, perhaps, but coming from the same place philosophically.)

So Revelle, Suess, and Craig carried on in the tradition of Callendar and Plass: they worked on nuclear waste disposal to pay their bills, while moonlighting on their real interest, which was carbon dioxide uptake. Meanwhile, a second collaboration was doing the same thing. Ernest Anderson at the Los Alamos weapons laboratory was funded by AEC, and James Arnold at Princeton was funded by the army's Office of Ordnance Research, the same organization that had funded the ENIAC computer. The two collaborations soon caught wind of each other and began corresponding, talking on the phone and paying visits to each others' labs, justifying the expenses to their granting agencies on the basis of their common interest in nuclear waste disposal.

In 1957, the five men published three back-to-back papers in the same issue of the Swedish geophysics journal *Tellus,* Arnold and Anderson writing one, Revelle and Suess the second, and Craig the third. All three groups took the next obvious step after Suess's earlier discovery and compared the radiocarbon age of contemporary wood with that of shells and other biological matter from the ocean, believing this would yield a more direct measurement of uptake.

Although they employed very different models of the ocean and atmosphere, the studies produced nearly identical numbers not only for the atmospheric lifetime of a carbon dioxide molecule but also for the time it would take surface waters to mix with the deep ocean—a concordance the five authors found reassuring. It is not unusual for creative minds to come up with completely different apparatuses to mimic a complex problem and, once they've plugged in their experimental numbers and turned the mathematical crank, to find that the different systems nonetheless produce similar answers. This is one of the more mysterious and beautiful aspects of mathematical physics.

The five concluded that the atmospheric lifetime of a carbon dioxide molecule was about ten years and that ocean mixing would take several hundred years more—short times compared to the estimates of Callendar and Plass. Thus, wrote Revelle and Suess, "most of the CO_2 released by artificial fuel combustion since the beginning of the industrial revolution must have been absorbed by the oceans."

This was a step backward, toward Arrhenius . . . but the odd thing is that Revelle and Suess have gone down in the history books for saying exactly the opposite. Even today, the Web site for Environmental Defense, a nonprofit environmental group, has them "report[ing] for the first time that much of the CO_2 emitted to the atmosphere is not absorbed by the oceans." And a certain resonant but, in truth, noncommittal passage in which Revelle (who probably supplied the rhetoric) referred to fossil fuel burning as "a large-scale geophysical experiment" has taken on the weight of prophesy, when in fact it was nothing of the sort. Revelle said later that he had not been particularly concerned about global warming when he wrote the paper.

On its face, Revelle and Suess's paper used a tiny bit of new experimental information to come to a misleading and previously demonstrated conclusion. As it turns out, however, the two scientists were expressing serious doubts in private. They were concerned that they had considered only the steady state, not the dynamic response of the ocean to a quick pulse of carbon dioxide into the air. Revelle had seen years earlier, after all, that the fallout from the Bikini blast had been absorbed very slowly. In a letter to Arnold, Suess wrote that he was "not too happy about the whole thing"; and Revelle, evidently, was telling them all that he actually believed about 80 percent of the emissions was staying in the air—a higher estimate even than Callendar's, whom Revelle had gone out of his way to impugn in the paper.

Science historian Spencer Weart, who has written a fine book and constructed an extensive Web site about what he calls "the discovery of global warming," has gone to great lengths to rationalize the strange legacy of this paper. After reading it many times, he had a realization about a minor incongruity concerning the "peculiar buffer mechanism" that Revelle had discovered as a grad student and resumed wrestling with after the Bikini bomb test. In a short paragraph near the end of the paper, Revelle and Suess suggest that buffering *might* slow the absorption of carbon dioxide by as much as 90 percent relative to their naïve main argument—in which case most of the gas would stay in the air (and the unusually astute reader would realize that Revelle and Suess agreed with Callendar and Plass).

Weart describes finding a late manuscript of the paper in Revelle's archives: "It was gratifying confirmation to find the paragraph . . . typed on a different kind of paper and taped onto an earlier version": it was an afterthought. On the basis of this forensic evidence—regarding an innocuous passage in a paper whose main conclusion was clearly the opposite—Weart attributes the discovery that the carbon dioxide produced by fossil fuel burning will stay in the air to Roger Revelle. I don't think so.

After reading Revelle's obfuscations many times myself, I developed a suspicion that his actual purpose was to show how little was known about this critical subject in order to justify more research; in other words, this famous paper was basically a grant proposal. Consider the last two sentences:

> Present data on the total amount of CO_2 in the atmosphere, on the rates and mechanisms of CO_2 exchange between the sea and the air . . . are insufficient to give an accurate base line for measurement of future changes in atmospheric CO_2. An opportunity exists during the International Geophysical Year to obtain much of the necessary information.

Forty-four years after the fact, I mentioned my suspicion to Charles David Keeling, who actually did make the discovery, in my opinion (and who expresses nothing but admiration for Roger Revelle, by the way).

"That's right," Keeling responded. "That's exactly what it was for. One way you can tell is because Revelle called it a great geophysical experiment when it was really a geochemical experiment and he was writing it for the

Geophysical Year. That's a secret I reveal for the first time. That's never been pointed out, but that's pretty much what happened."

The International Geophysical Year was undertaken jointly by almost seventy nations for about a year and a half, actually, from the beginning of July 1957 to the end of December 1958. Its goals were "to observe geophysical phenomena and secure data from all parts of the world; to conduct this effort on a coordinated basis by fields, and in space and time. . . . " Special emphasis was placed on the exploration of Antarctica, the upper atmosphere, and outer space, so it became a grand three-ring circus for Cold War competition, with space in the center ring. The Soviets launched Sputnik under its auspices.

It should come as no surprise that Roger Revelle was on top of this vast opportunity for research funding from the very start. He had been promoting the IGY for a few years by the time he submitted his "grant proposal" to *Tellus* and had already assumed the chairmanship of the program's U.S. Panel on Oceanography.

AS THE GEOPHYSICAL YEAR APPROACHED, DAVE KEELING WAS WORKING AS A POSTDOC at the California Institute of Technology, where Harrison Brown (who had extended the transatlantic offer that brought Hans Suess to Chicago) had recently founded one of the earliest research programs devoted exclusively to geochemistry.

As is not uncommonly the case, a casual remark by a great scientist set Keeling off on the quest of a lifetime. He hardly saw Brown during his first year at Caltech, for his advisor spent most of 1954 in Jamaica, writing a genuinely prophetic book, *The Challenge of Man's Future* (one of many sources from which Revelle later borrowed in his "grant proposal"). During a brief visit home, Brown happened to engage in an informal discussion with some students about the natural acidity of lakes and rivers. Keeling questioned his assumption that the bicarbonates in a body of water, which have a lot to do with determining the water's acidity, will sit in equilibrium with the carbon dioxide in the air above it (the same question that was irking Revelle and Suess, not far south on the California coast, at about the same time). Brown promptly challenged Keeling to find out experimentally—and soon left for Jamaica to work on his book, leaving Keeling to his own . . . well, devices.

Unable to find a commercial instrument that met his satisfaction, Keeling decided to build one himself. He eventually settled upon an old two-step method in which he first had to extract the carbon dioxide from an air sample by a chemical method, then determine its amount with a standard instrument for measuring gas pressure, known as a manometer. His vague and somewhat hazardous design specification was to make the method as sensitive as he possibly could. He would later admit that he made it about ten times more sensitive than it needed to be.

Since it took him more than a year to develop the method, he felt a little behind the eight ball when he finally succeeded. He began collecting air in glass flasks from the roof of his Pasadena laboratory once every four hours, often around the clock, sleeping on a cot in his lab and running home to attend to his pregnant wife in his few free moments. As his schedule happened to work out, he left her in labor in the hospital and missed the birth of their first child.

"His outstanding characteristic," Roger Revelle told the author Jonathan Weiner about twenty years later, "is that he has an overwhelming desire to measure carbon dioxide. He wants to measure it in his *belly.* Measure it in all its manifestations, atmospheric and oceanic. And he's done this all his *life.* Very single-minded."

Under the circumstances, however, Keeling's single-mindedness would turn out to be a good thing—although it would take a few years for anyone, including himself, to understand why.

Keeling loves the great outdoors. He still keeps a home in the mountains of Montana. In the spring of 1955, a few weeks after the blessed event, the new family embarked on a series of combined research/camping trips up and down the West Coast. Keeling took samples by the ocean at Big Sur, in the Sierra Nevada in Yosemite National Park (where a mule deer, ravenous from the winter, stole his lab notebook; Keeling miraculously recovered it, the deer having eaten only its green cover), and in the Inyo Mountains near Death Valley. In early September, the family took a trip to Washington State, where Keeling sampled the air in the Cascades and Olympic National Park.

Back in Pasadena he would run the samples through his procedure and add the results to a growing table that showed where and when they were taken, the temperature at the time, the wind speed and direction, and various other meteorological factors.

SINCE THE BEGINNING OF THE TWENTIETH CENTURY, THE WORLD CENTER FOR METEOROLOGY had been Scandinavia. The towering figure early on had been the Norwegian Vilhelm Bjerknes, and he had trained the next two: his son, Jacob, and Carl-Gustav Rossby, a Swede. The two younger men immigrated to the United States between the wars, Jacob Bjerknes to UCLA and Rossby, eventually, to none other than the University of Chicago. Both did military weather prediction during the war and eventually became heads of their respective meteorology departments. The mathematical frameworks they developed in the thirties and forties laid the foundation for the ENIAC program that von Neumann and his colleagues used to mimic the famous 1950 Thanksgiving Day storm, and for nearly all the computer models developed later to study global climate and the greenhouse effect. By the time Keeling entered the field, Rossby was splitting his time between Chicago and Stockholm, where he had just become founding director of a new Institute of Meteorology. Around then Rossby also founded the journal *Tellus*.

As plans for the IGY began to fall into place, Rossby suggested that his institute prepare the ground by monitoring trace atmospheric constituents from a series of stations around the globe. The participants at a 1954 conference then decided to set up a worldwide network of carbon dioxide monitoring stations during the IGY.

Current wisdom held that carbon dioxide levels would vary dramatically between different locations and that they would fluctuate even in a single location as each new weather system passed through. This belief was not entirely unfounded; it was based on the same measurements, performed mostly by Scandinavians, upon which Callendar had based his greenhouse calculations in the thirties and forties.

BUT AS KEELING'S TABLE OF MANOMETER READINGS GREW, HE BEGAN TO NOTICE TWO things that seemed completely out of step with current wisdom. For one, his numbers didn't vary all that much; and for another, what variation there was followed a regular, and remarkable, daily pattern.

In all of his wilderness locations, he found that the carbon dioxide level reached a high point each morning, just before dawn. The level would begin to drop at sunrise and would reach a minimum in the mid-afternoon. Then it would change course and rise until the next sunrise, to about the level of the previous morning.

He soon guessed that this diurnal cycle reflected the "breathing" of the local plants. During photosynthesis—which means, literally, "synthesizing with light"—plants use the energy from single particles of light, photons, to convert individual molecules of carbon dioxide and water into simple sugars. Sugars are the basic currency of biology; they store energy and life's building block, carbon, in transferable form, so they can be used later to build everything from DNA molecules to the more complex tissues and structures found in all living things.

As plants—and animals, for that matter—*spend* this currency, that is, as they break the sugars down, they "respirate" and give off carbon dioxide. While, in plants, respiration takes place continuously over the course of a day, photosynthesis takes place only in the presence of light. This explains why the levels dropped at Keeling's remote, forested sites during the day: trees "inhale" carbon dioxide while the sun is up. They take one breath a day.

As Keeling repeatedly scanned the neat columns of numbers he continued to obtain from a growing number of sites, another remarkable pattern was revealed: at every site, the samples taken at the minimum in the late afternoon produced almost the same number: 315 parts per million. He began to suspect that this was the background level of carbon dioxide in the atmosphere, independent of local effects, characteristic of the entire West Coast or, perhaps, the whole northern hemisphere—maybe even the entire planet.

Over the fall, he arranged for a colleague from Scripps to take samples from the forward deck of a research vessel on a cruise in the equatorial Pacific. The minimum in that set of numbers was again about 315 ppm.

In March 1956, on a hunch, Keeling packed thirty flasks into his car and drove to a field station at almost twelve thousand feet in the Inyo Mountains (within sight of Mount Whitney, where Langley had used his bolometer to answer a question posed by Tyndall about sixty years earlier). A westerly storm was approaching the arid mountaintop as Keeling set up camp, high above all sizable plants, where there would be nothing except his own breath and his campfire—both of which he made sure to keep downwind—to contaminate the pristine air. For five days he collected "Essence of Pacific Mountain Storm" following his usual procedure: he took a sample every four hours, day or night, wind or snow. According to his table, during one midnight sampling interval the wind was blowing thirty-nine knots, and the temperature was nine degrees Fahrenheit. The thirty flasks gave a tightly packed set of readings, all below 320 ppm. The minimum value was 315.

MEANWHILE, IN 1955, JUST BEFORE THE KEELING FAMILY SET OFF ON THEIR FIRST CAMPing trip, the Scandinavians had gone ahead with the monitoring program that had been planned at the conference the previous year. The program was run by a Finn named Kurt Buch, who had been measuring carbon dioxide from ships in the North Atlantic since the 1930s. Buch's group was publishing the results every three months in the back pages of *Tellus,* and although they did their sampling in open places where there should not have been much local influence, their numbers were all over the place.

Keeling, a lone and lowly postdoc in California who had yet to publish a single word about his work, seems to have been the only person in the world who had a suspicion that there might be something wrong with the

Scandinavian method. He says even "Rossby was looking for big variations that would correlate with weather systems, so he wasn't upset with what he was seeing in the data."

At the 1954 conference, responsibility for setting up the Pacific component of the monitoring network had been given to Wendell Mordy, chief meteorologist at the Pineapple Research Institute in Hawaii (which shows just how interdisciplinary Cold War research could be). Keeling got in touch with Mordy to express his concerns about the Scandinavian method, and things moved rapidly from there: two weeks after his blustery epic in the mountains, he found himself on a plane to Washington, headed for an eight A.M. meeting with Dr. Harry Wexler, the forceful director of the United States Weather Bureau.

For almost forty years Keeling remembered that meeting as taking only fifteen minutes, but upon looking through his diary recently he noticed that it had actually lasted an hour. Either way, it appears that this life-changing conversation happened so fast that it left him slightly dazed.

"Wexler asked a number of questions in rapid-fire, covering both the scientific and the practical," Keeling later wrote. "He was especially interested in costs." Wexler told him about a small hut that the Weather Bureau maintained near the summit of the Mauna Loa volcano in Hawaii and that plans were under way to expand it and make it a permanent research facility.

Keeling had two reasons to be nervous during this interview. Not only was he challenging the word of one of the greatest names in meteorology, Carl-Gustav Rossby, but Rossby and Wexler were friends. They had worked together at the Weather Bureau in the thirties. Wexler was an ardent supporter of the Scandinavian monitoring network and quite aware of the data that were coming out quarterly in *Tellus*. Keeling's fresh news from the mountaintop impressed him immensely, however—especially the stability of the numbers he had taken during a five-day storm.

The young man did have the poise to point out that his manual procedure, which took about an hour per sample, would not do for measuring the large number of samples that stood to be produced during the IGY, and to propose, furthermore, that Wexler purchase, from a company Keeling knew in Pasadena, a few ten-thousand-dollar infrared analyzers, which held the possibility of doing the measurement instantly and continuously. In essence, the new analyzer was identical to the one John Tyndall had designed in 1859 when he had discovered the molecule's heat-absorbing properties. It would measure the infrared absorption of the local air as it flowed through an optical chamber and produce a continuous readout on a strip-chart recorder. So, conceptually, Keeling's new idea turned Tyndall's on its head: he planned to use carbon dioxide's greenhouse properties to measure its concentration.

This was an uncharacteristic shot from the hip for a perfectionist like Keeling. He had hardly evaluated the instrument; he had only talked about it to a design engineer over the phone. He didn't hide this fact from Wexler. Nonetheless, Wexler decided, on the spot, to base the entire U.S. carbon dioxide monitoring effort on the preliminary work of this one young man. If and when the United States decided to join the IGY, he promised, he would allocate funds for the purchase of a continuous infrared analyzer on Mauna Loa and possibly a few other locations.

According to a table Keeling published later, from March 31 through April 1, 1956, just after his meeting with Wexler, he literally "took some air" at Assateague Island National Seashore, just east of Washington, D.C. It seems that carbon dioxide measurement was indeed "in his belly."

THE ALL-SEEING ROGER REVELLE WAS COMPLETING HIS GRANT PROPOSAL WITH HANS Suess at just that moment. He quickly obtained Wexler's commitment to base the U.S. component of the IGY carbon dioxide monitoring program at Scripps and, in late July, hired Keeling to lead it. (Thus, one specific purpose of the proposal seems to have been to hire Keeling.)

Ironically, however, once Revelle had Keeling under his thumb, he proceeded to show significantly less faith in him than Wexler had. His conservative side came out again. While Keeling felt, on the basis of the consistent numbers he'd already obtained, that it would be sufficient to take measurements in two or three strategic locations, Revelle held the traditional view that passing weather systems would cause unacceptable variations. He instructed Keeling to ship hundreds of flasks out to all the stations in the world that would be involved in the IGY, as well as to the various airplanes and ships that the U.S. military would be making available to the program.

However, Wexler had also come through on his promise to Keeling; so as the young man followed his boss's orders (and this would turn out to be a good thing), he also pursued his mandate to build a few infrared analyzers. The first shipped out for Little America, the U.S. base on Antarctica's Ross Ice Shelf,* on the day after Christmas 1956. In his rush to get the analyzer on the boat, Keeling had not tested it to his usual excruciating level of satisfaction, and it did not perform well the first year. It was refurbished over the next austral summer and performed well after that.

*Which Richard Byrd had established and from which he had made the first airplane flight over the South Pole.

Three months later, he had another instrument up and running at the end of a Scripps pier that jutted out into the Pacific. He'd validated this one well enough to know that he could trust its numbers, but he discovered that this was true only during a sea breeze: whenever the wind backed around north, the air from Los Angeles, the land of the automobile, sent the carbon dioxide level skyrocketing.

As Keeling set about getting a third instrument ready, for Mauna Loa, Revelle again intervened. There had been delays in getting the flasks to and from their various far-flung destinations—mostly due to the intransigence of various customs agencies, which was completely beyond Keeling's control. Revelle refused to sign Keeling's travel documents, thereby preventing him from installing his instrument in Hawaii. But as 1958 began and the end of the IGY loomed, a desperate Wexler sent Ben Harlin, who had operated the first instrument at Little America, to La Jolla to receive direct instruction from Keeling. In March 1958—scant months before the IGY came to an end—Harlin and Jack Pales, the first director of what was eventually christened the Mauna Loa Observatory, managed to install an infrared analyzer on a barren lava bed near the summit of the volcano, with an anxious Keeling in constant contact by phone. Keeling told Harlin and Pales that it would be a good sign if the analyzer gave the magic number 315 ppm. The very first reading was 314.

As auspicious as this seemed, it was only a coincidence.

WEXLER AND KEELING HAD FIGURED MAUNA LOA WOULD BE AN EXCELLENT PLACE TO take a representative sampling of the earth's air (or at least the air of the northern hemisphere, since they had little idea at that point how quickly the air north and south of the equator would mix), for Hawaii is the most remote archipelago on the planet. The trade winds blow west across a few thousand miles of open ocean before wafting up the volcano's eastern flanks.

For Keeling, however, the installation of his instrument was followed by about twelve months of nail-biting. In the analyzer's second month of operation, the carbon dioxide level began to rise. Then the electrical generators at the observatory broke down for a few weeks. When they came back up in July, the level had dropped below its March value. Keeling became anxious that he and Wexler had been wrong about the atmosphere, that the levels at Mauna Loa might be "hopelessly erratic." The level dropped again in late August. Then there were more power failures. Wexler finally found the cash for a better electrical system and fixed that problem. In November, Keeling found a way to work around Revelle and pay his instrument a visit. Then the level began to climb to the heights of the previous spring.

It slowly began to dawn on Keeling that they were watching a regular seasonal pattern—the level dropping in summer and rising in winter—which seemed to be the yearly, global analog of the daily "breathing" he had recorded in the forests by the West Coast. "We were witnessing for the first time nature's borrowing of carbon dioxide for plant growth during the summer and returning the loan each succeeding winter," he wrote. "The maximum at Mauna Loa occurred in May just before temperate and boreal plants add new leaves. The seasonal pattern was highly regular and almost exactly repeated itself during the second year of measurements at Mauna Loa. . . . Thus there was no need to wait for statistical studies to prove the reality of the oscillation as would have been required had less exact chemical methods been used."

So, forests take one breath a day, and the earth takes one breath a year.

BUT THE MOST IMPRESSIVE RESULT FROM THE FIRST YEAR OF MEASURING CARBON DIOXide at Mauna Loa was, as Keeling puts it, that "as soon as we had two Marches . . . we could see that it was going up." The average reading for March 1959 was one part per million higher than that for March 1958. The yearly oscillation seemed to be sitting on a rising baseline.

So that was the moment when mankind's "great experiment" provided an answer to one of Gilbert Plass's two questions: at least some of the carbon dioxide we were dumping into the sky was staying there. Now it remained to be seen if the rise in carbon dioxide would drag temperatures along with it.

It is testament both to the impartiality of good science and to the respect good scientists accord to it that one of the first champions of Keeling's work came from Scandinavia. Bert Bolin, who had become director of the Swedish Institute of Meteorology upon Rossby's death in 1957, visited Scripps in 1959. Understanding the value of Keeling's work immediately, Bolin invited him to speak at the 1960 meeting of the International Union of Geodesy and Geophysics in Helsinki and accelerated the publication process at *Tellus,* so that in March of that year, Keeling announced in a peer-reviewed journal that atmospheric carbon dioxide was on the rise.

Considering all the attention that has been paid to Mauna Loa over the decades, it is interesting to note that Keeling based that announcement not on the measurements from the volcano but on a series of flask samples he might not have obtained had Revelle not been so insistent about his "planes, trains, and automobiles." These samples came from the most remote IGY base of them all, the U.S. South Pole Station, the establishment of which during the IGY enabled the first people, indeed the first creatures of any kind, to begin living at that unique spot year-round.

Flask sampling at the pole had begun only a few months after the commencement of the IGY, so the pole's record was about six months longer than Mauna Loa's. And it was the perfect complement to the volcano, not only because it stood in the southern hemisphere but also because the air at the South Pole is the least contaminated in the world.

It was also a better place for measuring the slow rise in carbon dioxide, because the annual breathing of the earth's plant life has much less of an effect there than at Mauna Loa. Only deciduous plants in the zones that experience seasons participate in the annual give-and-take of carbon with the atmosphere—tropical plants contribute little if any—and since the vast majority of the earth's temperate and subarctic landmass is found north of the equator, most of the action in the annual carbon cycle takes place in the northern hemisphere. (Furthermore, since most of the earth's industrial countries lie in the northern temperate zones, the lion's share of the fossil fuel burning takes place north of the equator as well.) Annual oscillations are, therefore, smaller at the South Pole than they are at Mauna Loa. They are also exactly out of phase, because seasons are reversed in the southern hemisphere.

Based on just over two years of data from the South Pole, Keeling found that carbon dioxide was rising at a rate of about 1.3 ppm per year.

NOT ONLY AN EXACTING SCIENTIST BUT ALSO A THOROUGH ONE, KEELING SOON looked in on the Finnish monitoring program and discovered problems both with their method and with the sites they had chosen for sampling. Final proof came when Buch's group sent him a flask of air that they had measured at 350 ppm, and he got his magic number, 315. The Finns eventually switched to infrared analyzers.

Exhibiting open-mindedness in handsome complement to his industry, Keeling also got in touch with Guy Callendar, who had been following these events all along from the unique vantage point of the desk at his home in Sussex, at which he had been working on the carbon dioxide problem for more than twenty years. The two agreed that the art of direct measurement had been solid before the turn of the century and that the best estimate for the preindustrial carbon dioxide level was about 290 ppm. Together they deduced that Buch's group, in the process of streamlining the nineteenth-century method, had inadvertently compromised both its accuracy and its precision. (Keeling observes that this is "something government agencies sometimes do, called 'faster, better, cheaper,' in which only two out of the three are usually realized. The one that wasn't realized was 'better.' ")

Finally, in order to answer the all-important question of ocean uptake, Keeling turned his exactitude to the calculation of global carbon dioxide emissions. About fifteen years later, in 1976, based on those calculations and more than a decade's worth of data from the South Pole and Mauna Loa, he showed that roughly half the emissions were staying in the air—not far from Callendar's old estimate of three-quarters. (Callendar's calculations had been sound, by the way, but he had based them on Buch's atmospheric measurements, which had been systematically high.)

In the twelve years from 1959 to 1971, the level at Mauna Loa had risen from 315 to 330 ppm, about 3.4 percent, and the levels at the South Pole had tracked those at the volcano but always lagged behind. From the delay, Keeling estimated that it takes about six months for the disproportionate emissions north of the equator to mix completely into the air of the southern hemisphere.

HISTORIANS AND ECONOMISTS HAVE NAMED THE PERIOD SINCE THE EARLY EIGHTEENTH century, when coal burning first became widespread, the "Industrial Era." Geologists seem to take a longer view. When they sit around the lunch table they don't talk about what happened yesterday on the other side of town, they discuss what happened forty million years ago on the other side of the planet. It was over lunch, in fact, that I once heard Ellen Thompson refer to the present era as the Fossil Fuel Age. For, in the span of four or five centuries, give or take, we stand to burn the fossil reserves it has taken nature nearly a billion years to produce—and that burning stands to have a major and lasting effect on the climate we hand to future generations. Ellen is not alone in believing that the brief Fossil Fuel Age will propel this planet from the relative stability of the Holocene epoch, which has now lasted twelve thousand years, into the unknown territory of the Anthropocene.

When the industrial revolution hit America, at about the beginning of the nineteenth century, the vast majority of carbon dioxide emissions came from coal, and the sum total from around the world was about 32 million tons per year. Through that century emissions grew at the astounding rate of 4.5 percent per year, which means they doubled every sixteen years. The burning of oil, which commenced in 1870, did not have a noticeable effect upon this exponential trend, though it did falter in 1913 with the start of World War I, by which time emissions had multiplied one hundredfold, to 3.4 billion tons per year.

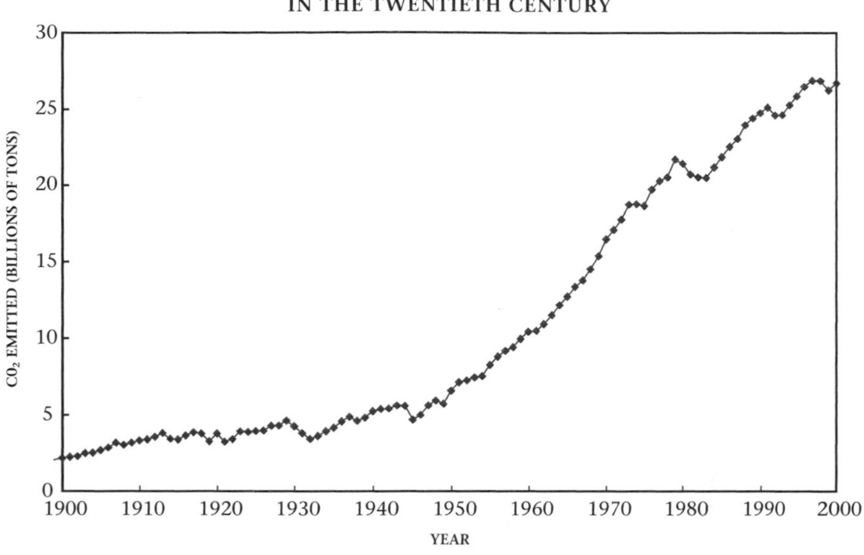

Economists often use the consumption of energy as a measure of economic output. Well, if we take their word for it, the emissions chart shows that the chaotic period from the start of World War I until the end of World War II was not good for business. Emissions did grow, but at an annual rate of less than 1 percent. There was a pronounced drop from 1929 to 1933, during the Great Depression.

Of course, much of the strategy of World War II was based on access to oil fields. Once it was over and that was all sorted out, it seems that unprecedented prosperity set in (for some people at least). In 1945 the growth rate shot up to 5 percent per year. It stayed there for almost three decades, until 1973 and the Mideast oil crisis, by which time annual emissions had quintupled again, to 18.6 billion tons. After two years in which they actually decreased, the blistering pace resumed—only to falter in 1979 with the revolution in Iran, which slowed that country's oil exports to a trickle. Iran promptly went to war with another major oil producer, Iraq, and emissions dropped for four years in a row, until 1983.

Since then, with the exception of the two-year drop initiated by the first Persian Gulf war, the growth has been less vigorous: about 2 percent per year. It stands to drop again with the economic slowdown and the second Iraq war, which are taking place as I write. But exponential growth is an insidious thing. Two percent may not seem like much, but it leads to a doubling every thirty years.

Over the first 250 years of the Fossil Fuel Age, then, human beings have sent roughly one trillion tons of carbon dioxide into the atmosphere; and such is the power of exponential growth that half that amount has been emitted in a tenth of the time: the last twenty-five years. Even if emissions continue to grow at the present, relatively lethargic, rate of 2 percent per year, we will send up another trillion tons in roughly the next twenty-five years. At that point we will have produced about as much carbon dioxide as the total amount that existed in the atmosphere when the Fossil Fuel Age began. As Dave Keeling has shown, about half those emissions will stay in the air; so, the way things are going, we stand to double the preindustrial level of carbon dioxide sometime in the second half of the present century.

I SUSPECT THAT IF WE COULD SEE THESE CLOUDS OF HEAT-TRAPPING GAS, AS WE CAN other forms of air pollution, we would have done something about them long ago. To convey just how massive they are, Lonnie Thompson points out that the carbon dioxide produced in the burning of a given amount of fuel actually weighs more than the fuel itself. A gallon of gasoline weighs about seven pounds, for instance, yet it produces twenty-two pounds of carbon dioxide. That's a quarter of a ton from a twenty-gallon tank. He sometimes takes his students on field trips to the Gavin power plant in southeastern Ohio, which receives its coal on a huge conveyor belt, directly from an underground mine ten miles distant. That single plant, just one of about five hundred in the country, burns 7.5 million tons of coal each year and sends up more than 20 million tons of carbon dioxide—almost as much as the entire global output in 1800.

Scientists constantly strive to imagine vivid ways to describe the scale of the burning. One creative recent observation is that a gallon of gasoline comprises the distilled remains of about one hundred tons of ancient plant matter, the equivalent of about forty acres of wheat. Put another way, all the vegetation that could possibly grow on this planet in four hundred years would not be enough to produce the fossil fuels we burn in one.

Jonathan Weiner makes a good point in referring to this as the "Industrial Eruption." Virtually all the carbon around—in the atmosphere, the oceans, and the soil; in the coal, oil, and natural gas below; in trees, plants, animals, you, me, the diamond in your ring—was originally spewed into the air by a volcano. Lonnie tells me that all the volcanoes on earth produce somewhere between three hundred million and a billion tons of carbon dioxide in an average year. Humans are now producing between twenty and ninety times that *every* year: twenty billion tons!

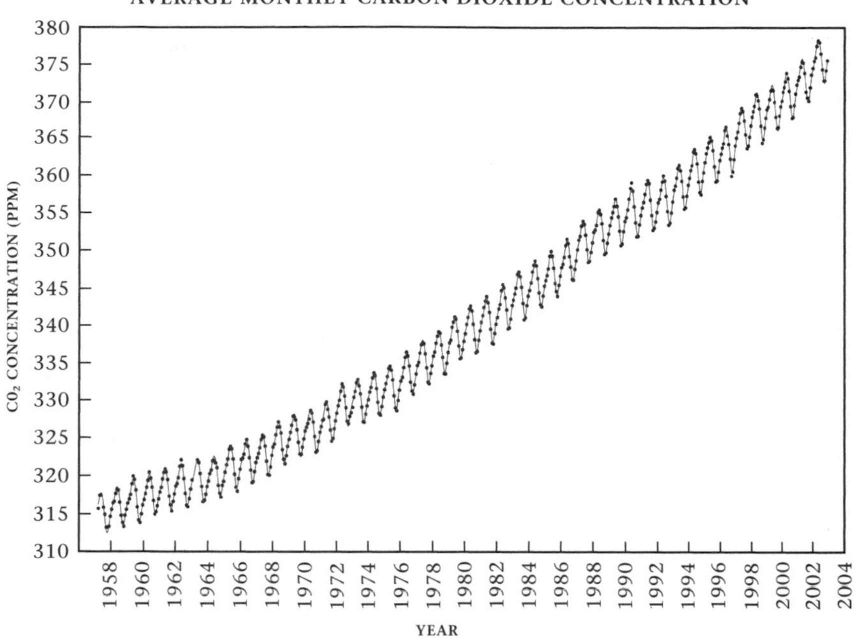

The seesaw pattern that rides the rising trend in CO_2 results from the annual "breathing" of the earth, as nature, in Dave Keeling's words, "borrow[s] carbon dioxide for plant growth during the summer and return[s] the loan each succeeding winter."

THE GRADUAL RISE IN THE LEVEL OF CARBON DIOXIDE AT MAUNA LOA, NOW KNOWN AS the Keeling Curve, has become one of the central icons of global warming. As of 2003, the level was about 373 ppm, 18 percent higher than it was in 1959. It's been rising at a rate of about 1.3 ppm per year, precisely the rate Keeling first measured at the South Pole during the International Geophysical Year.

I called Keeling in 2001 with a few questions about the early days ("It's always fun to talk about this stuff," he said) and one about ocean uptake. How long *will* it take all this stuff to come back to earth?

He said that the three famous papers in *Tellus* were right as far as they went, but that there was much more going on than even those estimable scientists figured at the time. There are many, many processes whereby carbon dioxide is dissolved into the ocean, circulated through it, and eventually deposited on its floor. Some take only six months; the ones Revelle and his colleagues studied take tens of years; some take hundreds of years, some tens of thousands of years, and some even hundreds of thousands of years.

"But I will make just one statement about it," Keeling said. "If you look at the whole Fossil Fuel Era, when we burn all the fossil fuel—which seems to be what we'll be doing if we can, whether it takes a thousand years or two hundred years—the one thing you can notice is that the curves [of carbon dioxide level over time] all look rather similar. The level rises to a peak, and then it declines over a *very* long time period. If we keep accelerating the use by two percent or something like that, the peak will be reached in a couple hundred years or so, and you can push it out to maybe three hundred years or four hundred years . . . and it's up there well over the double [double the preindustrial level]. Then it declines, and at *ten thousand years*—unless there's some process we haven't put in the model, and we've put in the things we know about—it's *still* more than double. So you might say that we've got a transient there that's maybe a hundred years and maybe two hundred years [while we push the level up to a peak], and after that it's *always* gonna be more than double, for a *lo-ong time!*"

You should have heard the way this patient man said those last two words. It seemed that it was a strain for him to be so imprecise.

"Will a doubling make a big difference?" I asked.

"It looks like it, doesn't it?" he answered. "Well, it was hard to tell until some things started to happen, but now they seem to be happening, and we're getting weather we're not used to."

· 8 ·

TEMPERATURE FOLLOWS SUIT

John Mercer was fully aware of these developments as they were unfolding, of course. Colin Bull says he was "the first person ever to use, in my hearing, the words *greenhouse effect*—and *climate distortion,* I think we called it at that stage," at an in-house symposium at Ohio State in about 1963.

Ohio State served as a collection center for data from the polar regions during the IGY, and the synergy this created among scientists from different disciplines led to the founding of the Institute of Polar Studies in 1960.

The scientists at the new institute naturally began to consider the implications of "climate distortion" for the regions they were studying, and Mercer chose from among his own myriad possibilities to focus on Antarctica. Aware of the notion first proposed by Arrhenius that a warming would be amplified at the poles, Mercer realized that a doubling of atmospheric carbon dioxide might lead to the melting of the Ronne and Ross ice shelves, which are the most important gates through which the enormous West Antarctic Ice Sheet calves into the sea. The bulk of these shelves floats on the ocean, so their melting alone would not affect sea level directly.* At the coast, however, where they are "grounded"—that is, frozen to the earth—they act as dams against the potentially catastrophic surge of an ice sheet that covers about half the continent. If these barriers were to melt away, Mercer argued, the West Antarctic Ice Sheet might become unstable, start to flow rapidly into the sea, and raise sea level by as

*Archimedes' Principle: the level of the water in a glass does not change as the ice cubes in the glass melt.

much as sixteen feet—enough to "flood all existing port facilities and other low-lying coastal structures, extensive sections of heavily farmed and densely populated river deltas of the world, major portions of the states of Florida and Louisiana, and large areas of many of the world's major cities," as Roger Revelle would put it in a report produced by the National Academy of Sciences many years later.

To show just how far ahead of the pack he was, Mercer waited fifteen years to publish the full analysis—in *Nature* in 1978—yet even that delayed publication is now seen as having been far ahead of its time. The *Nature* paper was the centerpiece of a feature article about the rising seas that appeared in *Scientific American* in 1998 and attracted much interest again in 2002 with the collapse of the Larsen B Ice Shelf, which borders the Antarctic Peninsula: a slab the size of the state of Rhode Island dissolved into a galaxy of icebergs in the space of about five weeks.

The question as to the stability of the polar ice sheets remains the most disconcerting and underappreciated wild card in the global warming game. Moreover, the apocalyptic warning Mercer sounded some forty years ago is actually seen as optimistic by many of today's experts.

BUT A RISE OF ONE PART PER MILLION DOES NOT SEEM LIKE MUCH, EVEN IF IT DOES come every year, and even those who realized half a century ago that there was a threat on the horizon also knew that the warming itself was a slow train coming.

More immediate threats rose to the fore in the 1960s. In 1962, Rachel Carson's *Silent Spring* warned that pesticides such as DDT were wreaking havoc on entire ecosystems; and the public, primed by warnings about nuclear fallout, or, worse, nuclear war—fears of which reached a crescendo just that year with the Cuban Missile Crisis—had no trouble seeing synthetic chemicals as just one more dire and all-pervasive side effect of industrial progress. Carson put fallout at the top of her list of global threats, but she made no mention whatsoever of carbon dioxide, even though Keeling had just shown that it was the most pervasive contaminant of all.

On the other hand, the notion that industrial by-products could spread to the most remote reaches of the planet was hardly news to the greenhouse scientists. When *Silent Spring* began to make waves, Secretary of the Interior Stewart Udall discussed it with his new science advisor, Roger Revelle. He later recalled Revelle's advice: " 'Well the evidence isn't all in, Stewart, but . . . she fits your department, you ought to back her.' We did back her, and President Kennedy backed her, and that gave her message a resonance. . . ."

Thus the environmental movement was born. In an appearance before the Senate Committee on Commerce in 1963, Carson suggested that a commission be formed to protect the public against the environmental threats posed by industrial development—a suggestion that eventually took the form of the Environmental Protection Agency.

In 1964, Revelle moved to Harvard University to become the first director of its Center for Population Studies; the following year, as chairman of President Lyndon Johnson's Science Advisory Panel on Environmental Pollution, he oversaw the first U.S. government report to recognize the carbon dioxide threat. A year or two later, Al Gore, the son of a U.S. senator from Tennessee, attended one of Revelle's classes at Harvard. Revelle's eloquence on the dangers of fossil fuel burning made a lasting impression on that particular young man.

Meanwhile, largely ignored by the public, greenhouse science marched on.

IN 1954, THE EVER-PERSUASIVE JOHN VON NEUMANN CONVINCED THE U.S. WEATHER Bureau and certain weather forecasting arms of the U.S. Navy and Air Force to establish the Joint Numerical Weather Prediction Unit, in Suitland, Maryland. JNWP began forecasting weather on a regular basis the following year, although it would take more than a decade for its predictions to show a clear improvement over the intuitive forecasting of human meteorologists.

Weather forecasting models simulate conditions over selected portions of the globe and are not really suited to the study of climate. They emphasize the circulation of the atmosphere. (The first were based on so-called dishpan experiments in Rossby's lab at the University of Chicago, where they literally studied spinning pans of water.) In 1955, von Neumann persuaded the Weather Bureau to take the next step and develop simulations of the whole atmosphere. This led to the creation of a special section for General Circulation Research, led by Joseph Smagorinsky, one of the heavyweights von Neumann had earlier recruited to his Meteorological Project at the Institute for Advanced Study.

Smagorinsky saw it as his mission to realize von Neumann's original vision of a full three-dimensional simulation of the atmosphere, incorporating not only the sort of heating and radiation effects that Gilbert Plass had put into the first computerized radiation model, but the circulation behind the weather models as well. Such a model would simulate the way different parcels of air rose as they warmed, fell as they cooled, and circulated horizontally, due to the spinning of the planet, the reflected heat or rising evaporation from the land or ocean, and the influence of neighboring parcels. This sort of program has become known as a "general circulation model," or GCM.

The basic idea is to divide the atmosphere up into many small boxes and design rules based on as realistic a set of physical laws as possible to describe how the air in the boxes will behave. One can also start the system off in a specific state and with a certain set of rules—one of which might be that carbon dioxide will rise at such-and-such a rate—and watch it evolve over time. General circulation modeling has become extremely sophisticated in the half century since its invention. The most significant advance since the early days has probably been the coupling of the oceans and their circulation to the atmosphere.

Climatologists often point out that Plass's great global warming experiment lacks an experimental control: for a number of practical reasons, including the fact that we have only one of them, we cannot compare what has happened to our planet over the past two centuries with what might have happened had we not been adding greenhouse gases during that time. The beauty of a GCM is that it represents a small working model of the earth on which controlled experiments may be run.

In 1959, Smagorinsky hired Syukuro "Suki" Manabe, fresh from his doctoral studies at Tokyo University. For almost forty years, Manabe would lead the most productive and enduring of the many GCM groups that took root around the world in the fifties and sixties.

Although Smagorinsky and Manabe had little interest in global warming at first, they had a great deal of interest in carbon dioxide, owing to the major role it plays in the atmosphere's energy budget. Thus, when Keeling showed that the level was rising, it was natural for the two computer experts to begin playing with the idea. Additional motivation was supplied by the inability of "simple" radiation models of the ilk of Gilbert Plass's to deal with water vapor feedback. Feedback caused these models to become unstable: depending on small changes in the details, a doubling of carbon dioxide could produce anything from no change at all to runaway warming.

When Manabe learned of this problem he took up the challenge and incorporated water vapor feedback into one of his simplest GCMs: a simulation of a single column of air at a particular latitude. The addition of circulation removed the instability, and, in 1967, Manabe and his colleague Richard Wetherald published the results of their first global warming study. They had found that a doubling of carbon dioxide produced a rise of about two degrees Celsius, or almost four degrees Fahrenheit. This led many of the working climatologists of the day to realize that global warming was a problem not merely for the ivory tower.

After another eight years, Smagorinsky and Manabe had incorporated water vapor feedback into a more realistic, three-dimensional GCM, in which the greenhouse effect was stronger: a doubling raised temperatures by 3.5°C, or 6.3°F. This result commanded more attention within the ivory tower, but caused hardly a ripple without.

Recall that when Lonnie Thompson entered the field, at around this time, a brief "global cooling" craze held sway. A climatologist from the U.S. Weather Bureau had set the craze in motion by announcing—on a suitably cold day in January 1961—that temperatures had been dropping for about twenty years. But this notion did not attract much attention until the early seventies, when a series of spectacular disasters carried the dark prospect of climate change to the level of the evening news: in 1968, a drought hit Africa's Sahel, the long, barely habitable strip of land along the southern edge of the Sahara, and contributed over the four years it persisted to the starvation of about one hundred thousand people.* Another drought hit the Soviet Union in 1972; the Soviets were forced to make massive grain purchases on the international market, and grain prices soared. That same year, an El Niño set in, shutting down the fisheries off equatorial Peru and Ecuador and causing the Indian monsoon to fail. Droughts in the Midwestern United States also made headlines.

The leading contributor to the public fears of an imminent global cooling was probably Reid Bryson of the University of Wisconsin, who believed that the apparent (and largely illusory) trend had been caused by the parasol effect of airborne dust particles, or "aerosols." Bryson employed the phrase "human volcano" in the opposite sense of Jonathan Weiner: to describe not our greenhouse emissions but our pernicious habit of sending dust and smog into the air. Both metaphors work, because human activity mimics both the greenhouse warming and aerosol cooling effects of an eruption; Bryson's belief in the possibility of a cooling did have some scientific basis, for large eruptions often do cool the atmosphere for a few years.

The most dramatic historical example is probably the massive eruption of Indonesia's Tambora in 1815, which was followed by the infamous "year without a summer." Lord Byron and his guests Mary and Percy Bysshe Shelley spent that nonsummer on the shores of Lake Geneva, where they found the weather so deliciously gloomy that they sought additional stimulation in the reading of evening ghost stories. Mary Shelley was inspired to write her novel *Frankenstein,* and Lord Byron composed the bleak poem "Darkness."

*Recent work demonstrates that the drought in the Sahel was caused not by a cooling but by the warming of the Indian Ocean, which may, in turn, have been caused by global warming.

In 1974, when Lonnie Thompson and Colin Bull revealed the link between dustiness and cold temperatures in the Byrd and Camp Century ice cores, they referred to simulations of the parasol effect that had been carried out a few years earlier by Stephen Schneider and S. Ichtiaque Rasool—which had led to Schneider and Rasool's premature warning of an oncoming ice age. In the meantime, Schneider had moved to the National Center for Atmospheric Research in Boulder, Colorado, and tweaked his computer model for a few years—its greenhouse aspects in particular. These improvements had changed his thinking: in 1975, he predicted that greenhouse warming would begin to dominate aerosol cooling "soon after 1980."

The confusion in this one scientist's mind seems to have exemplified the confusion that obtained all through the climate science community at the time: when journalists and newscasters asked the experts what might have caused the recent string of droughts and other disasters, they received conflicting answers. Some scientists believed in the cooling; others thought the parasol effect might work the other way, because the soot from smokestacks and so on is black and would, therefore, absorb sunlight and heat the air; and in the background were the greenhouse scientists, who had always expected global warming to win in the end. Since the media tend to follow the temperature trend of the moment, only the first camp received much notice.

Paradoxically, however, the confusion on the surface represented consensus on more fundamental matters. Whether they expected temperatures to rise or fall, most climate scientists now believed that mankind had become a significant player in climate change and that if change were to occur, it might occur very rapidly. This was a dramatic departure from the Victorian belief in divine stability. Climatic balance seemed more like that of a tightrope walker. Temperature might swing wildly in either direction, with serious and mostly negative consequences for society either way.

Proof that past changes had been extremely abrupt was just beginning to emerge in the peer-reviewed journals. Willi Dansgaard had noticed the first strange indications of flickering and sudden warming in the deepest ice layers at Camp Century, Greenland, in the late 1960s.

In the same year that Schneider announced his about-face, an influential paleoclimatologist named Wallace Broecker, from Columbia University's Lamont-Doherty Earth Observatory, also predicted a warming. Broecker based his thinking not on computers (which he distrusts to this day) but on field data: Dansgaard's record of oxygen isotopes at Camp Century. Broecker saw a natural cycle in that record, which would soon bottom out, and deduced that on the rebound the cycle would add to the "expected exponential temperature rise . . . from the carbon dioxide buildup" so that we would soon "experience a substantial warming."

AT THIS JUNCTURE, AN ASTRONOMER AND PHYSICIST NAMED JAMES HANSEN ENTERED the picture. Principal investigator on a Pioneer space probe project exploring the atmosphere of Venus, Hansen was quite content to be studying a planet other than his own. He knew, however, that the laws of physics are no different on Venus than they are on Earth—and that both our planetary next-door neighbors happen to provide illustrative examples of the greenhouse effect at work.

Venus is almost nine hundred degrees warmer than the estimate based on radiation balance. Most of its carbon has somehow escaped the planet and entered the atmosphere, to boost carbon dioxide levels to roughly 350,000 times those on Earth. This makes the Venusian atmosphere so dense that the air pressure at its surface is equivalent to that found about one mile deep in Earth's oceans, and it elevates temperatures enough for the planetary surface to glow red, like a hot coal. Lead would melt there.

Mars, on the other hand, has a thin atmosphere. The temperature at its surface is only eleven degrees higher than the estimate from radiation balance. Temperatures sometimes drop below –112°F—even in summer by the equator—cold enough for a significant portion of what little carbon dioxide there is in Martian air to fall to the ground as "dry ice."

As Arrhenius pointed out, the greenhouse effect on Earth elevates temperatures just enough to make our planet hospitable to life as we know it. This is sometimes called the Goldilocks Effect: Mars is too cold. Venus is too warm. Earth is just right.

There is a cautionary aspect to this tale, for both Venus and Mars may once have had climates like our own. Since the mid-1970s, when the Viking spacecraft first sent back photographs of what appeared to be dried-up rivers and lakes on the Martian surface, scientists have suspected that the planet may have once been as lukewarm as Earth—perhaps capable of supporting primitive forms of life. The fact that water did once exist on the surface of Mars was finally confirmed by the Spirit and Opportunity rovers, which NASA placed on the planet in January 2004. There is also evidence that Venus was once cooler, that its present, searing temperatures may have been brought about by a runaway positive feedback, similar to the fry cycle of Snowball Earth.

IN 1975, HANSEN WAS ASKED TO JOIN AN ONGOING STUDY OF EARTH'S GREENHOUSE EFFECT by a Harvard postdoc named Yuk Ling Yung. Agreeing, Hansen convinced his superiors at NASA's Goddard Institute for Space Studies (GISS) to allocate the staff and computer time he would need to design a global circulation model.

The first thing Hansen did with Yung was to review the constituents of our atmosphere more comprehensively than the earth scientists had before him. Among other things, he considered trace greenhouse gases that had mostly been ignored: nitrous oxide, ammonia, ozone, methane, the freons, and others.

Freons—known more commonly today as chlorofluorocarbons, or CFCs—are highly stable synthetic compounds that happen to be the most powerful greenhouse gases known. Molecule for molecule, the greenhouse potency of most freons is more than a thousand times that of carbon dioxide. They were used extensively in refrigerators and spray cans at the time of Hansen and Yung's first study, and it so happens that right around then scientists began to realize that they also destroy the ozone layer. Within a decade, their connection to the growth of the ozone hole over Antarctica would be proven, and a successful international effort to ban their manufacture would be joined. Few people realize that this has had the added benefit of slowing the growth of greenhouse forcing, that is, the change in Earth's radiation balance that has been caused by the greenhouse buildup.

His first "earthly" study told Hansen that human activity was increasing the atmospheric abundance of all the "lesser" greenhouse gases. By 1976, he and Yung had developed a primitive GCM, which showed that the increase of carbon dioxide and the lesser gases was likely to warm the planet by the mid-twenty-first century. They suggested that the other gases be monitored as well.

It did not take long for Hansen to realize that it was more interesting to study a changing climate than a stable one. (He has always maintained that he is motivated by a fascination with science rather than a love for the environment.) So, in 1978, he resigned from the Pioneer project and began to devote his efforts exclusively to Earth. The following year his group revealed the results from their first true GCM. They found climate to be almost twice as sensitive as Manabe had: a doubling of carbon dioxide raised temperatures by about four degrees Celsius, or seven degrees Fahrenheit.

In 1977, a National Academy of Sciences committee chaired by Roger Revelle issued a report on "Energy and Climate," which concluded that if emissions continued to grow at their present rate, temperatures could rise by as much as eleven degrees Fahrenheit by the middle of the twenty-first century—more than the difference between now and the depths of the Wisconsinan glacial stage. Either by media savvy or simple luck, they made this announcement in the midst of the hottest July the United States had experienced since the Dust Bowl, and it "caught fire" with the press. (Lonnie would have been working on Quelccaya at precisely that moment.) The droughts and ice age fears of the past few years had tuned the media in to the notion of climate change, and now that scientific opinion seemed to be shifting toward a warming—and the weather seemed to be agreeing—they did an about-face with a charming lack of irony and short-term memory. The environmental movement also added global warming to its agenda. Thus the issue began to make a faint impression on the public mind.

The new level of consternation prompted Frank Press, science advisor to President Jimmy Carter and a geophysicist, to ask the National Academy for an opinion on the credibility of the computer models. The academy chose MIT's Jule Charney, a former student of Jacob Bjerknes and one of the original members of von Neumann's Meteorology Project, to chair a panel of experts. The Charney panel concluded that the predictions of both the GCMs and the simpler radiation models were believable, and set a standard that has held to this day for communication on the global warming issue. The panel gave a *range* for climate sensitivity, rather than a single number: between 1.5 and 4.5°C (3 and 8°F).* And that range has held up amazingly well. Only in 2005 did a serious study call it into question (stretching out the high end, unfortunately). The Charney panel foresaw serious consequences for society even at the low end of the range, while at the high end they saw climatic zones shifting dramatically, to change the very face of the planet. "A wait-and-see policy may mean waiting until it is too late," they wrote.

Of course, the seventies were also the decade in which our ability to burn oil finally began to rub against some limits: first with the Mideast oil crisis, then with the revolution in Iran and the ensuing Iran-Iraq war. Faced with high oil prices and double-digit inflation, Congress came up with the bright idea of producing synthetic oil and gas from domestic shale and coal reserves; but when Jimmy Carter then proposed the formation of the United States Synthetic Fuels Corporation (Synfuels), Roger Revelle, Dave Keeling, and two other eminent scientists pointed out that synthetic fuels take a double whack at the greenhouse, for fossil fuels are burned both when they're made *and* when they're used. Thus, when Congress established Synfuels as part of the Energy Security Act of 1980, it mandated yet another National Academy study. Revelle and Keeling had finally put global warming on the government's radar screen.

*The general belief is that they took Manabe's estimate of two degrees Celsius as the low bound and Hansen's of four degrees Celsius as the top bound and added half a degree Celsius on either side as a hedge.

• • •

JIM HANSEN WOULD KEEP IT THERE. IN A SENSE, HE IS THE MODERN, LESS QUIRKY COUNterpart to Guy Callendar, with better science and many more resources at his disposal, plus the advantage of working in the academic mainstream. In place of Callendar's hand-calculated radiation model, Hansen has a computerized GCM; he has the solid measurements of Keeling and others to tell him what's in the air; and he keeps very good track of global temperatures. Once a month, his group at GISS collects the reports from about ten thousand meteorology stations around the globe and adds them to a database that extends back to 1880—about the time temperature was first measured with scientific instruments.

In their first analysis of temperature trends, published in 1980, Hansen's group revealed that the cooling trend that had been announced by the Weather Bureau almost twenty years earlier had been largely mythical: the bureau's data had been biased toward the northern hemisphere, where the vast majority of meteorology stations are located. The GISS group found that temperatures north of the equator had dropped slightly between about 1940 and 1965, but that they had held steady south of it. And, ironically, both hemispheres had begun to warm in the mid-sixties, just as the global cooling craze was revving up. Earth's average temperature had risen about 0.4°F since then and about twice that since the beginning of the instrumental era. In other words, the mid-century cooling had not been as strong as previously believed, and temperatures had resumed their inexorable rise.

Hansen had added some elegant mathematics to his GCM, so that it ran faster and more efficiently than Manabe's, and with this nimbleness he could run different historical scenarios and search for a cause. By adding or removing what was known about volcanic eruptions, changes in greenhouse and aerosol levels, and solar luminosity over the past century, he showed that random, year-to-year temperature variations came mostly from eruptions and changes in luminosity, while the warming from decade to decade had probably been caused by rising carbon dioxide.

After almost a year of checking, writing, and revising (during which he was promoted to director of GISS), Hansen's group revealed these results in *Science*. "Anthropogenic carbon dioxide warming should emerge from the noise level of natural climate variability by the end of the century," they predicted, "and there is a high probability of warming in the 1980s. Potential effects on climate in the twenty-first century include the creation of drought-prone regions in North America and central Asia as part of a shifting of climatic zones, erosion of the West Antarctic ice sheet with a consequent worldwide rise in sea level,* and opening of the fabled Northwest Passage." Furthermore, "the continued increase of fossil fuel use would lead to about 2.5°C [4.5°F] global warming by the end of the twenty-first century, making the Earth warmer than it has been in millions of years—in fact, approaching the warmth of the Mesozoic, the age of the dinosaurs."

Walter Sullivan, the highly regarded science reporter for the *New York Times,* got that story onto the front page. A few days later, the paper's lead editorial argued that, despite the uncertainty of the science, the idea of a total change in U.S. energy policy was "no longer unimaginable." Similar editorials appeared in the *Washington Post* and other newspapers.

But when the traditionally conservative scientific community settled down to its collective morning paper, a few nearly spat out their coffee. Some impugned Hansen's integrity, accusing him of employing scare tactics to attract research funding—an oft-heard and unfortunate refrain that has been leveled by both sides in the greenhouse debate. If that *was* Hansen's goal, however, his tactics certainly backfired.

I AM SITTING WITH HIM IN A SMALL CLEARING IN HIS LARGE, PAPER-FILLED OFFICE AT the Goddard Institute for Space Studies, which is housed in a cramped and slightly dingy office building belonging to Columbia University, on the corner of Broadway and 112th Street on the Upper West Side. The more famous tenant is Tom's Diner, on the first floor, where Jerry and his friends supposedly hung out on the TV show *Seinfeld.*

Hansen speaks evenly, as usual. "There was a change of administration at that time, in 1981. The [Reagan] administration wanted to cut back on that sort of research. . . . That paper managed to get us in trouble with the Department of Energy, because they had by that time changed their perspective on this problem. The guy who was in charge of the carbon dioxide program [under Jimmy Carter] was replaced by a guy who really wanted to wind the research down."

"A hatchet man?" I asked, and Hansen nodded. It was too strong a phrase for him to use himself.

Shortly after Hansen made headlines, the Department of Energy reneged on a promise to fund his research. "The DOE position was further clarified," Hansen later wrote, "when they told an independent researcher that his funding would be terminated if he used results from our climate model." Hansen had to lay off five people and trim back his research

*A reference to the seminal work of John Mercer.

program. For about a year he worked forty instead of his usual eighty hours a week. Then he managed to win funding from the Environmental Protection Agency and returned to his productive ways.

THE PREDICTIONS HE MADE IN THAT CONTROVERSIAL PAPER HAVE PROVEN TO BE UNnervingly accurate, and the first came true within a year. In 1980 a record was set for the highest average world temperature in the history of instrumental measurement—substantially higher than the previous record set at the height of the Dust Bowl. So, although it would take the rest of the scientific community about twenty years to realize it (the first UN report to attribute the warming of the second half of the twentieth century to rising greenhouse levels would appear in 2001) Hansen provided the answer to the more important of Gilbert Plass's two thirty-year-old questions in 1980: rising carbon dioxide was indeed dragging temperatures along with it.

Of course, Stephen Schneider's and Wallace Broecker's previous predictions were confirmed as well. Thus, 1980 marks the start of a run of accurate predictions by the greenhouse scientists, which they continue to build upon a quarter of a century later. Jim Hansen would make a significant fraction of them.

AND AS LONNIE THOMPSON, BRUCE KOCI, KEITH MOUNTAIN, AND THEIR FRIENDS plugged away on Quelccaya, racing to the summit each morning and scrounging for leftover food in base camp at night, the warming hung on. Nineteen eighty-one set another record. Temperatures then dropped slightly in 1982—to about the heights of the Dust Bowl—and in 1983, the year they drilled to bedrock on the summit, temperatures rose almost to the level of 1981. It was the second hottest year on record. Looking back, it is no wonder that Qori Kalis began to shrivel up.

But Lonnie had no time to dwell on it.

·9·

CHINA OPENS UP

At the Sees' pension in Lima, at the start of the 1984 expedition on which he would find Qori Kalis in full retreat, Lonnie received word that the National Academy of Sciences desired his presence at a briefing about research opportunities in China. Richard Nixon had normalized diplomatic relations with Mao Zedong's communist government in the early 1970s, and now, more than a decade later, the two countries were planning their first, limited scientific exchange. The academy flew Lonnie back to a resort in Virginia for a week, then returned him to Lima.

The meeting was mostly about protocol, how not to "make mistakes and get our Chinese colleagues in trouble," according to Lonnie. Photography would not be permitted; travel would be greatly restricted. He would be prevented from crossing certain high mountain passes that the military viewed as strategic even if interesting glaciers happened to cross them.

But Lonnie's interest in China did not spring from the government. Ever since graduate school, he had been hoping to visit the ice caps rumored to exist on the far reaches of the Tibetan Plateau, when he'd heard about them in a course taught by Richard Goldthwait, the now deceased founder of the Institute of Polar Studies.

Colin Bull relates that Goldthwait had taken part in a glossy commercial expedition "in '46 or thereabouts," to look for a mountain, allegedly higher than Everest, that had been sighted by the pilots flying bombing missions over the Himalayan "Hump" from India to Japan near the end of World

War II. "I don't know how much 'research' he was doing," says Bull. "[The expedition] was sponsored by the people who make Biro pens."*

However, Goldthwait did capitalize on the connections he made on the Biro jaunt to initiate a true research program on China's Loess Plateau. Lonnie recalls a particularly vivid story Goldthwait told in class of being caught doing fieldwork on the plateau on the day in 1949 when Mao and his Red Army seized power. The communists actually fired at the airplane carrying Goldthwait and his colleagues out of the country, as it was leaving the airstrip. They radioed the pilot to return, but he kept on going. It was one of the last planes to make it out.

And while the Biro expedition came up dry in its search for the world's highest mountain, Goldthwait uncovered a few papers written by Western geologists in the 1920s that reported the sight of distant ice caps on the Tibetan Plateau. Sixty years later, these claims remained unsubstantiated.

AS SCIENTIFIC DÉTENTE WITH CHINA COMMENCED, BULL HAPPENED TO BE CHAIRMAN of a working group on glaciology associated with the Scientific Committee on Antarctic Research (SCAR), an international body charged with coordinating research at the bottom of the world. "The Chinese were absolutely desperate at that stage to become noticeable in science," Bull recalls. "They were going to do three things: (a) transplant a human heart, (b) set off an atomic bomb of some sort, and (c) establish a research station in the Antarctic. Those were their stated three methods of achieving greatness."

Chinese science had been decimated by the Cultural Revolution, but what remained of the country's premier institute for glacial studies still existed in Lanzhou, the city at the eastern edge of the Tibetan Plateau that once marked the start of the legendary Silk Road. Bull says that when he visited the Lanzhou Institute in 1983 he engaged in "some highly interesting conversations" in which the Chinese authorities—assuming that a chairman such as he had the power of command—tried to persuade him to send a few of their people to Antarctica. He informed them that Western science didn't work that way; he didn't tell people what to do, he asked them, and a proposal such as theirs would have to be weighed on its scientific merits by a committee. He gave a series of lectures about climatology and established a good relationship with the director of the institute, Shi Yafeng. "So," Bull points out, "I did a little bit of spade work for Lonnie there."

*Being British, Bull employs the generic name for the ballpoint used in most parts of the world. This new, convenient alternative to the fountain pen had been invented in 1938 by two Hungarian brothers named Biro. It became a big fad after the war, and Bic later bought the Biro patents.

Lonnie had corresponded with Shi Yafeng as early as 1978 and had met him at a conference in Australia the following year. He says, "Shi reminded me of Yoda in *Star Wars*. He looks like Yoda in many ways, and he's wise, very wise, because he's been through the Cultural Revolution and he's still active. He's in his late eighties now."

SO, IN 1984, NOT LONG AFTER CASTING A LAST BACKWARD GLANCE AT THE SHRIVELING tongue of Qori Kalis, Lonnie visited China himself, on a trip that he remembers as the loneliest of his life. He traveled for three months in the company of twenty Chinese scientists and their assistants, none of whom spoke English.

"Everyone in China was wearing the blue or gray Mao suits," he says, "and there were still horses in downtown Beijing. If any country in the world has changed at lightning pace, it's China."

His host was the new director of the institute, Xie Zichu, another of the rare scientists to have survived the Cultural Revolution (during which, according to Lonnie, he had been "put out to raise hot peppers somewhere in western China").

Although Goldthwait's fabled ice caps may have held the status of myth among Western geologists, the Chinese were quite familiar with them. They first took Lonnie to about as remote a place as one can imagine: the eastern half of the Tien Shan Range, which spills east into China's northwest Sinkiang Province from neighboring Kazakhstan. The Tien Shan lie west of the town of Ürümqi, about two thousand miles from Beijing.

Lonnie had taken along Bruce Koci's composite hand auger, sounding equipment to measure the thickness of the glaciers, and a large stash of plastic water bottles. His extreme loneliness might be explained by the amount of spare time he had on his hands: it took him only three days to realize that this half of the Tien Shan had no glaciers worth drilling, yet his hosts' rigid travel plans kept him in the range for six weeks. He took samples anyway, to keep busy, and befriended Yao Tandong, a graduate student who was completing his Ph.D. on the water discharge in the Ürümqi River, which issues east from the Tien Shan. They took long walks together, conversing mostly in sign language.

After returning to Lanzhou, the group traveled northwest to the Qilian Shan, the range that defines the rough, historical border between ethnic Tibet and Mongolia. The Gobi Desert stretches to the north, the highest portion of the Tibetan Plateau, known as the Qaidam Basin, to the south.

THE MINUTE HE SAW IT, LONNIE KNEW THAT THE QILIAN SHAN'S DUNDE ICE CAP WOULD be a good place to drill: vivid horizontal strata were evident on its vertical margins; it was about the size of Quelccaya, twenty-three square miles; and it was mostly flat, with a 17,500-foot central dome—high enough to remain frozen year-round.

Domes and saddles are known as "origination points" in the ice drilling trade. They tend to be the best places to drill, because, owing to their one- or two-dimensional symmetry, they tend to be places where glacial flow lines originate, and where horizontal movement is thereby minimized. The ice tends not to move en masse, but simply to thin out, as opposite streams flow symmetrically downhill in two or four directions. Consequently, there is less chance of a break in the strata, and it is more likely that a core from an origination point will archive a continuous climate record all the way to bedrock.

Again Lonnie accomplished his work in a few days but remained in the mountains for six weeks; however, this time the director, Xie Zichu, had joined the expedition. He now accompanied Lonnie on his walks, and the two became fast friends.

Not one simply to hang around with the boss, however, Lonnie also joined the Mongol guides and horsemen who supported the expedition in the target practices they held to while away the hours after dinner; his experiences as a youth in West Virginia, shooting squirrels off the tops of beech trees with a .22, served him handsomely there. (It had also qualified him to train as a sharpshooter during his short stint in the military.) The Mongols hunted for the expedition's meat—wild sheep and horses mostly, a welcome change from the guinea pig at Quelccaya—and Lonnie particularly enjoyed the horsemeat.

When he'd returned to Ohio, he received a letter from Xie Zichu saying how pleased they'd been to have him, how well he fit in, that he was a "real Chinese." The letter still hangs in a frame on Lonnie's mother's living room wall in Gassaway.

SO HIS SIGHTS WERE SET ON DUNDE. THE SHALLOW CORES AND SOUNDING DATA FROM his solitary reconnaissance provided enough information for a proposal to NSF (which included the letter from Xie). He encountered a serious diplomatic hurdle, however, in a State Department rule that forbade the spending of U.S. tax dollars in the People's Republic of China. He could use NSF money for science, but not for trucks, porters, food, nor other logistical expenses on the expeditions themselves. That money had to come from a private source.

Lonnie soon discovered that even his next, relatively small expedition would cost more than the most obvious alternative, the National Geographic Society, ever spent on such things. But the society had a great deal of interest in China and "some sort of loan arrangement with the Chinese," so that Lonnie was "soon able to arrange for National Geographic to pay for the logistics and NSF to pay for the science." National Geographic even bought a few large six-wheel-drive trucks for the next two Dunde expeditions, as well as the subsequent multiyear effort on Guliya, which Lonnie still sees as the greatest logistical challenge of his career (so far).

Owing to the delays induced by these details, 1985 was the first year in a decade in which he did not take a major expedition.

IN JULY 1986, LONNIE, DAVE CHADWELL, BRUCE KOCI, AND KEITH MOUNTAIN FLEW TO Beijing with about twenty-five huge duffel bags checked in as excess luggage. They were met by Ms. Wu Xiaoling, the new co-leader with Xie Zichu of the Chinese delegation.

The very presence of Madam Wu, as the Americans came to call her, indicated just how far Chinese science had been set back by the twelve-year break in the training of new scientists that had accompanied the Cultural Revolution. Wu was an associate professor at the Lanzhou Institute of Glaciology and Geocryology, yet she had been trained in Russia about mudflow. The Americans soon learned that she knew next to nothing about ice and was an insecure and difficult individual as well. Most of them still shudder at the sound of her name.

The scientific aim that year was to conduct a more complete reconnaissance than had been possible on Lonnie's solo visit. And he was not necessarily committed to returning a third time; he wanted to make absolutely sure that there *would* be a record at Dunde before expending considerable time and effort—not to mention the goodwill he had only recently won back from NSF—on a costly deep-drilling operation on the other side of the world.

The trip got off to a slow start, as they were delayed in Beijing for ten days. The Americans suspect that Madam Wu contrived to confuse the customs process in order to spend extra time in her favorite city.

This time they weren't allowed to approach the Qilian Shan from the south, as Lonnie had in 1984, for the route passed through a long-range nuclear ballistic missile base near the town of Delingha. This meant they had to circle west around the range and approach it from the north, which delayed them another four or five days.

After the last large town, Golmud, the ethnic mix changed from Han to Mongol, and the letters on street signs from Chinese to Arabic. They passed camels carrying people and their belongings, and yaks and clusters of yurts surrounded by grazing livestock, owned by Mongol nomads. They enjoyed their last good meal in what Dave calls a "one-room truck stop," wallpapered in newsprint, with a kitchen consisting of a single large wok on a stove fashioned from a fifty-five-gallon oil drum.

IT WAS A LONG WAY IN FROM THIS DIRECTION, SO THE CHINESE ARRANGED FOR THE team to ride on small Mongolian horses, mounted with what are universally remembered as "hard" wooden saddles. That horse ride has generated more stories than any other single event in the annals of Lonnie's forty-odd expeditions. Wretched travel experiences usually do. "Good life experience, but miserable while you do it," says Lonnie.

Bruce says the Mongols expressed their respect and affection for the Americans by providing them with "what they called 'very strong' horses. That meant they'd be breathing fire and standing on their hind legs and everything else."

And they all remember Keith, the largest of the group, looking especially ludicrous in the saddle, because his horse was so small and his stirrups so short that his knees nearly touched his chin. I once asked him if his feet could touch the ground.

"I tell you, I *wish* they were touching the ground. I'd have got off and walked the damn horse. . . . First of all let me state, unequivocally, that I do not respect the horse as a primary means of transportation. They have huge heads, comprised mostly of bone; they smell—even worse when wet—and they have chronic gas. Being there behind a horse . . . there's only one place to be, the lead dog. There's a reason they say the lead dog needs to be there. . . . Above all, when mounted on one of these things, you quickly realize that they possess neither clutch nor brake. This is inherently a bad idea. The things would just take off."

Thirty yards from the road head they encountered a flood-swollen stream that was moving so fast they couldn't go straight across; the current dragged them sideways.

"So here we are in the middle of goddamn China, sitting on this horse," Keith continues. "We're crossing a river, bloody sitting on this horse, and the water's up around the saddle. I really did think I was going down for the count this time. . . . The horse is not happy; I'm not happy; and it's heading into deeper water. . . . Chadwell's horse isn't happy; Chadwell isn't happy. . . . So what happens? The horses decide there's security in numbers. They start bumping into each other. We're just about falling off these things. Finally they get out of this creek, and we're off for three days through the mud and muck of this vast, treeless plain that stretches into the distance like a bad dream.

"And then it started to snow. There was no escape from this trial. Snow was building up on the horse, on me; it was cold, wet, and my ass was both sore and itchy. This was akin to the retreat from Stalingrad, and it was just the beginning. It was a mess. You're out there in the mud, out there way in the middle of goddamn nowhere. The goddamn snow's starting to build up in your lap. The horse has got gas. It's buried in the mud up to its knees, and I'm thinking, 'This was not a good idea. There's just something very wrong with this picture.'

"These horses were amazing, though. The more it snowed, the more the animal couldn't see where it was putting its foot. And you're going up these steep cliffs. I thought, 'Shit. Either it's going to fall and break its ankle or it's going to fall off the cliff.' And you certainly had no real sense of where you were; you were in the middle of this big whiteout. It was snowing. The good news is it stopped snowing. Then it started raining. And trying to set up camp in the mud with all these animals and all this equipment . . ."

Dave remembers seeing a group of camels crossing the plain in the distance, veiled by blowing snow: "That was just a bizarre concept for me, to see camels in the snow."

And to add to the confusion, in the middle of the snowstorm, according to Keith, "one of the guys jumps off, hauls out his damn gun, and shoots this, like, wild mule—just lays it out. Goes over there, cuts this damn thing up; guts it right there—skinned, gutted, sliced, and diced in a heartbeat; and then throws the dead carcass—three or four hundred pounds—across the back of one of the packhorses. Well, the horse smelled nothing but death. It clearly sensed that this was a very dead but close relative it was expected to carry for the next two days. The poor creature was terrified.

"These Mongols were genuinely tough folks. One was unwilling to complain or show any degree of discomfort around them. You had a distinct sense that they didn't really see a lot of difference between you and the thing they just shot and gutted. On the other hand, they seemed to like us, and, given my social standing, it gave me a sense of comfort and security to know that somewhere in the world I would always have friends like these to call upon."

As is often the case on the first day of a trek, the team hadn't had a chance to sort through their gear, so they couldn't find what they needed when darkness quickly fell. They were caught in the cold and wet wearing blue jeans and cotton shirts, and they slept on the hard ground without pads: "one of the more miserable sleeping nights," in Dave's opinion. They stuffed their wet clothes into their sleeping bags in order to dry them with body heat as they slept.

• • •

BUT NOTHING COULD KEEP KEITH DOWN FOR LONG, OF COURSE; SO WHEN THEY EVENtually got their act together and settled in for the ride, he graciously offered to tutor Yao Tandong in English. (Yao's first few phrases must have been quite colorful.) In the two years since Lonnie had joined him on their walks in the Tien Shan, Yao had completed his doctorate and a postdoc in Grenoble and had now returned to a permanent position at the Lanzhou Institute. This was the start of his meteoric rise to the directorship of the institute. He would also become Lonnie's steadfast partner on all of his future expeditions to China—and one of his closest friends.

While the Chinese, with the exception of Yao, were no help with the physical labor, the Mongols and their long-suffering horses were magnificent. Dave remembers the Mongols' "calm way of approaching things. . . . They did all the portering and the heavy, heavy work." They erected a large tent the moment they reached base camp, and Dave recalls that the cooks "had spent all night butchering the horse. We came in the next morning, and there was meat strung everywhere. The food was very good—quite a contrast from Quelccaya. There were fresh dumplings and that kind of thing every day."

THE SURFACE OF THE GLACIER PROVED HARD ENOUGH FOR THE HORSES TO WALK ON, SO they were employed to carry the equipment to the drill sites or to drag it on one of the plastic children's toboggans that Bruce had had the good sense to pack. Nevertheless, Keith remembers that trip as being "very demanding physically," because even though they took in less equipment on such reconnaissance missions, they "had considerably less logistical support, too, which meant more shit for us to haul up and more onus on us to get the samples, get them processed, and get them back." They had not yet made the transition to camping on the summit at this point, so they faced "a five-mile walk every damn morning. The first couple days were just agony."

Part of the reason Keith remembers the trip as being so demanding may be that he did nearly all the hand drilling, which is about as tiring and repetitive an activity as weight lifting, especially at an altitude of about 17,500 feet. Chadwell, who mainly surveyed, describes the drilling as "a killer. I can't think of anything worse to do at altitude. It just zaps you. It laid me flat very quickly, so I only helped at the start. When we got fairly deep, I wasn't strong enough. Keith did it. Physically, he is probably the strongest of the bunch."

Lonnie wanted to test the idea of bringing some of the ice back frozen that year—for the first time ever—so Bruce had found some insulated boxes and Cryopaks of the sort one freezes for use in a picnic cooler. They had taken along two boxes with enough room for about half of one of the thirty-five-meter hand-drilled cores.

As it turned out, they did uncover a glitch in this process, but it had nothing to do with the integrity of the boxes. After they had carried them "essentially all the way across China," in Lonnie's words, United Airlines promptly lost them somewhere between Beijing and Columbus. In a scenario familiar to many, persistent calls to the lost baggage department at United produced only frustration; after two days, Lonnie received "final" word that there was no record of the core boxes anywhere in the airline's system. "And these are *big* boxes," he points out. "You don't just lose them. They're huge!"

He finally put in a call to the president of United, in Chicago. Within half an hour they had found the core boxes in the basement of the Beijing airport. "They put them on dry ice and shipped them, and we finally got them and were able to analyze them. But it raised all the red flags about what might happen the next year."

LAB TESTS SHOWED THAT THERE *WAS* AN ANNUAL SIGNAL IN THESE CORES, SO LONNIE and Ellen decided it would be worthwhile to return and drill to bedrock. Ellen describes the conservative and somewhat ad hoc approach they took on the grant proposal: "We thought, 'Well, we got fifteen hundred years at Quelccaya. Wouldn't it be great if we could get five thousand at Dunde?' because Dunde was about the same thickness but receives less snow. That turned out to be serendipity."

Still, the idea of drilling ice in subtropical Asia (thirty-eight degrees north latitude) points to the same instinct for exploration that had sent Lonnie to Quelccaya at the beginning of his career. He and Ellen had no *specific* questions to ask about the midlatitudes. He was going to Dunde just to see what he could see.

Since the Quelccaya cores had reached nowhere near as far back as the last major change to encompass the globe, the Younger Dryas, they had done nothing to alter the prevailing view that climate was driven mainly from the poles. And the mainstream ice coring community remained very much in the grip of its polar obsession. The core they had finally recovered five years earlier at Dye 3, in Greenland, had provided tantalizing hints of multiple, dramatic, and abrupt climate changes there during the late Wisconsinan, so scientists on both sides of the Atlantic now lusted after Greenland's ultimate prize: a deep core at the island's summit, where the ice is almost two miles thick. But the infighting among the various funding agencies persisted, unfortunately, so there would be no movement in that direction for another three years. In any event, the North Atlantic

School did not rally to the Thompsons' support. As Bruce puts it, "When it came to deep drilling at Dunde, we got the same kind of reviews, 'Oh, Quelccaya was just a fluke. It's not going to work.' "

But the grant came through anyway.

ELLEN, BY THE WAY, HAD ESTABLISHED A STRONG FIELD PROGRAM IN GREENLAND AND Antarctica by this time, specializing in short cores with high temporal resolution; but while she *did* work at the poles, she was definitely *not* a member of the polar mafia. She shared both drilling equipment and drilling crews (most notably Bruce and Keith) with her husband and—in contrast to the usual drawn-out polar project—generally accomplished her field objectives in a single season with a small team.

To see that she was no slouch either, consider the trip she took with Bruce over New Year's 1987 (less than six months after he returned from Dunde) to a spot they named Plateau Remote, roughly 150 miles from Antarctica's Pole of Relative Inaccessibility, the geometric center of the continent and, presumably, the coldest place on earth. Russia's Vostok Station, which should be a bit warmer since it lies about four times farther away from the same pole, holds the record for the lowest temperature ever recorded: –129°F.

Ellen was the leader and only woman in a group of six who were airlifted to the middle of the oceanlike East Antarctic Ice Sheet to drill for twenty-one days without what is known as "station support." Alone on the ice, they set up their own camp and cooked for themselves in temperatures that ranged, according to their thermometer, from –40°F to –15°F Only their cook tent had heat, and then only at mealtimes. They were about two miles (the thickness of the ice cap) above sea level, which did not improve their ability to withstand the cold; and their only drilling shelter consisted of two air force pallets lashed together and propped up as an ersatz windbreak.

The safety protocol called for them to make radio contact with South Pole Station once a day. Sometimes atmospheric conditions made that impossible, so they would relay a message through Siple Station or make no contact at all. Their weather reports were so consistently dire that the residents at "Pole" set up a betting pool to guess when the team would finally give up and call for a flight home.

"They didn't realize there were no whiners on our team," says Ellen. "I don't put up with whining. If you go to do a job, you do it—unless, of course, you've got a life-threatening situation. We were all in good health, but we were losing weight. The altitude wasn't the issue; it was the cold—just fighting the cold and the hard work all the time, and the fact that our cooking facilities were inadequate. None of us could eat enough to sustain weight. We were eating six thousand calories a day, and it wasn't enough."

In these challenging conditions, they retrieved two two-hundred-meter cores with an electromechanical drill; thirteen more ten- to twenty-meter cores by hand; and excavated a pit three meters deep, three meters long, and two meters wide, taking samples on a ten-centimeter grid from three of its walls.

"In the nineties, when I was going to Pole every year for another project, people would still say to me, 'God, you should have seen how you guys looked when you got off that plane from Plateau,' " Ellen continues. "Yes, toward the end it was getting a little rough."

At "Pole" they discovered that their thermometer had been reading warm and temperatures at their field site had actually dropped as low as –60°F. Bruce, who has been on one or two difficult trips in his time, remembers Plateau Remote as one of the more difficult of his career.

He left for the deep drilling on Dunde roughly six months later.

CHADWELL COULDN'T MAKE IT. HE WAS STILL AT OHIO STATE, BUT HE HAD GRADUATED the previous spring and was now pursuing a graduate degree in geodesy. He was replaced by Henry Brecher, the mapping specialist who had taken the first metric photographs of Qori Kalis back in 1978, and Mary Davis appears on the roster for the first time. She had been hired to do analytical work in the dust lab four years earlier, the first permanent addition to the Thompsons' group.

Lonnie's longest-standing friend on the Chinese side, Xie Zichu, was missing and would never join Lonnie again. It seems that the Communist Party was putting him out to pasture. Lonnie points out that Xie "was enthralled with the Western way of life and democracy" and later lost his job for leading a march in Lanzhou during the events at Tiananmen Square. Thus, Madam Wu alone led the Chinese contingent of twenty-one, which included Yao Tandong. Counting Mongol porters and guides, there were fifty participants all told. The trip began inauspiciously.

As they unpacked their fleet of taxicabs at the entrance to their Beijing hotel, two scam artists took advantage of the confusion to steal Lonnie's small carry-on briefcase, containing twenty-four thousand dollars in travelers checks, all of his cash, and his field books from the first two Dunde expeditions. When American Express balked at the idea of replacing the checks, Lonnie was obliged to remain in Beijing for a few days, while the rest took the seventy-two-hour train trip to Lanzhou.

It seems that someone was at least trying to compete with Lonnie in the area of high-altitude ice core drilling even that long ago, for Bruce remembers meeting a very well funded, "big, strong, burly" Japanese group at the Lanzhou Institute: "They had brand-new Land Rovers—everything—and they sort of looked down their noses at us. They were going to go to this ice cap and drill fourteen million meters of ice; and here were Henry Brecher, myself, Mary, Keith Mountain—an unlikely-looking crew if ever there was one. We looked like a bunch of janitors. It was a funny contrast."

The janitorial staff managed to avoid a horse-riding adventure this year by taking a train through Delingha at night, when the missile base would be shrouded in darkness. Approaching Dunde from the south, they made it all the way to base camp by truck, just in time to celebrate Lonnie's thirty-ninth birthday (the first of July) with what Mary remembers as a "twenty-course" Chinese meal. She was unable to enjoy it, however, having been laid low by acute mountain sickness for the first time in her life.

Mary was awed by Dunde's immensity: "At the time it seemed like an insurmountable task to me. But when I look back at the places we've been since—like Guliya and Huascarán—it's a small ice cap, comparatively. If we had drilled Dunde in '97 instead of '87 it would have seemed like nothing.

"I've got a soft spot for Dunde because it was the first place, and I thought it was very striking scenery up there, even though it's not very remarkable compared to Dasuopu [in the Himalaya]. I didn't bond with the Himalaya on an emotional level the way I did with the Qilian Shan. At Dunde I felt like I'd come home for some reason. I just felt—once I got over the altitude sickness—I felt really comfortable. I felt really happy. I didn't want to go home. I didn't want to leave Dunde. And I've never felt that way about another mountain. By the time the expedition is over, I usually just want to get out of there; I just want to go home. I didn't feel that way about Dunde. I was sorry to leave.

"Dunde and Guliya and Dasuopu and Gregoriev, which was up in the [Kyrgyz] Tien Shan—in one way they all looked the same: glacial erratics and the brown earth and the blue sky and the white ice and these mountain chains off in the distance—monochromatic in a lot of ways. For some reason I always felt a lot more comfortable and, I don't know, *at home* there than I did in the Andes. I always felt more at ease on the Tibetan Plateau.

"I think if you're suited to this kind of work, you're going to bond with the first place, because it hits you at a deeply emotional level. It was the first time I'd been out of the U.S., the first time I'd been in the mountains, first time I'd been on the ice. Everything seemed kind of magical because it was new; it was the first time.

"I've seen other people go into the field for the first time and they think they've landed in hell. They're uncomfortable; they can't stand it; they want to go home. Some people are suited to this kind of work, and some people aren't, and you don't really find out until you're there."

HOWEVER, IN THE INITIAL FOG OF HEADACHE, NAUSEA, AND LISTLESSNESS THE TEAM ENcountered a glacier covered in sugary hip-deep snow. The horses that had so valiantly dragged their equipment to the summit the previous year "got halfway up and sunk in to their bellies," according to Lonnie. "There was no way they could pull those sleds."

The Americans soon found themselves in a makeshift camp, halfway to the summit, at the end of a long line of equipment that had been abandoned along the route. The Chinese had been little help, and their leader, Madam Wu, was already declaring it time to give up. Even Lonnie admits that "for a couple of days there we weren't sure we were going to be able to drill this ice cap."

But at that all-time low, with the success of the expedition in jeopardy and even Lonnie's optimism sagging, Keith remained indomitable. Mary remembers him regaling them with "a story that made me laugh so hard I almost got sick from that." So they sent him down to base camp to try to raise the Chinese morale.

As he descended, Keith says, "Christ, there was shit everywhere. Stuff was scattered all over the glacier. We were moving it up bit by bit, but it was damn clear the Chinese folks weren't going to do it. And you couldn't really blame them. They were in a difficult environment. They weren't always well dressed or prepared for the conditions. I mean they were eating watermelon up there. The temperature's in the teens, and their primary food is watermelon, for Christ sake. These people were just decomposing, just falling apart. They get extremely low salaries for what they do; they're eating dog shit; they're miles away from home with no chance of going home for the next two months; and they get the same pay—the same sense of reward—hauling shit up the mountain as they do sitting down in the sand.

"Now, on our side, we are very driven to accomplish the task. It doesn't matter what the Christ you've got to do. You will put it on your back. You will drag it. You will do *anything* to get it up there. This is a group of people that just don't back off lightly. It is ugly. We're not pretty. We smell a lot and we have horrible bodily habits; but the fact of the matter is, we can take a disaster and turn it into something that can be a success. And in order to get it through to these people that this was not a game, that we had not come this far to end it here unless we had done our very best—well, you have to convey that.

"I did things like divide up the loads so that they could be moved on up and make them feel like things were being accomplished. I convinced them that it could be done without any major overhaul of the way they were working. People were carrying awkward loads. We changed that."

At about six one morning, the Americans on the glacier were awakened by yelling from below.

"I crawl out of my tent, and there's an army heading for us," says Mary. "They had these red-and-white pennants."

"Here came the whole Chinese party up over the ridge with flags flying in front of them," adds Lonnie, "everyone roped to a sled! It was a man-haul! We all joined in, sometimes six people to one sled. It was an awful job getting all the drilling equipment up there, but it was done, and it was done because everybody worked together and had a common goal. It was very focused. I will never forget the Chinese coming over the ridge with the flags flying—the human side of it, you know? They turned the tide."

"They became very patriotic about it, and that was great," muses Keith. "Whatever it takes. Whatever it takes."

FROM HIS SEAT IN THE CRATER ON KILIMANJARO, BRUCE OBSERVES THAT CONDITIONS ON this African mountain are similar to those they encountered at Dunde: "Warm during the day—hot sun and all that; so it was kind of a bare-bones thing. We went with generators, because we were in the middle of the monsoon season, and solar panels don't work in thunderstorms. We were right out in the open all the time—no dome in those days—so the drill would heat up. We learned *that* on the first core. We could drill until a certain time in the middle of the day, depending on the wind and the cloud conditions. So what I would do is get up at the crack of dawn—as soon as you could see any light at all. I'd go and start the generator and start drilling, and we would drill like mad until—you could tell when the core barrel was getting too warm; it would start to stick a little bit. Then we would turn everything off, run the drill down the hole [to keep it cold], and take off until later in the afternoon. We would have two or two and a half hours later on, and we would drill until we couldn't see the gauges anymore. We didn't have any lights.

"We hit bedrock at night the first time. I couldn't see the gauges, but it was pretty obvious when we hit because the drill bounced and twitched and practically flew out of the hole. We absolutely destroyed a set of cutters doing that."

"We always joked about Bruce getting started early every morning before the sun was up," Lonnie says, smiling. "Pow! You'd hear the generator starting, and everybody would groan. But we were all on the same page. We were going to drill three cores to bedrock, and we needed to get the job done so we could get home. Get it done, accomplish what we came there to do, and leave."

Since this was the first time they lived on the ice, it was also the first time they experienced the "trickle-up" effect that I would later encounter on Sajama: the drillers, who were working the hardest and at the highest altitude, tended to get the least nutritious food. There was plenty to eat at base camp, but very little was getting to the top. Everyone was hungry, so the porters carrying food up the glacier ate most of the provender before it got there. The drillers finally arranged for the Mongols to bring food that had been precooked in pressure cookers up to the top by sled once a day. "Some sort of mixed-up goulash, but very welcome," recalls Lonnie. "We would bring the pressure back up on the stove up top and get some warm food each day."

Once they got the drill going and had established a pattern, Mary descended to base camp to help melt the first core with a fellow they nicknamed "Smiley" for his friendly manner and beautiful white teeth. These two split the core, layer by layer, into two sets of bottles, one for the Chinese and one for the Americans. But Mary's most vivid memory from that week or so in base camp was witnessing perhaps the most memorable of Madam Wu's many antics. Believing it was her right (more than her duty) to see what was going on at the drilling site, but unwilling or unable to walk up the now well-beaten track, Madam Wu demanded that the Mongols pull her to the top in a sled.

"Now, fuel had to be man-hauled up regularly to keep the drill going," Mary recalls. "The drillers were always sending notes down, 'Send more fuel! Send more fuel! Send food!' " Mary laughs. "So Mrs. Wu went up that day instead of the fuel. And the next day Lonnie sent down a very angry note, 'What happened to our fuel?' and I had to say, 'Well, you saw Mrs. Wu come up. That was the fuel, basically.' And all she did up there was scream at the Chinese and get everyone all worked up and then get back on the sled and get man-hauled back down."

This was another reason the Mongols and other ethnic minorities who would porter and guide on later expeditions worked so hard for Lonnie. He treated them with respect. He points out that the Americans "looked pretty good compared to Wu Xiaoling. We didn't insist that the Mongols pull us in sleds to the summit. They hated her for that."

• • •

DESPITE THESE AND OTHER DISTRACTIONS, INITIATED MAINLY BY THE CHINESE (A FEW bloodcurdling cases of animal abuse among them), the Americans succeeded in drilling three cores to bedrock, all about 140 meters long. The deal called for them to leave one frozen core in China and melt a significant fraction of any they took home, so their take from this expedition consisted of the bottled core that had been split by Mary and Smiley and a second complete core—its upper fifty-six meters, corresponding to the last 150 years, in liquid form, the rest as ice. The Americans refer to the core they left behind as the "sacrificial core," because nothing was ever published about it outside of China.

Back in Lanzhou, they reencountered the big-time Japanese drillers only to discover that they had not had quite as successful a season. In fact, the Americans had retrieved more ice by hand the previous year than the Japanese had retrieved with all their fancy equipment this year. "We didn't say much," observes Bruce, "but they didn't want to talk about it at all."

Lonnie went ahead with the ice under his watchful eye and left the others to wrap things up. Thanks to a heads-up letter written beforehand to United Airlines (and copied to NSF and National Geographic), both he and his cargo were treated like VIPs all the way. He was taken out to inspect the boxes at every stop and even treated to dinners in town on one or two of the longer layovers. The core arrived "in beautiful condition" after traveling more than halfway around the world.

"DUNDE WAS, AGAIN, ONE OF THOSE FOUNDING EXPERIENCES," SAYS KEITH. "GOING TO China; traveling in that part of China; seeing that part of the world . . . It was great. I wouldn't say I like any place better than any other, but certainly I do believe that there's something about a really seminal experience that stays with you.

"The other good thing was the Chinese boys never left themselves short of beer—world's worst beer, but it was beer nonetheless. That always helped out. Those were good times."

THEY DISCOVERED PERHAPS A MORE VALUABLE BONUS WHEN THEY EVENTUALLY ANAlyzed the cores: Dunde's record reached back one hundred thousand years, not the mere five thousand Ellen and Lonnie had guessed in their grant proposal. (This is the reason Ellen spoke of "serendipity" earlier.) To the Thompsons' way of thinking, this was the main point of the paper they published in *Science* in 1989—although it is also worth noting that Dunde's ice displayed the same change to dusty conditions at the transition to the

Wisconsinan glacial stage that had been observed in every ice core reaching the glacial that had been recovered at the time.

But the press, including the *New York Times* and the *Wall Street Journal,* seized upon a minor point at the end of the paper, regarding a "striking feature of the [oxygen isotope] record" over the previous sixty years. Dunde's isotope ratio had risen dramatically in that period, indicating that temperature had risen as well. "The only comparable values are for the Holocene maximum, 6,000 to 8,000 years ago," the Thompsons wrote. "Climate model results of Hansen and others suggest that the central part of the Asian continent might be strongly affected by the anticipated greenhouse warming."

This statement was more than powerful enough to galvanize the press at the time; but, in fact, it represents a conservative interpretation of the data. In conversation, Lonnie now plainly states that the last fifty years at Dunde have been the warmest in about forty thousand years.

Meanwhile, Jim Hansen had been making headlines, too.

IT IS AN ODD AND PERHAPS DEMORALIZING COINCIDENCE THAT JUST AS LONNIE WAS fighting through to his high-altitude breakthrough on Quelccaya only to discover that high-altitude glaciers were beginning to disappear everywhere on earth, and as Hansen was pointing out that warming had resumed and fossil fuel burning had probably caused it, the prospects for a solution to the greenhouse problem took a decided turn for the worse. Ronald Reagan's eight years in office, which commenced in January 1981, transformed environmental protection from an issue everyone could love into a partisan battleground.

Recall that Hansen had lost the support of the Energy Department when Reagan took office but returned to his hardworking ways by winning funding from the Environmental Protection Agency. In retrospect, this can be seen as the first salvo in an interagency battle, signifying grave differences in mandates and philosophies, that has persisted to this day. EPA has continued to play the watchdog against the environmental threats posed by industry even as it has been gutted time and again, first by Reagan, then by George Bush, the father, and now, as I write this, by George Bush, the son. Rachel Carson's eloquence before Congress in 1963 has had a profound and lasting effect—although I must admit that the second George Bush's strategy of stacking scientific and health-care committees all through the government with his ideological comrades has come very close to neutralizing that effect altogether.

Since Reagan could see no evil in private enterprise of any kind, EPA's very existence was a thorn in his side, and it nagged him all through his presidency. In relation to the greenhouse issue specifically: in 1983 the National Academy of Sciences published the report that Congress had commissioned with the setting up of Synfuels (before Reagan took office). The NAS panel, which was influenced by the presence of one global warming skeptic and two economists, came up with the same range for climate sensitivity that had been proposed by the Charney panel a few years earlier, but presented this old news with a new, reassuring spin: "Overall, we find in the CO_2 issue reason for concern, but not panic." As scientists will, the NAS panel recommended more research, and the foot-dragging executive branch was happy to oblige—in an early example of what is now a time-honored stalling technique.

But EPA upstaged the NAS by releasing its own report just three days later, suggesting not only that "substantial increases in global warming may occur sooner than most of us would like to believe," but that they might change "habitability in many geographic regions," leading to potentially "catastrophic" results within decades.

This made the front page of the *New York Times,* and when Reagan heightened his pitch by labeling the EPA view "alarmist," he promoted it to the TV news. The media, the public, government agencies, and even senators and congressmen quickly developed an interest in global warming; and this, in the end, worked against Reagan's purpose, for the underlying message of both reports, of course, was that, no matter what the consequences might be, hotter times *were* on the way.

But Democrats and environmentalists also bear some responsibility for turning the environment—and global warming in consequence—into a partisan, litmus-test issue. For they took every opportunity while Reagan was in office to wave the environment like a matador's cloak before his eyes, knowing he could always be counted upon to charge in with an outrageous and eminently quotable response. Al Gore was a leading partisan, having followed in his father's footsteps upon graduating from Harvard and entered politics (though Lonnie, who came to know him at the end of the decade, believes Gore has a sincere interest in the environment and, thanks in large part to Roger Revelle, an impressive grasp of greenhouse science). First as a congressman, then as a senator, Gore held hearings about global warming all through the eighties.

Hansen testified regularly, issuing ever-more-serious warnings with each passing year—as, indeed, temperatures remained high. Apart from the occasional TV sound bite or short article in the back pages of a newspaper, however, the only practical effect seems to have been that in 1987 the Office of Management and Budget attempted to toe his official statements more to the Reagan line. Hansen said what he wanted to say anyway, as a private citizen rather than a government employee.

Environmental discussions were considerably less politicized in other parts of the world, however—and at the United Nations in particular. An international agreement to phase out the production of chlorofluorocarbons and other ozone-depleting compounds, the Montreal Protocol, was signed in the same year as OMB's muzzling attempt; and during the decade that it took for this landmark agreement (some call it "miracle") to be reached, progress was made on greenhouse policy as well.

The First World Climate Conference was held in Geneva in 1979. As the first international forum devoted exclusively to global warming, it resulted in no calls for action, but it did lead the World Meteorological Organization (WMO), the United Nations Environmental Programme (UNEP), and the International Council of Scientific Unions jointly to create the World Climate Programme, which held a series of workshops in Villach, Austria, during the first half of the eighties. The attendees of the last Villach workshop, in 1985 (speaking only for themselves, not for their respective governments), made a significant attempt to influence policy by declaring in a joint statement that the first half of the twenty-first century would probably see "a rise of global mean temperature . . . greater than any in man's history" and recommending that "scientists and policymakers should begin active collaboration to explore the effectiveness of alternative policies and adjustments."

There are many parallels between the ozone and greenhouse issues. (Some people still confuse the two.) The leaders of the ozone effort—Mostafa Tolba in particular, the Egyptian director of UNEP, who spearheaded the effort and gained much visibility in the media as a result—hoped to take advantage of the momentum from Montreal to forge an agreement to limit greenhouse emissions.

As the greatest greenhouse emitter by far, the United States had a huge stake in the outcome of any negotiations. As the largest financial patron of the UN and the nation with the most expertise of any in climate science, moreover, it held a very strong hand. When it became clear that the game would be played with or without U.S. participation, the Reagan administration—despite its intimate ties to the automobile and fossil fuel lobbies, which virulently opposed even the idea of negotiation—felt obliged to join in.

It has been observed that while "politics caught up with ozone, climate change was born in politics." The panels that negotiated the agreement in Montreal had based their decisions almost exclusively on the *science* of ozone depletion. It could be argued that the scientific case for limiting greenhouse emissions was at least as strong by the mid-eighties; but while science may have gotten the drop on industry the first time around, that first battle taught industry how to fight the next, which involved significantly higher economic and political stakes.

When global warming entered the global spotlight at the end of the eighties, however, the same balance of power that had caused Jim Hansen to be "fired and rehired" at the beginning of the decade led to what proved to be a crucial ambiguity in U.S. policy. DOE and the White House favored stonewalling; the National Academy wanted to "wait and see"; and EPA, and to a lesser extent the State Department, were in favor of action. The compromise solution, proposed at a WMO congress in May 1987, was to "establish an intergovernmental mechanism to carry out internationally coordinated scientific assessments of the magnitude, impact and potential timing of climate change"—the key word being *intergovernmental*. According to climate policy expert Shardul Agrawala, "[The] intergovernmental assessment mechanism which the U.S. finally proposed addressed DOE concerns regarding the [exclusive] involvement of 'official' experts. At the same time it precluded immediate action and provided an opportunity for the administration to buy time ('let's study the problem more'). Yet, by encouraging participation it also made an eventual climate convention more feasible, consistent with the goals of EPA and the State Department."

In March 1988, WMO's executive council sent a letter to its member governments, asking if they would like to participate in an Intergovernmental Panel on Climate Change (IPCC). Backroom negotiations then ensued about the precise form the panel would take, with the competing U.S. agencies again playing leading roles. In the end it was decided that the IPCC would comprise three working groups: one devoted to science, a second to impacts, and a third to response strategies. A peer-review process was implemented that would grow to be incalculably more cumbersome than anything ever applied to a scientific issue before and would invite endless opportunity for political shenanigans. Executive summaries, for example, would be subject to line-by-line approval by all member governments. This seemed to play into the hands of the fossil fuel lobby: as Agrawala observes, "Through this innovative set-up the founding fathers of the IPCC sought to advance what many thought was an oxymoron: quality scientific assessments by democratic consensus."

Few realized how clear the science would be.

THEN, WITH THE "HEAT" BUILDING ON THE SCIENCE AND POLICY FRONTS, CLIMATE ITself decided to make a statement. In 1988, the very summer the IPCC was established, as Lonnie and Ellen Thompson analyzed ice cores from the center of Eurasia that would demonstrate half a century's warming there, a memorable wave of heat and drought settled in on the centers of both that continent and North America—the exact places Jim Hansen had predicted global warming would begin.

Farmers in the American Midwest experienced their worst growing season since the Dust Bowl; southern cotton shriveled on the vine. Barges were stranded by the thousands in the Mississippi River. Civil War vessels last seen when Confederate troops scuttled them on their retreat from Vicksburg rose above the surface of the Big Muddy, a Mississippi tributary. The West experienced the worst forest fires in recorded history; more than six million acres burned. Almost half the forests in Yellowstone National Park turned from deep, fulsome green to smoky and skeletal black. President Reagan signed a four-billion-dollar farm relief bill.

On June 23, Senator Tim Wirth of Colorado called a hearing of the Committee on Energy and Natural Resources and warned the Washington press corps that something interesting might happen. Jim Hansen was among the witnesses, and he was armed with fresh news: his latest paper had just been accepted by the *Journal of Geophysical Research*.

In the fifteen minutes that one is given to speak in such venues, Hansen made three main points, which he vividly recalls and stands by to this day: first, that he was "ninety-nine percent confident" that the earth was getting warmer and that it was not a chance fluctuation; second, that the warming was sufficiently large to "ascribe with a high degree of confidence . . . to the greenhouse effect"; and third, that his circulation model showed that the greenhouse was already strong enough to increase the frequency of extreme events such as summer heat waves and droughts in the United States, especially in the Southeast and Midwest. He pointed out that the 1980s had already seen the four warmest years of the century and that the first five months of 1988 had been warmer than any comparable period in the instrumental record. He then racked up what would prove to be another in his unbroken string of accurate predictions: that 1988 would set a new global temperature record.

In the question period that followed, one senator, hoping to evoke a juicy quote for the headlines, asked if the greenhouse effect had caused the present drought. Hansen replied that the effect only alters probabilities; no specific drought could be linked to it directly. (This led to endless confusion. The senators kept trying to lure the experts—Manabe was one—into linking the drought to the greenhouse effect, but none would bite. Nevertheless, sometime later, a question on the game show *Jeopardy* indicated that Hansen *had* linked global warming to that specific drought.)

While he had prepared his written testimony in advance, Hansen had found no time until the night before, in his hotel room, to think about the words he would use in his brief time slot. There, and at an unrelated meeting at NASA headquarters on the morning of the hearing, he had struggled to come up with ways of presenting his findings to the nonscientist. He finally found an opportunity when he was surrounded by reporters just after the hearing: "It's time to stop waffling so much and say that the greenhouse effect is here and is affecting our climate now."

This apocalyptic warning, delivered in a monotone by a matter-of-fact, almost bland Midwesterner, on a day when the temperature broke one hundred degrees in downtown Washington, flashed across television screens and made newspaper headlines around the world.

Hansen was called before the House of Representatives two weeks later. "I just xeroxed the testimony I presented to the Senate," he says. "I decided . . . [laughs] . . . I had a lot of phone calls and pressures, including political pressures, you know? 'Do you really mean this or that?' Because the administration was not very happy about that testimony. It became a bit of an issue. So if I changed anything . . . [laughs again] . . . So I decided to use exactly the same words for the second testimony."

The summer of 1988 turned out to be the hottest on record in Washington, D.C. In August, Michael Oppenheimer, the senior atmospheric scientist at the Environmental Defense Fund, told the *New York Times,* "I've never seen an environmental issue mature so quickly, shifting from science to the policy realm almost overnight." Interest increased again in September, when Hurricane Gilbert wheeled through the Caribbean, causing many deaths and more than a billion dollars in property damage. By the end of the year, thirty-two climate-related bills had been introduced in Congress.

But the scientific community again demonstrated its innate conservatism by going even more ballistic than it had over Hansen's first headlines, in 1981. A fall climate workshop in Washington devolved into what one participant called a "get Jim Hansen session."

It didn't phase him a bit. In May of the following year, the first George Bush's first year in office, Al Gore invited Hansen to testify before the Senate Committee on Commerce, Science and Transportation. Before the hearing, the Office of Management and Budget asked Hansen, as a government employee, to submit his written testimony for review. OMB proceeded to change the text, Hansen's main point in particular: that the greenhouse effect was changing climate. His superiors rewrote it to say the cause was unknown.

Hansen then made his most spectacular headlines ever by disavowing his own written testimony in person before the committee. He also revealed more recent work indicating that global warming would increase the chance of extreme climatic events: hurricanes and floods on the one hand, droughts and forest fires on the other.

Having learned in advance of OMB's editing job, Al Gore took the revelation public and accused the Bush administration of "science fraud." It was the lead story on all the major television networks that night. White House spokesman Marlin Fitzwater attempted to distance Bush from the controversy by blaming the edits on an OMB bureaucrat "five levels down from the top."

But the minds of scientists were finally beginning to change. Hansen went straight from Washington to a climate workshop already in progress in Amherst, Massachusetts. *Science* magazine reporter Richard Kerr summarized the meeting in an article entitled "Hansen vs. the World on the Greenhouse Threat," but the text shows that "the World" did not disagree all that much with Hansen's science; what bothered his colleagues was the nonscientific way in which he had presented his findings. Some quibbled with his exact wording; others disagreed on sophisticated technical points—small matters that did not take away from his main argument. On the other hand, they all objected to his simplification, his lack of caution, his disregard for the formal, highly qualified—one could even say codified—manner in which scientific conclusions are stated in the peer-reviewed journals. A scientist from the Woods Hole Oceanographic Institution put it most succinctly: "This kind of giving the result and not telling the whole story, that's what I'm criticizing." But, as a GCM expert from California told Kerr, "If there were a secret ballot at this meeting . . . most people would say the greenhouse warming is probably there."

This was a trying time for Jim Hansen. As he faced the glares of both the media and his colleagues, his family experienced a wave of sickness and death. He turned inward and began reading fiction for the first time in his life, mainly because he had difficulty sleeping. He gravitated toward stories of heroes who flouted convention (odd for a man who remains quite conventional in most aspects of his life) or were persecuted for sticking to their principles. Two favorites were *Pride and Prejudice* and *The Grapes of Wrath*.

And as testament to his fine character, he never accused his colleagues of pettiness. In the thirty years that he has been working on the greenhouse issue he has been the object of endless scorn and personal attack, yet he has never responded in kind. He has always focused resolutely on the science and the facts and, moreover, gone out of his way to point out the shortcomings in his own arguments. When I spoke to him twelve years after that most difficult passage in his life, he seemed honestly to bear no grudge: "Mostly it was legitimate disagreement or concern that the statements were stronger than they should have been," he said mildly. "The scientific method does require that you continually question the conclusions that you draw and put caveats on the conclusions—but that can be misleading to the public. It seems to me that when we talk to the public we have to try to give such a summary. And it's not easy for most scientists to do—and not easy for me."

And all through that period he continued to produce outstanding work.

What many still consider to be the most definitive piece of evidence linking changing temperatures to changes in carbon dioxide had been revealed in 1987 (just as the IPCC was being born) in three exquisite back-to-back papers in *Nature,* written by the Franco-Russian collaboration that had recently recovered a 2,083-meter (one-and-a-third-mile) ice core at Vostok Station, Antarctica.* Among other things (which shall be covered later), that particular Vostok core demonstrated that carbon dioxide and temperature have tracked each other closely for the past 160,000 years. Furthermore, the greenhouse effect calculated from the fluctuations in carbon dioxide was about the right magnitude to account for the temperature swings. Thus, trapped air bubbles and isotopic temperature measurements from the very same record demonstrated the greenhouse effect to be the primary governor of temperature on geological timescales.

As Michael Oppenheimer—who has now moved from Environmental Defense to Princeton University and who has led a long and distinguished career in both the science and policy of climate change—observes, "The ice core work is the single most compelling set of evidence we have on climate history and what's likely to happen in a greenhouse world. The discovery at Vostok grabbed a huge amount of public attention. It crystallized things for both the scientific community and the average person."

Those first Vostok studies, however, did not go very far in examining the physics of the greenhouse–temperature connection. As Hansen knew, carbon dioxide is not the only important factor in the greenhouse.

He has always had a knack for seeing all facets of a problem. In 1990, realizing that Vostok furnished rather a complete picture of the many factors that have affected the earth's radiation budget over the past few hundred thousand years, he joined four French members of the Vostok collaboration in a deeper look at the physics. The ice core provided records not only of

*This was the second of three cores recovered at Vostok in a monumental eighteen-year effort. The third, completed in 1998, would be the longest ever: 3,623 meters, or 2.25 miles.

carbon dioxide but also of methane (a strong greenhouse gas that also happens to have dropped during cold spells and risen during warmings), dust, and other aerosols. Hansen and his colleagues referred to astronomical tables to determine how intense the incoming sunlight would have been at particular times; and aware of the strong positive feedback of the reflective polar ice caps, they incorporated estimates of their size that had been inferred from oxygen isotopes in deep seabed sediments (another topic to be covered later).

Since they now knew the extent to which all these factors had changed since the Last Glacial Maximum, eighteen thousand years ago, when the earth was colder by about eight degrees Fahrenheit, they had everything they needed in order to calculate climate sensitivity. Vostok's empirical evidence told them that a doubling of carbon dioxide (or an equivalent forcing change resulting from an increase in any combination of greenhouse gases) would change global temperatures by between five and seven degrees Fahrenheit. The beauty of this approach was that it accounted for all climate feedbacks implicitly. They found, in fact, that only a third of the sensitivity arose directly from the greenhouse; the rest resulted from feedbacks.

Furthermore, this measured climate sensitivity landed right in the middle, though a bit toward the high end, of the range that had been proposed more than a decade before by the Charney panel—which range had been based, of course, on the GCM studies of Hansen and Manabe. This actual measure calculated from real-world data, therefore, validated the GCMs.

I believe it is fair to say that this *proved* global warming to be a clear and present danger and that the small piece of climatology that is global warming research fell away from the cutting edge of science just then. Since 1990 it has mainly been a matter of dotting the *i*'s and crossing the *t*'s. And watching it happen.

SO IT WAS IN AN INTENSELY POLITICAL ATMOSPHERE THAT LONNIE AND ELLEN THOMPSON found themselves preparing their Dunde report. When Lonnie visited NSF to give his administrators a preliminary look at the news he was about to deliver from China, he received some unsolicited advice.

"You get a 'briefing,'" he says with a wan smile. "I must have talked with the program manager at NSF for an hour on this subject. He said, 'The ramifications are much larger than your program, Lonnie. The NSF budget could be cut.' They thought Congress might cut their funding if we published reports that made it look like global warming was really happening. 'You have to be sensitive to the unknowns' [said the administrator] and, 'You know, it's only one site.' . . . I mean, all true, but at the same time . . ."

Did it change what they wrote?

"No. No. I have never backed off an interpretation of anything, ever since that time at Jay Zwally's house, when I was a graduate student. You have to make your best interpretation of what you see in the data. That's what we did on Dunde."

Bruce Koci believes Lonnie's conservative Appalachian background inclines him to the skeptical side of the global warming debate. (Indeed, Lonnie speaks highly of Ronald Reagan to this day.) Well, if Bruce is right, the scientific overrode the political mind here.

"All through the eighties, temperatures were high," Lonnie continues. "We set a few global records. . . . There was evidence building on that end of things that the system was getting warmer. But the thing is that there's natural variation. If at any stage the temperature of the earth turns, your credibility goes out the window. But every year has confirmed and added strength to the arguments.

"A lot of scientists gave Jim Hansen a hard time, because we're never ninety-nine percent sure of anything in science. You could take him to task for that, but at the same time he took a stand. And so far the system has vindicated him."

PART IV

THE ESSENCE OF LIFE

CIVILIZATION EXISTS BY GEOLOGIC CONSENT, SUBJECT TO CHANGE WITHOUT NOTICE.

—WILL DURANT

Lonnie's opening statement in a lecture to an undergraduate class in environmental geology.

·10·

A CITY BY A LAKE

> Technology is a blessing to be sure, but every blessing has its price. The price of increased complexity is increased vulnerability.
>
> GARRETT HARDIN

At the same time, largely unnoticed by his colleagues as they accumulated evidence for a human origin to today's warming, Lonnie's work on Quelccaya was quietly demonstrating that ancient, natural climate changes had led to the collapse of a number of sophisticated pre-Incan civilizations in and around the Andes. This is a cautionary tale not from Venus or Mars, but from our own planet.

It began with a 1988 report in *Nature,* written by the Thompsons, Mary Davis, and an anthropologist from Louisiana State University named Kambiu Liu. Disarmingly brief at only two and a half pages, this fine study would initiate a minor revolution in the archeology of pre-Incan South America.

Quelccaya's ice bore evidence of two stretches of about 130 years—one centered around A.D. 600, the other around A.D. 920—in which the burden of airborne dust and aerosol salt particles in the air above the Altiplano had been unusually high. The Thompsons and their colleagues had sought a climatic explanation, but the other indicators in the dusty layers painted a confusing picture.

Elevated levels of airborne dust and salt would normally be associated with drought, but the peaks of both events coincided with periods of higher-than-average precipitation; in fact, the peak of the second matched a precipitation maximum. And the events didn't even resemble each other: the first began warm and dry and ended wet, while the second was wet all along. The chemical composition of the dust ruled out volcanic eruptions—a few of which, such as Huaynaputina, were present at other depths—and, besides, the effects of a volcano persist for half a decade at the most. The chemistry of the dust itself indicated that it had probably blown onto the ice cap from the Altiplano, but there was a higher fraction of very fine particles in the dust events than in the rest of the core. This led the Thompsons to guess that it might have come from a lake bed—Titicaca perhaps, which lies to the southeast of the mountain, on the border between Peru and Bolivia. However, twentieth-century records of Titicaca water levels show that they generally rise and fall in concert with snowfall on Quelccaya, so the lake was probably full during the dust events; they could not be explained by retreating shorelines.

Quelccaya stands at the edge of the Altiplano, about 125 miles north-northwest of Lake Titicaca. During the dry season, from April to October, winds come mainly from the south; hence, during the harvest and the times of year when agricultural plots would tend to be bare, dust, pollen, and whatever other biological particles might become entrained by the wind will blow from the lake in the direction of the mountain. This is the season in which the thin layers of dust are laid down between the fluffy annual snow layers. In the rainy season, when the Altiplano receives 80 to 90 percent of its precipitation, winds blow from the east, carrying moist air from the Amazon Basin and the Atlantic Ocean beyond.

In the decade during which he had been visiting Quelccaya, Lonnie had become fascinated by the dozens of vivid cultures that had flourished and collapsed in Peru and Bolivia during the long prehistoric period before the conquistadors arrived. So it was natural, once he had exhausted all the climatic factors he could think of, to look to humanity for an explanation. Andean research being a very small world, he also knew exactly where to look: Alan Kolata, an anthropologist from the University of Chicago. Kolata is arguably the leading expert on the mysterious people known today as the Tiwanaku, who once ruled a vast empire from an ancient city at the sharp, 12,500-foot edge where the brown Altiplano meets the azure waters of the lake.

THE HEART OF THE CITY OF TIWANAKU—THE SACRED SITE OF ITS MOST IMPRESSIVE ARchitectural monuments, its places of ritual and worship, and the dwellings of the priest-kings who ruled the empire—was once surrounded by a wide moat, now completely dry. The centerpiece of this sanctuary is the Akapana, a terraced mound with a footprint in the shape of a pre-Christian Andean cross, its four corners aimed in the four cardinal directions.

Archeologists believed that the Akapana was a natural hill until the late 1900s, when excavations undertaken by Kolata's group demonstrated that it was entirely man-made. The excavations revealed seven terraces, each of which had once been retained by a wall of huge cut stones, irregular in shape and exquisitely crafted and faced, with rounded, beveled edges at the joints where they fit perfectly together like the pieces of a jigsaw puzzle. The structure is roughly 650 feet on a side and 60 feet high, and its summit is the only point in the extinct city from which it is possible to view both the glacier-capped peaks of the Cordillera Real—among them Illimani, which I stopped frequently to admire during the week I spent in La Paz on my way to Sajama, above which the sun first appears every morning—and the sacred islands of Lake Titicaca, behind which the sun sets every night. These islands—more precisely, mountains rising from the sea—were the crucible of creation in Tiwanaku myth, and the sanctuary within the moat was probably built to resemble one.

Interspersed between the thick layers of structural clay that fill the seven terraces of the Akapana—and completely covering the topmost surface—were thin layers of bluish green pebbles taken from the beds of the streams that flow down the sides of the Quimsachata* and Chila mountain ranges, which are visible across a wide plain to the south. The same pebbles line the rivers that debouch onto the plain from wide canyons and valleys. They give a distinctive color to all the streams and rivers in the area, a color that would naturally be associated with water—an essential to human life anywhere and an especially critical resource in this high desert environment.

At the height of its significance, when the Tiwanaku "metropolitan area" is believed to have comprised some 350,000 people, the Akapana seems to have served as both a monumental altar and an arresting display of the priests' ostensibly magical power. Kolata's excavations have revealed a hidden network of sophisticated stone-lined and copper-joined drains and tunnels. Rain falling on the topmost terrace of the pyramid—or carried from Lake Titicaca during the dry season by a long line of llamas ascending a huge axial stairway to the east and descending its counterpart to the west—would have streamed across it into a series of drains at the pyramid's four outer edges; then down, in vertical, stone-lined channels open to the air, to

*These are not the same Quimsachatas I saw from Sajama. *Quimsachata* simply means "three peaks."

a pool comprising the inner "ring" of the next terrace; into subterranean channels again, from which it would have emerged at the edge of the second terrace, to flow down the outside of the structure again; and so on and so forth, until at ground level the water entered the moat, which was joined to the vast drainage system that served the city and its environs.

"It was fundamentally stone," says Kolata. "There's no indication that it was frescoed or painted, but in some cases there are large stone blocks with small drill holes in them, and these drill holes suggest some kind of banner—it could have been textile; it could have been metal. There are still golden nails in some of the holes, which suggests that gold was hammered into the side, so the structure could have potentially even been shimmering gold. When there was water flowing, it would have been very dramatic. I mean it would literally have been hemorrhaging water—streams and pools flowing out from multiple sides and down to the bottom."

The Akapana's alternating layers of flowing water also happened to mimic the streams on the mountain ranges in the distance, which continue to flow intermittently, both in space and in time, today. These natural streams appear only during the rainy season, when dark clouds climb the mountainsides nearly every day to produce spectacular thunderstorms and violent rain—snow up high—nearly every afternoon. As they cascade down the hillsides, the natural streams—like those that once ran on the Akapana—disappear into subterranean channels, reemerge lower down, flow along the surface for a while, form pools on saddles and ledges, spill from the pools to form new streams, disappear into the earth again, and eventually converge, either above or below ground, with the larger flows that finally reach the plain.

Most of the water in Peru and Bolivia originates on mountains, and most of the myth and ritual of their aboriginal peoples—even up to the Aymara, who just a few years ago poured the blood of a white llama on the sand to placate Sajama—involve the supplication of mountain *huacas* to make rain. Sacred places in the landscape—mountains, unusual rock formations, rivers, almost all of them associated with water—are still known as *huacas* to many Andean peoples. They are seen as openings in the body of Pacha Mama, Lady Earth, through which sacrificial offerings are made to the ancestors who live inside her.

TIWANAKU EMERGED AS THE RITUAL CENTER OF A LOOSE-KNIT COLLECTION OF LAKESIDE fishing villages sometime between A.D. 100 and A.D. 300. In about the fifth century, just before the first of Quelccaya's two dust events, the communities by the lake began to coalesce into the urban center and capital of an empire that held economic and political sway for about seven hundred years: across the Altiplano, down into the Bolivian *yungas* (lower, wetter subtropical valleys intermediate in altitude between the high plain and the Amazon cloud forest), and as far down as the shores of the Pacific in present-day Chile and Peru, more than two vertical miles below the city.

By mirroring the islands at the center of his cosmogony and by mimicking the mountains that made water for fields at the very margin of agriculture—which were nevertheless capable of feeding hundreds of thousands of people—the monument in the center of the capital city must have seemed an immense display of wealth and technological wizardry, even magic, to the common man of Tiwanaku, the enduring symbol of an oasis in the desert.

I MAINLY SLEPT, OF COURSE, AS I CROSSED THE ALTIPLANO ON MY WAY TO SAJAMA WITH Javier and Manuel, but I spent quite a few hours staring at it through a bus window on my return to La Paz. Visually, the Altiplano is spectacular, with its brilliant high-altitude sunlight, uniform sere flatness, and the majestic white-topped mountains floating like clouds in the distance. But it seems an impossible place for a farm. The people are very poor, the dusty plain evidently barren, the towns desolate and windblown.

And the experts would seem to agree with that first, naïve impression. Bolivia's few agricultural development projects, funded mostly by foreign governments and international agencies, are located almost exclusively at lower altitudes. Agronomists from the best universities in the United States and Europe characterize the Altiplano as "unfit for agriculture, suitable only for extensive and temporary grazing"—which has indeed been its primary use for the last thousand years.

Odd, then, that one of the greatest empires in the ancient world thrived in this place and that its wealth sprang—as that of all preindustrial societies did in the end—from an abundance of agricultural produce.

IN THE WETLANDS ALONG THE EDGE OF LAKE TITICACA, SURROUNDING THE SACRED CITY of Tiwanaku, the earth undulates in motionless waves, like frozen sea swells. Until recently, even the descendants of that ancient people—the Aymara, who live today in the poor farming and llama- and alpaca-herding villages at the edge of the lake—believed these undulations to be natural forms on the land. Today's Aymara farm higher plots, because the potatoes and other tubers that comprise most of their diet would rot before reaching maturity in the moist lowland soil—the roots of any vegetable crop would rot there.

In the late 1960s, William Denevan, a biogeographer from the University of Wisconsin, realized that these undulations were remnants of the most extensive network of man-made "drained fields" ever to have existed in the prehistoric Americas. In his landmark paper, which reviewed similar massive earthworks throughout North and South America, Denevan pointed out that the undulations were often difficult to discern and that the fields at Tiwanaku, in particular, had been "overlooked until 1966—even by me when I drove through them between Puno and Juliaca in 1965." Two friends later brought them to his attention.

The Tiwanaku planted their crops in raised beds laid out in a checkerboard pattern, surrounded by irrigation ditches that received water from Lake Titicaca and the rivers that fed it by means of a monumental network of reservoirs and aqueducts. The earliest beds were constructed sometime before the birth of Christ—and before the villages on the lakeshore gave rise to a city. Through a process of experimentation that evolved for more than a thousand years, first their ancestors, then the Tiwanaku themselves developed the raised bed into a food-growing technology for that "unfit" environment—subject to withering drought in winter, flooding rain in summer, and killing frost at any time of year—that surpassed the inventiveness even of today's agricultural experts.

"In the case of the pre-industrial state," writes Kolata, "economic well-being was synonymous with agriculture. . . . This essential truth holds even more rigorously in the context of the Pre-Columbian Andean world where markets and mercantile activities were, on the whole, non-existent, or at least, severely restricted in geographic and economic scope. . . . Not surprisingly, some of the greatest public works of Andean civilization are its monumental agricultural and hydraulic structures which reshaped deserts, mountains and high plateaus into economically productive landscapes."

The abandoned city of Tiwanaku sits on a platform fashioned from a former lake terrace. It contains a number of walled compounds representing former barrios, each related to a specific far-flung colony, which paid taxes to the empire with its particular form of wealth. Among the more important were those in the *yungas,* which grew the most important ceremonial crops: maize and coca. A few raised-bed fields existed within the city walls, and at the height of the empire, the raised beds surrounding the city covered more than two hundred thousand additional acres—three hundred square miles. The capital was encircled by a series of satellite townships in the marshlands, "surrounded by raised fields and linked by roads and elevated causeways," according to Kolata. The Tiwanaku metropolitan area would have qualified as a great public work in any age.

UNABLE TO COME UP WITH A SATISFACTORY CLIMATIC EXPLANATION FOR THE TWO-century-long dust events at Quelccaya, the Thompsons corresponded and met with Kolata, and he told them some of what he knew about the history of the Tiwanaku. In their short *Nature* paper they then proposed that the dust they had found in Quelccaya's ice may have been stirred into the air by massive earth-moving operations associated with the construction of the aqueducts and raised-bed fields around the city. This may seem like a wild guess, but it was probably a brilliant one. A quarter of a century later, with a rather complete picture of the sequence of Tiwanaku florescence and decline now firmly established, the notion has been widely accepted.

Kolata has studied virtually every aspect of Tiwanaku civilization in mind-boggling detail, and one thing he had learned by the time of the Thompsons' brilliant insight was just how sophisticated—and appropriate to life on the Altiplano—the raised-bed technology really was.

His excavations in the beds themselves showed that the laborers piled soil into mounds between the irrigation ditches as they dug them, sometimes on courses of rocks or gravel that they had laid down in advance. Along the edges of the mounds, they built low, straight walls of sod bricks, which retained the soil and imparted an elegant architectural appearance. The soil in the mounds was aerated—as it needed to be for gardening—and loose enough to permit nutrient-rich irrigation water to flow by the plant roots, something like a modern hydroponic garden. The soil was elevated so that tubers or roots would not sit in the water and rot.

And when Kolata's team went so far as to construct a few raised beds themselves, they were amazed at the vitality of the ecosystem that quickly emerged in and around them. Algae and other nitrogen-fixing aquatic plants thrived; small fish from Lake Titicaca found their way into the irrigation channels; bugs proliferated; birds built nests nearby for easy access to the bugs. A close look at the soil showed that the Tiwanaku had used fish and other biological matter from the ditches as fertilizer and had supplemented it with livestock dung. They had also planted densely, both to maximize yield and to squeeze out the tenacious weeds that blanket the Altiplano today.

In Kolata's hands the raised-bed method can produce up to ninety-three tons of potatoes per acre, seven times the yield of the dry-field methods used today by the Aymara and twice that of dry fields treated with modern, and expensive, chemical fertilizers. He estimates that the network of raised beds that existed at the height Tiwanaku imperial power could have fed the metropolitan population of 350,000 almost twice over.

One of the few foreign agricultural projects to have been attempted on the Altiplano utilizes a simple greenhouse technology involving polyethylene sheets stretched across small wooden cribs. But plastic sheets and lumber require cash, which the Aymara don't have much of; they are among the poorest people in the world. The ancient technology of the Tiwanaku not only requires no foreign materials and no cash money, it also produces more food.

In the late 1980s, believing the raised-bed method might be useful today, Kolata and a Bolivian colleague managed to convince the Aymara in the lakeshore village of Lakaya to try it themselves. The first year proved to be a wet one, so the program got off to a good start. Even the traditional fields did well. The two anthropologists thought their main problem would be to demonstrate that the old method produced more food.

Then, on successive nights in late February 1988, the Altiplano experienced unusually early killing frosts. Kolata happened to be staying in Tiwanaku, which is lower in altitude and warmer than Lakaya, but awoke to severe frost even there. Concerned after two frigid nights, he drove to Lakaya, where he found the Aymara elder who had championed his cause salvaging what was left of the crop in his traditional dry fields on a hillside. Nighttime temperatures had stooped to twelve degrees. More than three-quarters of the plants in the traditional fields had been destroyed.

But when they looked to the marshy plain below, they were surprised to see the experimental raised beds looking as green and healthy as ever. Walking across the *pampa* to get a closer look, they encountered a young man, very excited and slightly drunk, running toward the village bursting with some news. Upon taking his cattle out to graze near dawn, he had noticed that a warm mist had risen from the irrigation ditches to blanket the raised beds and protect them from the frost. The raised-bed crop had sustained only superficial damage to very few leaves.

Kolata had run computer simulations of the heat flow around a raised bed (see what I mean by mind-boggling detail?) and found that it *should* enhance warmth, but this was an extraordinary confirmation in the real world; and the effect was stronger than he had expected. Essentially, the water in the irrigation ditches acts as a passive solar collector: it absorbs energy in the intense high-altitude sunlight of day and meters it out in the form of heat all through the night.

It might make sense for the people of the Altiplano to take up raised-bed farming again, and indeed, the publicity the method received as a result of the success in 1988 did enable it to catch on for a few years. But Kolata observes that "it's a bit of a hard sell given its labor requirements, and there's not much of a market for produce from the Altiplano. There has also been a lot of political upheaval in Bolivia. Many of the community leaders who were involved in this project have gotten involved in coca production and migrated out. So some communities are still doing it, but others, whose leadership has been lost, have walked away from it. And there has been continuing migration to El Alto [the slum by the airport above La Paz]."

MIGRATION TO A MODERN CITY WAS NOT AN OPTION IN THE FIRST MILLENNIUM A.D., and the Tiwanaku also had the aid of a hierarchical political system, buttressed by blood-soaked religious belief, to keep workers in the fields. Those factors, along with enough water to keep their aqueducts and irrigation ditches full, ensured them reliable harvests for generations. "The (Malthusian) response to abundant surplus production," notes Kolata, "was a logistical capacity to sustain large populations." And as its population grew, Tiwanaku edged closer to what he calls an "environmental threshold": thanks to one very specific technology, based on one critical—and, in their land, extremely limited—resource, they had developed a complex set of social, agricultural, and economic habits, which had allowed their civilization to flourish for about five hundred years. Having encountered a certain range of environmental conditions as they evolved, the Tiwanaku had developed habits that were robust to small, short-term changes in the environment; but as their population approached the absolute carrying capacity of the land, they left themselves more and more vulnerable to a change outside their range of experience—especially if it were to last more than a few years—and such a change assumed the growing possibility of pushing them across a deadly threshold. Indeed, the archeological record shows that the Tiwanaku disappeared almost instantly in the eleventh century A.D.

ONE IMMEDIATE BY-PRODUCT OF THE THOMPSONS' NEW INTEREST IN ARCHEOLOGY WAS a reciprocal interest by archeologists in the history of climate in that part of the world—simply, for the most part, because the brief *Nature* paper had put Quelccaya on their radar screen. Some, including Alan Kolata, had long suspected that climate change may have had something to do with the unusual volatility of the many prehistoric civilizations that had thrived in the Andes at various times, and Quelccaya now represented a crystal-clear climate record that might help determine if these suspicions were justified.

While Lonnie had been drilling ice on Quelccaya, in fact, Kolata had been drilling sediment cores in the bed of Lake Titicaca, seeking an explanation for Tiwanaku's abrupt collapse. But lake bed sediments don't have the resolution of ice cores; they see decades rather than individual years. All he needed was a quick glance at the Thompsons' paper, however, to find the explanation he was seeking: in about A.D. 950, roughly a century before that high, dry city of nearly half a million was completely abandoned, a drought more severe than any in the previous experience of the Tiwanaku set in on the Altiplano, and it lasted for more than four hundred years. The ice cores show a pronounced drop in annual snow accumulation from about A.D. 1000 to 1500. Since the Thompsons' paper also showed that accumulation correlated well with Titicaca lake levels, Kolata quickly deduced that the level of the lake must have fallen as well.

But this was only a hint. Kolata got in touch with Lonnie, who gave him the raw data. Charles Ortloff, a hydrological engineer on Kolata's team, then ran a statistical study of annual snow accumulation on Quelccaya and showed that the average amount of precipitation in an average decade dropped by about 15 percent during the drought, compared to the two centuries preceding it.

This was more compelling, but Quelccaya stands more than a hundred miles from the Tiwanaku heartland. If this were the only evidence, writes Kolata, "we might remain justifiably skeptical." But his limnological evidence (limnology being the science of lakes) supported the inferences from Quelccaya. He had drilled in spots that were underwater but close to the present shoreline. By evaluating the mix of deep- versus shallow-water plants in the different layers of the sediment cores—or plain dust at times when a particular spot had been dry—and then carbon dating the biological material, he found that lake levels had reached a maximum in about A.D. 500, just as Tiwanaku first began to expand; and—as precisely as he could tell with carbon dating—that they had begun to recede in about A.D. 1000, a few decades after the ice cores showed that the drought had begun.

This was no small drought. Kolata estimates that the level of the lake dropped more than fifty feet in fifty years. And Quelccaya showed, furthermore, that during the period of Tiwanaku expansion, A.D. 500 to 1000, accumulation had been unusually high relative to the entire 1,500-year record. As the lake rose, more wetlands would have become available for raised-bed farming, and a larger population could have been sustained. At the turn of the first millennium, however, when the water level fell so low that an entire basin abutting the Tiwanaku heartland (now known as Lake Wiñaymarka) completely disappeared, the population that had burgeoned during the previous, relatively wet four hundred years either starved or was forced to leave. With no water at all, even raised beds couldn't feed them.

Some survivors made the long trek north or east across the Altiplano to the *yungas,* where the ceremonial maize and coca had been grown. In those tilted environments the difficulty wasn't water so much as the availability of arable land. Thousands of horizontal terraces, hacked out of the rocky slopes and fed by canals from mountain streams, rise above the handful of towns that were built by people of Tiwanaku descent in the centuries following the abandonment of their city. Other settlements from that time are clustered in a few mountain valleys *above* the Altiplano, where a mountain aquifer or a stream and a sunny exposure made existence possible for an especially hardy few. But large-scale raised-bed farming and the urban development it made possible have not been seen on the Altiplano since about A.D. 1150. The sophisticated but significantly smaller Aymara city-states that emerged at the end of the drought relied primarily on llama and alpaca herding and the usual sort of terraced or irrigation-based agriculture.

Kolata's later work also demonstrated that the very first agriculture on the shores of Lake Titicaca appeared in conjunction with a climate change. Additional lake bed cores showed that a very long dry period commenced in about 5700 B.C. and ended in about 1500 B.C., while the earliest agricultural plots date from the subsequent moist period, and the raised-bed technology seems to have been invented sometime after the fifth century B.C. Thus life on the Altiplano literally ebbed and flowed with the rhythm of climate.

AT THE SUMMIT OF THE AKAPANA, KOLATA ONCE FOUND A ROOM THAT WAS LITERALLY filled with evidence that the high priests of Tiwanaku had conducted a last ritual—similar to but considerably more flamboyant than the rite Lonnie would witness in the shadow of Sajama almost a thousand years later—before leaving their sacred island forever. The central potion in that ancient ritual was also blood, the magical essence of human life, representing water, the magical essence of the land. By spilling this essence, by filling their sacred pyramid with it and by pouring it on the ground, the priests made one last desperate attempt to induce the distant mountain huacas to give water and life back to their dying land.

> Sometime around A.D. 1150 [writes Kolata] the descendants of [Tiwanaku's] creators performed a desperate ritual on the summit of the Akapana pyramid. On the day of the ritual, the sun illuminated a land clad in the doleful colors of parched grass and lifeless soil. The horizon was hazy with a powdery mist of dirt relentlessly scoured from the dry earth by the winds of the Altiplano.
>
> The celebrants ascended the seven eroding tiers of the sacred mountain-pyramid laden with riches, leading a small herd of young llamas. The great pyramid no longer sparkled and roared with the life-giving waters of the rainy season. The secret tunnels within were long choked with debris, forgotten remnants of an earlier time.
>
> Gaining the summit, the celebrants proceeded quickly with their sacrifice.

One by one the native priests bent the long necks of the llamas backwards toward the west, toward the land of the dead. Then, with sure, practiced strokes, the priests struck deep with blades of translucent obsidian. Great bursts of blood spurted from arteries pumped taut with the terrible pressure of the animals' dying hearts. The priests caught the bloody torrent in elegant ceramic cups. Striding to the four corners of the temple, they cast the liquid in viscous, crimson streams to the ground below.

Having fertilized the dry earth, the priests returned with sacred, blood-smeared hands to butcher the dead animals. With great effort, they cut through sinew and tendon, muscle and bone, to decapitate the llamas. They placed the severed heads and upper jaws of the sacrificed beasts in the north and west corners of a room that had once been the priests' quarters. They carefully removed the animals' lower jaws and piled them in the southeast corner of the room. Then, with the foundation of the sacrifice properly laid, the priests began to assemble the offering to the gods who had abandoned them.

First they placed copper pins, plaques, and figurines in the northeast corner of the room. A magnificent three-dimensional sculpture of a miniature fox, the *tiwula,* stood on the lime-plastered floor of packed earth, eternally frozen in the posture of a canine howl. Layered on top of the copper objects were countless sheets of pale, hammered silver. Then the headless bodies of the sacrificed llamas, still oozing blood, were packed inside the room. The last bodies were draped over the cut-stone threshold, blocking access to the dark interior where no one would ever live again.

In the rising heat of the day, the stench of spilled blood and the dull drone of torpid, black flies enveloped the room. Just outside the door, the priests burned pungent resins in an incense burner of painted clay shaped in the form of a powerfully muscled mountain lion. They lifted their eyes to the heavens and to the great, glistening mountain peaks over the horizon, intoning the long litany of the [huacas], supplicating the ancestors. As white smoke rose up to a cloudless sky through the open fanged mouth and perforated anus of the clay puma, the priests poured libations of maize beer from lyre-shaped cups onto the ground. The *keros* they held were painted with a resplendent crowned figure whose impassive face, vacant of human expression, stares out eternally from the Gateway of the Sun.* This was the manifold image of Viracocha and Thunupa: the creator-god in his guise as Lord of the Atmosphere. This Lord was a violent, unpredictable spirit holding thunderbolts in one hand and a spear-thrower in the other. He brought

*The Gateway of the Sun, perhaps the most well known ancient sculpture in the Andes, frames the eastern entrance of the Kalasasaya temple, which sits next to the Akapana, within the sacred moat.

Annual layers of snow and dust on the sharp, vertical margin of the Quelccaya ice cap, Peru, back when it was healthy, in 1977. (*Lonnie Thompson*)

Bruce Koci and Lonnie Thompson (holding a segment of ice core) at their drilling camp on the 18,600-foot summit of Quelccaya in 1983. Solar panels to power the world's first high-altitude ice coring drill, which these men invented, stand in the background. (*Lonnie Thompson collection*)

The 1983 drilling team descending to Quelccaya base camp under threatening clouds. (*Bruce Koci*)

The infamous horse ride into the Dunde ice cap in China's Qilian Shan Range in 1986. "Off for three days through the mud and muck of this vast, treeless plain that stretches into the distance like a bad dream," according to Keith. (*Keith Mountain*)

Mary Davis, Henry Brecher, Bruce Koci, Lonnie Thompson, and Keith Mountain at the 17,500-foot drilling camp on Dunde in 1987. "We looked like a bunch of janitors," observes Bruce. (*Keith Mountain*)

Traffic on the road to the Guliya ice cap in Islamic, far western China, 1992. "Like being back in biblical times," says Bruce. (*Bruce Koci*)

A thunderstorm approaches the 20,300-foot drilling site on the Guliya ice cap in China's western Kunlun Shan, 1992. (*Bruce Koci*)

Looking up from High Camp at porters carrying empty core boxes up the north-west ridge of 21,500-foot Nevado Sajama, the highest mountain in Bolivia. Needles of *nieve penitente* bristle from the rocky ridge in the foreground. (*Mark Bowen*)

Bolivian porters gathering around a large plastic bag of coca to stuff the leaves into their cheeks in preparation for our hike to High Camp in 1997. Their leader, Gonzalo, left, smiles as usual. (*Mark Bowen*)

The porters who did most of the high-altitude work on Kilimanjaro, standing around our campfire on the Shira Plateau on the evening before our fateful return to the summit. Their leader and "elder," Julias Minja, stands at the far right. His dutiful and indefatigable half brother, Stephen, sits in the center, preparing *ugali* for our dinner. (*Mark Bowen*)

Yak carrying core boxes on the Dasuopu ice cap in the Tibetan Himalaya, 1997. (*Bruce Koci*)

The drilling dome on Kilimanjaro's northern ice field, during the deep drilling expedition in 2000. Our camp sits on the sand of the crater, beside the ice. (*Mark Bowen*)

A view of Kilimanjaro's dusty summit crater from Uhuru Peak. Lonnie's tiny drilling rig stands on the small Furtwängler glacier, which drains into the Great Western Breach. The northern ice field spans the far crater rim. (*Mark Bowen*)

Assembling the dome on Kilimanjaro's northern ice field. The rock formation to the right, on the opposite rim of the summit crater, is 19,344-foot Uhuru, the highest peak in Africa. (*Mark Bowen*)

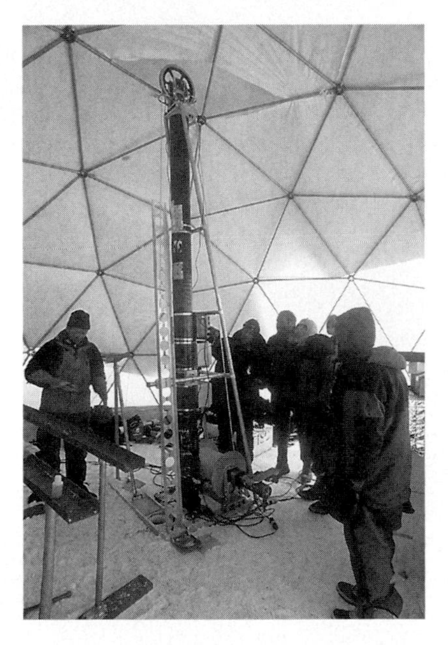
Victor Zagorodnov running his drill in the dome on Kilimanjaro. From left to right: Vladimir Mikhalenko, Victor, Keith Mountain, Lonnie Thompson, Bruce Koci. (*Mark Bowen*)

Lonnie and Mary inspect a core segment, and Victor and Vladimir fine-tune the drill's inner "screw," while Bruce and Keith look on. The drill's outer housing lies on the stand between Vladimir and Keith. (*Mark Bowen*)

Above: The retreat of the "Snows of Kilimanjaro." Top: 1912 (*E. Oehler*) Bottom: 1999 (*Mark Bowen*)

Left: Sajama's ice cores in the Thompsons' cold room in Columbus, Ohio—which may soon be the last place on earth in which tropical mountain ice will be found. (*Mark Bowen*)

wind and rain, lightning and thunder, life and death, all with sublime indifference.

After drinking deep from their ruby-colored keros, consecrated to the Lord of the Atmosphere, the priests, in a sudden, whirling movement, shattered the ritual beakers on the ground. Then they slowly removed the jewelry with which they adorned themselves before the ritual, placing the sacred objects, imbued with the spiritual status of their owners, next to the smoking incense burner. A pile of gleaming polished stone and bone plugs drawn from their lips and septums lay next to earrings of gold, silver, and copper. From bundles of exquisite woven cloth, the priests drew amulets and razor-sharp flakes of quartz crystal and obsidian, scattering them before the gore-spattered threshold. The jewels and fragments of crystalline rock glittered brightly in the sun. But the great pools of llama blood spilled inside the sacrificial room had already begun to blacken and congeal. The dark contents within remained concealed for over eight hundred years.

Given the current climate change, it is troubling to learn that the drought that drove these priests from their sacred dwelling happens to have arrived in conjunction with a mild Andean warming. This was evident in the graph of Quelccaya's isotopes that the Thompsons included in the *Nature* paper, although they didn't pin any numbers to it there. Kolata and Ortloff later showed, statistically, that the average temperature at Quelccaya had risen roughly two degrees Fahrenheit between A.D. 1000 and at least 1400. (A few years later, Lonnie would find signs of a concurrent warming in his ice cores from Huascarán, more than a thousand miles to the north.)

This mild warming also coincides well enough with the recently discredited notion of a so-called Medieval Warm Period. Until very recently, there has been a general belief among climatologists that most if not all of the earth experienced some sort of a warming in a vaguely defined time slot centered around the twelfth century. Popular books and newspaper articles—not to mention scientists themselves—have gone out of their way to point out, for instance, that northern Europe experienced unusually mild winters and increased rainfall in the twelfth century; that Norsemen established farming communities in Greenland in about A.D. 1000 and abandoned them in about 1350; that abundant wheat and barley grew in Iceland; and most picturesquely, that the *Domesday Book,* a census of British counties commissioned by William the Conqueror after he won the Battle of Hastings in 1066, lists thirty-eight vineyards, which, it is also pointed out, are not a common sight in the British countryside today.

Unfortunately, however, a closer look shows that these entertaining anecdotes are almost all false. The dates are true for the Norsemen's comings and goings to Greenland, but their movements probably had more to do with economics than with climate change; and the English have been growing wine constantly for the past thousand years. An Englishman invented sparkling wine during the comparatively cold seventeenth century, in fact, about thirty years before champagne was invented in France; and there are ten times as many vineyards in England now than there were in the twelfth century. It is becoming increasingly clear that what little warming did take place in High Medieval times was restricted to Europe, the Andes, and other isolated regions, while most other areas remained unchanged or even cooled—though the hot spots do seem to have brought the average temperature of the northern hemisphere (not the earth as a whole) up very slightly from around A.D. 800 to 1400.

Questions as to the reality of a "natural" medieval warming rose (or perhaps descended) to the political level at the beginning of 2003, when two ubiquitous global warming contrarians, Willie Soon and Sallie Baliunas, funded partially by the American Petroleum Institute, published a paper in the journal *Climate Research* arguing not only that there *had* been a Medieval Warm Period, but that the earth had been warmer then than it is today. Five editors of the journal resigned in protest over the publisher's subsequent refusal to change what they saw as the shoddy and easily manipulated peer-review process that had let the paper through in the first place, and the methods employed in the paper were shown to be faulty. (One patently specious method the authors employed was to assume that any change in precipitation—to either wetter or drier conditions—indicated a rise in temperature.) Nevertheless, a few conservative politicians, especially James Inhofe of Oklahoma, chairman of the Senate Committee on Environment and Public Works, and higher-ups in the Bush administration, attempted to use that single paper as evidence that the majority view on global warming was "junk science." The paper was even read into the Senate record.

But the very fact that Soon and Baliunas found so many climatic changes available for their manipulation demonstrates that *something* big happened during the first few centuries of the second millennium; the question is, what was it? Since the changes seem to have had more to do with changes in air circulation and rainfall than simply in temperature, Scott Stine, a paleoclimatologist from California State University in Hayward, has suggested the name "Medieval Climate Anomaly."

Stine has wandered—and paddled—all over his home state, collecting samples from dead trees rooted as deeply as seventy feet in Mono and Tenaya lakes, the Walker River, and other inland water bodies. The radiocarbon "death dates" of the trees fall into two distinct groups, both during the "Medieval Climate Anomaly." Some died in A.D. 1100, others in 1350. Since these trees were Jeffrey or lodgepole pines and cottonwoods for the most part, which die within weeks if their roots are saturated with water, Stine infers (quite plausibly) that two distinct California droughts must have caused the levels of these lakes and rivers to fall below the levels of the stumps—that is, as much as seventy feet below their present levels—for at least enough time for the trees to grow, and that subsequent returns to moist conditions filled the lakes back up and killed the trees. Estimating the age of the trees (again quite plausibly) by counting growth rings, Stine shows that the first drought lasted more than 200 years and the second about 140. Both would dwarf any drought that has occurred since Europeans landed in the Americas. The Dust Bowl, for instance, which is remembered as "biblical" by those who lived through it, lasted only 7 years.

Evidence for both droughts also appears in the rings of ancient bristlecone pines in the Great Basin region of Nevada, southwest Utah, and northwest Arizona; and Stine himself has found signs as far south as Patagonia of an extreme drought coinciding with the earlier of the two.

IT HAS ALSO NOW BEEN PRETTY WELL ESTABLISHED THAT STRANGE WEATHER ASSOCIATED with the Medieval Climate Anomaly caused significant human hardship in North America, as well. Both the construction and subsequent rapid abandonment of the spectacular cliff dwellings in the Four Corners area of the southwestern United States, for example, the homes of the so-called Anasazi or "Ancient Ones"—Mesa Verde, Colorado; Chaco Canyon, New Mexico; and Canyon de Chelly, Arizona—have been ascribed to some very strange weather that occurred in that time period.

To simplify only a little: in about 1500 B.C., many different tribes began to try their lot at farming in the typically arid American Southwest. They tended to be seminomadic, sometimes employing irrigation, but rarely dwelling in the same place for more than a few decades. During an optimum in rainfall that occurred from the eighth to the early twelfth centuries A.D., the population increased to the saturation point, and nearly all the arable land was converted to agriculture.

Regional tree-ring records show that a severe, two-decade drought commenced in 1130. In response, the people moved closer to the sources of water: to higher areas of the Colorado Plateau in the vicinity of the Four Corners itself, into the canyons where the rivers and streams originated and where it was possible to build dams and terraces to control and conserve water. As a result, the upland population reached a density that prevented movement in search of more water or richer land, and it became necessary for people to live in larger groups—hence the cliff dwellings, the largest of which were built during and after this relatively brief but life-changing drought.

So, new technologies and economic and cultural patterns allowed the Anasazi to capitalize on the same critical and limited resource that the Tiwanaku had based their spiraling growth upon less than two centuries earlier. The upland population reached the carrying capacity of the land, and the Anasazi edged closer to the same sort of environmental threshold that their counterparts had recently crossed in the southern hemisphere. None of the cliff dwellings were occupied for more than two centuries—some for only one.

But it wasn't drought this time. An impressive study of tree-ring records from twenty-seven sites in the Four Corners area shows that the seasonal cycle of rainfall in the uplands shifted into a chaotic mode in roughly A.D. 1250 and stayed that way for about two hundred years (coinciding, more or less, with the second of Stine's droughts). A pattern was disrupted that had been relatively predictable and around which agricultural methods had evolved for more than five centuries. "For those of us whose cultural history includes notions of the biblical forty years of wandering in the desert or the chaos of revolution lasting twenty years, the 200 years of chaotic, and hence, unpredictable, patterning in precipitation in the Southwest is an event of unimaginable magnitude," writes Linda Cordell, the anthropologist from the University of Colorado who conducted the study. The archeological record shows that building ceased and the cliff dwellings were largely abandoned between 1250 and 1400—quite rapidly in the case of Mesa Verde. And to support Cordell's thesis, the cliff dwellers seem to have migrated to lowland areas, primarily in the south, where rainfall remained as meager as ever—but predictable at least.

COULD SUCH A THING HAPPEN TODAY?

Consider the circumstances of modern city dwellers in Los Angeles, Las Vegas, and even notoriously wet Seattle. Scott Stine and his fellow climatologists happen to know that the last century and a half, during which Californians have established what he calls "the most colossal urban and agricultural infrastructure in the entire world" (to which I might add "and all of history"), has also been the third- or fourth-wettest period of that duration that California has enjoyed in the last four thousand years. Not unlike the Tiwanaku and the Anasazi, this modern society has taken advantage of a climatic optimum to construct its own sophisticated infrastructure and has placed itself, similarly, at the brink of its own environmental threshold.

Water has been a key ingredient in the California success since the state's beginnings a century ago, when, in a series of intrigues that formed the basis for the movie *Chinatown,* the nascent city of Los Angeles quietly bought 320,000 acres of land in the Owens Valley, along with water rights to the Owens River. In 1913, after leading an army of five thousand workers for five years, William Mulholland, chief engineer of the Los Angeles water department, completed construction of a 223-mile aqueduct to deliver Owens River water to the city; at the grand ceremony commemorating the release of the first trickle, this laconic Irishman is said to have uttered the five most important words in the city's history: "There it is. Take it." If not for that water, Los Angeles would still be a small, dusty town in a coastal desert valley.

Most of California get its water from snowpack high in the Sierra Nevada Mountains—the Owens River relies on it entirely—and even in this climatic optimum the state has been drawing down the principal in that bank account for decades. Stine had a much easier time doing his fieldwork than he might have, for example, because by the early eighties, when he visited the Owens Valley's Mono Lake, Los Angeles had already diverted enough water from the lake's tributaries to lower its surface by fifty feet, exposing thousands of acres of new shoreline and dozens of medieval tree stumps.

Since snowpack acts as a natural reservoir, storing water as it accumulates in the winter—when California receives more than 90 percent of its precipitation—and metering it out over the long, dry summer, global warming promises to push the state nearer or even across its environmental brink. Higher temperatures may not necessarily mean less snow or rain, but they will certainly mean less water in summer for agricultural use, recreation, the sprinkling of lawns and gardens (and the making of cocktail ice) in Los Angeles. The reason dams and aqueducts were built throughout the West in the early 1900s was to supplement the storage capacity of mountain snowfields by saving some portion of their spring runoff for later summer use. But as monumental as Mulholland's network of ditches and tunnels and the Glen Canyon, Grand Coulee, and Hoover dams may seem, they are as nothing compared with the hundreds of thousands of square miles of mountaintops, which rise literally like water towers above the vast, dry flatlands that characterize most of the West.

Scientists at the Scripps Institution of Oceanography have recently shown that the peak in annual spring runoff from mountain streams in the Sierra Nevada now comes about three weeks earlier than it did in 1948. Since this effect is most pronounced at lower elevations, where the snowfields are closest to the melting point, the likely cause is the average 1.4°F-degree warming that the American West has experienced since the 1950s. This explanation is supported by the observation that the earliest springtime blooms of Sierra Nevada honeysuckle and lilac, which respond primarily to early spring warmth, have advanced a similar one or two weeks.

The latest circulation models suggest that Western temperatures will rise between four and thirteen degrees this century. The same Scripps team, plugging the conservative lower bound of four degrees into a model simulating Sierra snowpack, estimates that the reserves in California's mountain water towers will drop by a third by 2060 and by half by 2090. As this process unfolds, the seasonal distribution of available, flowing water will shift from spring and summer to the winter months, increasing the frequency of winter floods on the one hand and deadly summer forest fires on the other. (And the latter would seem, from the newspapers, to be on the increase already.)

A survey of nearly six hundred snowfields in the Sierra Nevada, the Rocky Mountains, and the Cascades of Washington and Oregon shows that 85 percent have lost volume since the 1950s. Since the Rockies lie in the interior and tend to be colder, they have been less affected than the other ranges, but the Cascades, which are the warmest for being closest to the ocean, have experienced the greatest change of the three. Since even "rainy" Portland, Oregon, receives 90 percent of its rain in winter, the expected decrease this century of 60 percent in Cascade snowpack—even in the most reassuring of global warming scenarios—stands to have devastating consequences, believe it or not, in the Pacific Northwest.

The Columbia River's Grand Coulee Dam and other Western monuments to human ingenuity and perseverance, most of which date to the 1930s, have made it possible for the Pacific and Mountain states to absorb more than sixty million people since that time, roughly one-fifth of the U.S. population. It will be difficult to build enough new dams to substitute for the West's natural mountain water towers as snowlines slowly climb up their sides and disappear.

Furthermore, all of this ignores the prospect—indeed the growing reality—of long-term drought. Snowpack may be doing comparatively well in the Rockies, but the Rocky Mountain states are about to enter the sixth year of a serious drought. According to a May 2004 article in the *New York Times,* "The period since 1999 is now officially the driest in the 98 years of recorded history of the Colorado River, according to the United States

Geological Survey." The *Times* article also points out that "[c]ontinuing research into drought cycles over the last 800 years . . . strongly suggest[s] that the relatively wet weather across much of the West during the 20th century was a fluke. In other words . . . the development of the modern urbanized West—one of the biggest growth spurts in the nation's history—may have been based on a colossal miscalculation."

A small part of that miscalculation seems to have been Arizona's Glen Canyon Dam. Construction commenced in 1956, Ladybird Johnson dedicated it ten years later, and it took seventeen years to fill Lake Powell, which it created. The lake now contains less than half the water it did at its fullest (about the amount that it had as it was filling up during the Watergate hearings), and hydrologists are saying that if it continues to lose water at the present rate it will be incapable of generating electricity by 2007, and reduced to the size of the original riverbed—mocking the awesome wall looming above it—very soon thereafter.

Part of the problem stems again from conventions adopted in a time of wealth. Water was relatively plentiful and far fewer people lived in the Southwest in 1922, when the Colorado River Compact was signed, requiring 8.23 million acre-feet of water to be released every year to Nevada, California, and Mexico, which lie downstream from the present lake. (It also happens that the Hoover Dam and Lake Mead, which supply Las Vegas, lie downstream.) Upstream, however, Denver is siphoning quite a bit more water now from the Colorado watershed than it did as a small frontier town eighty years ago.

And through the former decades of water surplus, southern California—including not only Los Angeles, but also the farmers of the Imperial Valley and the citizens of San Diego—has become accustomed to using even more water than it was allotted in the original pact. The present deficit, therefore, brought about a crisis in 2002, pitting the farmers, who own most of the rights to Colorado River water, against the city folk. So fierce was the acrimony that the federal government found it necessary to step in and broker a deal, which now requires Californians not only to stick to their allotment but to divide it equitably among themselves. But the deal involves only "excess" water from the Colorado, and the Department of the Interior is now projecting that there will be no "excess" at all in the very near future.

Sounding something like the anthropologists studying the vanished civilizations of the Andes, Professor Daniel McCool, director of the American West Center at the University of Utah, says, "The law of the river is hopelessly, irretrievably obsolete, designed on a hydrological fallacy, around an agrarian West that no longer exists. After six years of drought, somebody will have to say the emperor has no clothes."

SCOTT STINE HAS FOUND PARALLELS BETWEEN RECENT MEMORABLE, BUT RELATIVELY short-lived, California droughts and the two longer, natural ones he has uncovered in his research. All seem to have been caused by shifts in the position of the jet stream (also known as the "storm track" or "polar front"), which steers Pacific storms in to the West Coast. When the jet stream shifts north, storms hit Alaska; hence, dry winters in California tend to be wet winters farther north. The winter of 1976–1977, for example, which brought the only drought in the instrumental era to have rivaled those in medieval times—in strength if not duration—was one of the wettest on record in Alaska. Glaciers in southern Alaska are retreating rapidly at present (due mainly to global warming), and dead stumps that were once uprooted by medieval advances are being revealed in their moraines. Since the uprooted stumps date from the times of the great California droughts, it seems that Alaskan glaciers advanced and uprooted certain trees just as California lakes evaporated and permitted other trees to grow. In other words, California's medieval droughts may also have been caused by shifts in the jet stream.

That these shifts were "natural" is no real comfort, for there is virtually no doubt that they will happen again—as with earthquakes, it is only a question of time. On the other hand, some believe that global warming increases their likelihood. Peter Gleick, president of the Pacific Institute for Studies in Development, Environment, and Security, who was awarded a MacArthur Fellowship in 2003 for his efforts to build awareness of the global freshwater problem, calls global warming "the down card in the poker game—you can't see it, but you know it's going to be a factor."

Climate modeling is better at predicting changes in temperature than changes in air circulation and rainfall, precisely because differences in temperature tend to drive the winds that will smooth those differences out. Therefore, modelers can't yet predict the effect that a warming might have in a region as small as a single state. But it is nearly certain that it will change California's rain patterns somehow. If, as with the Medieval Climate Anomaly, it were to shift the jet stream north, the San Joaquin Valley, which is presently the richest irrigated land in the world and among the most profligate in its waste of water—which streams again almost entirely from the High Sierra—might very well dry up, just as the raised beds on the Altiplano did a thousand years ago.

The San Joaquin provides a superb example of the absurd web of "economic and social habit" that can evolve in the wealth of a climatic optimum. The federal government originally helped divert four rivers carrying Sierra Nevada meltwater to help irrigate the land owned by the giant agribusinesses that now operate there. One of those, the JG Boswell Company, for example, has drained the valley's Tulare Lake—which was once the largest body of freshwater west of the Mississippi—to grow water-intensive cotton in its bed. It takes fifteen thousand tons of water to produce one ton of cotton—about fifteen times that required for a ton of grain. Meanwhile, the United States subsidizes cotton farmers and the American textile companies that buy their cotton to the tune of $12.9 billion per year.

This has been a major sticking point at recent meetings of the World Trade Organization (WTO), where Brazil has led an emotional battle to end the U.S. cotton subsidies, arguing that they deflate the price of cotton in world markets and hurt not only Brazil's economy but those of a few West African nations that rank among the poorest in the world: Chad, Mali, Benin, and Burkina Faso. The WTO ruled the subsidies illegal in 2004, but there has been no change in the U.S. policy as of this writing.

WE ARE USED TO HEARING ABOUT WATER SHORTAGES IN CALIFORNIA AND THE WEST, BUT they have been cropping up with increasing frequency in less likely places over the last few years—sometimes as a result of heedless overuse by farmers, but more often as a result of thirsty suburban sprawl. A regional planning commission reports that six Chicago counties on the shores of Lake Michigan—the largest lake in a system that contains one-fifth of the earth's open freshwater—may face serious water shortages in twenty years. In the verdant suburbs of Atlanta, population growth and its attendant development have drained the Chattahoochee River to the point that a "water war" is brewing over allotments between Georgia and its downstream neighbors, Alabama and Florida. Georgia is also fighting over the Savannah River with South Carolina and the Tennessee River with Tennessee. Maryland and Virginia have a disagreement over the Potomac. South and North Carolina are in a dispute about the Pee Dee. Texas and Mexico are involved in an international dispute over the Rio Grande. The list goes on and on.

And that's just the water we can see. The level of the Ogallala Aquifer, the largest in the world—it lies beneath eight states: Nebraska, South Dakota, Kansas, Wyoming, Colorado, Oklahoma, New Mexico, and Texas—has been dropping for decades. The farmers most responsible have finally realized that they're dealing with a limited resource and have begun conserving and employing less wasteful irrigation methods; while this has slowed the rate of depletion, it has not stopped it.

Rice farmers in Arkansas have drained an aquifer named the Alluvial nearly dry. The U.S. Army Corps of Engineers (demonstrating that it hasn't learned much from the California experience) has drawn up a plan with the state of Arkansas to divert the White River, the lifeblood of two important wildlife refuges, in order to replenish the aquifer. The estimated cost? A staggering $319 million, about $300,000 per farmer. And their crop, like cotton, is additionally subsidized. The government guarantees rice farmers more than double the present market price per bushel. As a result, the amount of U.S. land devoted to rice farming has grown even as the market price has dropped to a fifteen-year low.

And seeing a future in privatized municipal water distribution, corporations have quietly begun to invest in water rights. Fully aware of the dry prospects for Texas cities, T. Boone Pickens, an oilman, has been buying rights to the Ogallala from nearby farmers. And before its spectacular collapse, the infamous Enron Corporation had formed a water division, hoping to lay claim to a prospective global industry with a worth the company estimated at four hundred billion dollars.

Some believe water will be to the twenty-first century as oil was to the twentieth. Turkey has already started work on a thirty-billion-dollar irrigation project in the headwaters of the Euphrates, where farming was invented twelve thousand years ago. Once it is completed, Turkey's downstream neighbors, Syria and Iraq, largely desert already—and the latter suffering enough for other reasons—will simply have to make do with less water than they have now. Water may soon eclipse oil as the most divisive liquid resource in the Middle East.

It seems obvious that the underlying pressure today is population growth—as it was for the upland Anasazi 750 years ago, and as it also is, parenthetically, for the spiraling greenhouse emissions of today. But with the global population having just passed six billion and projected to reach nine or ten billion by the middle of this century, moving to greener pastures, as the Anasazi did, is becoming less of an option. While the United States seems to be wealthy (and wrongheaded) enough to believe it can forestall its looming environmental threshold with monumental technology, most of the world decidedly cannot. Forty percent of the global population—mostly the very poor, of course—already suffer from serious water shortages. More than a billion lack safe drinking water, and almost two and a half billion lack adequate sanitation. Water-related diseases resulting from these failures already cause between two and five million deaths annually. A report issued in 2002 by the United Nations Environmental Programme predicts that if market forces continue to "drive the globe's political, economic and social agenda," 55 percent of the earth's people will be experiencing severe water shortages in thirty years. In West Asia, which includes the Arabian Peninsula, the estimate is an astounding 95 percent.

JOURNALIST JACQUES LESLIE, WHO CUT HIS TEETH IN VIETNAM AND IS NOW WRITING A book about the global water crisis, sounds an oblique echo of the ancient blood rites of the Tiwanaku:

> When I was a war correspondent twenty-five years ago, I paid more attention to blood than to water. . . . Now when I envision the globe, I try to see beyond political boundaries to the world as it really is: a collection of watersheds, lakes, rivers, and aquifers that together maintain the earth's biota—which is to say, us. Now the world's quotidian skirmishes and conflagrations are mere background noise. Now it is water that scares me.

·11·

THE SEESAW

> The overwhelming lesson from history is that most irrigation-based civilizations fail.
>
> SANDRA POSTEL

The next anthropologist to take an interest in Quelccaya was Izumi Shimada, then of Harvard University, whose main interest at the time was an especially bloody people, as silent as the Tiwanaku, whom anthropologists have named the Moche.

They appeared at about the beginning of the Christian era on the very dry northern coastal plain of Peru, about a thousand miles northwest of Tiwanaku and two miles lower, and dominated that region from the second to the early eighth centuries A.D. As we shall see, it is probably no coincidence that the Moche fell just as the Tiwanaku rose.

In many ways conditions are harsher on the coastal desert than they are on the Altiplano, for there the reliance on water is starkly binary: when it flows, the crops grow; when it stops, the people starve. This may explain the intensity of Mochica mythic life.

The central figure in their iconography is the Decapitator, a warrior-priest sun king who holds a human head by the hair in one hand and a sacrifice knife in the other. His image is found on the exquisite ceremonial vessels in the Moches' royal burial chambers—which have been called the richest in the New World—on the crumbling friezes and murals on the walls of the enormous stepped pyramids (still known as huacas) they left behind, and is fashioned in gold on the bells and armor adorning the corpse, for instance, of the "Lord of Sipán," who evidently performed this ritual in the northernmost Lambayeque Valley, where his burial chamber was found.

The Moches' taste for human blood was not unique in the Andes. Nearly every pre-Incan civilization engaged in some sort of human blood rite. Dozens of dismembered male bodies were interred near the base of the Akapana, for example; and according to the accounts of the conquistadors, Incan emperors used the skulls of vanquished foes as drinking bowls. But it would appear that the Moche spilled—and drank—human blood more regularly and more frankly than any of the others.

At the height of their power they ruled the twelve northernmost watersheds that still cross the coastal plain, each fanning east to west from the Andes to the Pacific Ocean. The Moche myth cycle comprised just a few epic themes, similar to those of the Arthurian legends, which are repeated on walls and ceremonial objects all through their former domain. The images indicate that Moche priests performed elaborate rituals on the summits of the many man-made huacas still standing in these watersheds—one of which, the Huaca del Sol (Huaca of the Sun), in the town that is now named Moche, is the largest adobe construction in the New World. Until 1602, when a group of gold-crazed Spaniards diverted the Moche River against its western walls in a futile attempt to mine its base hydraulically, the monument comprised nearly one hundred million bricks of adobe, the holy food of Pacha Mama. Anthropologists estimate that it took thousands of workers nearly two centuries to build it.

In the "Sacrifice Theme," richly clad warriors bearing maces and spears prod a line of prisoners up the steps of a huaca. The prospective victims—naked and showing signs of having been battered about the head—are connected neck to neck by ropes. The head priest, in the role of Decapitator, either sits on a throne or stands with a raised knife at the summit of the huaca, which will be the prisoners' final destination. The priest's crown emits squirming rays terminating in snakelike heads with forked tongues, and his belt also resembles a serpent. Gloriously arrayed at the top of his man-made mountain, he resembles the true sun—the first god—which still casts its first rays every morning from the summit of one or another of the stupendous Andean peaks that tower to the east.

The relics buried with the Lord of Sipán invite comparison to the Pharaohs of ancient Egypt, some of whom were explicitly regarded as sun gods. His eyes and nose, chin and cheeks are protected with gold covers. He wears one necklace of gold plates and two necklaces of beads in the shape of peanuts: one gold, one silver. His skull lies in a large gold saucer. He wears a rayed, crescent-shaped headdress. His right hand holds a gold ingot and a gold rattle, and a copper ingot and a copper sacrifice knife rest in his left. Other gold and copper relics are placed to the right and left, respectively, as if the two sides of his body represent the rising and setting suns.

The avowed purpose of these theatricals was probably shamanic: they were probably a way to cross into the spirit world, enter the space and time of myth, and magically influence the processes of nature. The rattles and drums seen on Moche friezes and found in their corpses' hands are common to shamanic rituals everywhere. An animal medium or familiar, most often a dog, sat at the sun king's side during the rituals and was later buried with him. Some scenes show the psychotropic San Pedro cactus, used by *curanderos* in northern Peru to this day; so the priests may have entered the spirit world on hallucinogenic wings. Finally, ritual violence introduced the most powerful medicine: blood, the life force, the essence of vitality and fecundity. In drinking it, the priests may have believed that they were incorporating the vitality of their victims into themselves. One kingly corpse wears a necklace of gold beads in the shape of spiders. And to ensure that the blood would flow all through the rituals (and remain drinkable), the priests kept supplies of *Ulluchu* juice—a strong anticoagulant—close at hand. In some scenes *Ulluchu* fruits hang from the priests' belts.

Since they performed these astounding ceremonies in broad daylight on the summits of their awe-inspiring huacas, the priests probably used these rituals both to proclaim and to assert their power. There are few signs of physical control in the layout of old Moche. It was essentially a random collection of huts scattered around the bases of the Huaca del Sol and its smaller, more sacrosanct sister, the Huaca de la Luna (Huaca of the Moon). Thus a common—if inequitable—belief system seems to have united two very distinct classes: godlike rulers and worshipful commoners.

IN NORMAL YEARS, THE NORTHERN HALF OF PERU'S COASTAL PLAIN IS ONE OF THE DRIest places on earth. Therefore, in addition to the huacas for their masters, the commoners were required to build massive public aqueduct and irrigation systems for the growing of food. A typical Mochica, who lived only a few decades on average, experienced perhaps two or three significant rainfalls in his lifetime. Thus the drinking of the magical essence by the sun god must also have been aimed—like the sacrifice of so many llamas on the summit of the Akapana five hundred years later and of a single white llama at the base of Sajama a thousand years after that—at sustaining the life force of the land. Indeed, the huacas of the Moche were usually aligned with the mountains that were the most important natural huacas in their world: from the deep canyons of the Andes flowed the rivers that fed the irrigation ditches that watered their fields, while on their sacred mimetic summits, the priests released rivers of blood.

And virtually all the infrequent rain that falls in the Moche heartland is associated with El Niño. Positioned near the equator, on the Pacific coast, it sits right at El Niño's ground zero. But El Niño is unpredictable. It visits every four years on average, while it may stay away for as many as ten or return in as few as two. One can see why the Moche priests might have believed it took very strong magic to invoke rain. They walked a razor's edge, however, because El Niño is both a blessing and a curse in their land. It brings not only rain, but torrential rain, frequently accompanied by floods and landslides.

In non–El Niño years, the surface of the Pacific by the coasts of Peru and Ecuador is surprisingly cold. The frigid Humboldt Current—also known as the Peru Current, which splits from the Antarctic Circumpolar Current south of Cape Horn and flows north, at the bottom of the sea, along South America's Pacific coast—normally wells to the surface when it reaches the equator. One of the first signs of an approaching El Niño is a warming of the coastal surface waters, as the Humboldt Current recedes to the depths, taking the anchovy that subsist on the rich plankton content of its cold water along with it. Fishermen generally take El Niño seasons off, staying at home with their families and mending their nets.

This episodic "sea change" has made this area one of the most ecologically dynamic on earth. Millions of aquatic birds, mainly the Guanay cormorant, Peruvian booby (known locally as the *piquero*), and gray pelican, live and nest on small rocky islands near the coast, feeding on the anchovy and producing guano in such quantities that it had reached a depth of 150 feet in some places by the time Europeans realized its value as a fertilizer and began exporting it. The Moche used it on their fields, and they ate anchovy and other species from the local fishery, which is still one of the richest in the world. Moche sacrifice victims have been found buried in the guano, perhaps in ceremonies that were meant to sustain that resource as well, for it would also have come and gone with El Niño.

The birds disappear with the anchovy, of course, often leaving recent hatchlings behind in their desperate search for food. The seals in the area also subsist on the anchovy and happen to breed at about the time of year that El Niño begins, so the lack of food has an especially strong impact on expectant females and their young. Nearly all the fur seal pups in the Galápagos Islands, which lie right on the equator to the west of Ecuador, were lost during the El Niño of 1982–1983. This mysterious climatic seesaw is more than a change in the weather in this land; it turns the world upside down. And it has an equally strong effect nine thousand miles away.

British meteorologists first noticed in the nineteenth century that monsoon failure in India usually coincided with drought in Australia, and that both were related to changes in air pressure above the tropical Pacific. A failure of the monsoon brought unspeakable misery not only to the subcontinent but to Asia as well. After the disastrous monsoon failure of 1877, which caused an estimated forty million deaths in India and China, British colonials founded the Indian Meteorological Observatory, for the main purpose of understanding the monsoon and predicting its strength.

Sir Gilbert Walker, who was appointed director-general of the Indian Meteorological Service in 1904—not long after another disastrous monsoon failure, in 1899—obsessed about the monsoon for decades. A mathematical physicist from Cambridge University, Walker sought global patterns in climate through the statistical analysis of weather station data from around the world. In the 1920s, this empirical rather than theoretical approach led him to the discovery of the "Southern Oscillation," a seesaw in air pressure between the eastern equatorial Pacific and the neighborhood of Australia and Indonesia. When pressure drops in the east it climbs in the west, and vice versa. Walker knew as early as 1923 that high pressure in Australasia was usually accompanied by drought in Australia, India, and parts of sub-Saharan Africa, warm winters in southwestern Canada, and cold winters in the southeastern United States. Oddly however, Walker never considered the implications of El Niño, even though it was well known at the time and both 1877 and 1899 were documented El Niño years.

It was another forty years before Jacob Bjerknes—who along with Carl-Gustav Rossby had previously laid the mathematical foundation for global circulation modeling—connected the Southern Oscillation to the Peruvian El Niño. As with carbon dioxide monitoring, it took the massive push of the International Geophysical Year—which saw another visit by El Niño, providentially—to produce the comprehensive set of atmospheric observations that Bjerknes used to make that link. His was a more theoretical understanding of the dynamics underlying the combined phenomenon, which he gave the name "El Niño/Southern Oscillation," or ENSO.

Mark Cane of Columbia University's Lamont-Doherty Earth Observatory calls Bjerknes's work "[t]he towering landmark in the understanding of ENSO," and Cane should know, for he and his student Stephen Zebiak based the first computer program to be successful in predicting an El Niño on Bjerknes's work.

The Southern Oscillation Index, a measure of the difference in air pressure between Tahiti in the central Pacific and Darwin on the northern coast of Australia, is one of the simpler tools enabling scientists to watch an El Niño as it forms.* Normally, the pressure is lower at Darwin than it is at Tahiti. This causes the "trade winds" to blow east to west, thereby peeling the sun-heated surface waters away from the coast of South America. In the normal state of affairs the Humboldt Current rises to replace them, and a strip of cold water stretches west along the equator. And, since warm water is less dense than cold water, the surface of the Pacific Ocean actually tilts; it is normally about eighteen inches higher around Indonesia than it is at the coast of South America.

As an El Niño forms, the Southern Oscillation Index falls. The pressure drops at Tahiti and rises at Darwin; the trade winds falter and sometimes reverse direction; warm surface water rolls back in "Kelvin waves" toward Peru and Ecuador; and the Humboldt Current withdraws—with its anchovy—to the deep sea.

After an El Niño plays itself out, the pendulum usually swings too far back: the pressure difference between Tahiti and Darwin rises above normal, the trade winds blow more strongly than usual, and the equatorial strip of cold water stretches farther than usual across the Pacific. This is El Niño's sister, La Niña. When she comes, the land of the Moche is drier than ever.

And while coastal Peru and Ecuador are normally dry, the sun's heat on the warm surface waters of the western Pacific pumps massive cumulus anvils into the skies above New Guinea and the Philippines, making them among the wettest places on earth. This bank of water vapor, the so-called Indonesian Low, which constitutes a significant fraction of the earth's total atmospheric water budget, supplies the Asian monsoon. ENSO's direct swath, therefore, extends in a thin line along the equator from Peru and Ecuador, across the Pacific, through the Indonesian Archipelago, across the Indian Ocean, and as far as Africa.

Bjerknes's great insight was to realize that in normal years—and more intensely during La Niña—the greenhouse effect of the moisture in the Indonesian Low will heat the local atmosphere and act something like a radiator standing against the wall of a room: the warm air rises, and new air moves in from the east, along the ocean's surface, to replace it, thus generating the trade winds. When the rising arm of air reaches the upper troposphere, having shed most of its water vapor in the form of clouds and rain, it curls back east. Nine thousand miles later, in the vicinity of Peru, it drops back to the surface, now dry, to fill the vacuum that was left by the departing trade winds in the first place. In the late sixties, Bjerknes named

*I first heard of the SOI, remember, from Bernard Francou, on my way to Sajama, when he accurately predicted that an El Niño was on the way.

this circular air current, which stands above the equator like a huge Ferris wheel, the "Walker Circulation," or Walker Cell, after Sir Gilbert. He also understood that the cooling of the surface waters to the east and the rising air mass to the west reinforce each other in a positive feedback, or in Bjerknes's words, "chain reaction." As he wrote, "An intensifying Walker Circulation also provides for an increase in east–west temperature contrast that is the cause of the Walker Circulation in the first place."

But his most penetrating insight was to realize that the Walker Circulation could run the other way. He saw no reason why it couldn't have two modes: the predominant one, in which air rises in the west and falls in the east, and an alternate, in which air rises in the east and falls in the west. This linked it to El Niño, which is the periodic attempt by the atmosphere and the surface waters of the Pacific Ocean to switch to the alternate mode. As the trade winds diminish and warm water rolls east, the low over Indonesia moves to the central Pacific, stealing water vapor from the monsoon and making it available for the storms that lash the west coasts of North and South America once an El Niño begins.

The reason for the oscillation is that neither of the two modes is stable. Mark Cane's computer models show that if the Walker Circulation settles into either one, a very small perturbation will send it tilting the other way. It is the spinning of the Earth that predisposes it to its usual mode. The Kelvin waves that carry the warm water east from Indonesia during an El Niño are speedy enough to cross the Pacific in about two months, whereas the Rossby waves* that carry the water back west may take as long as a decade. So something is always out of phase; it's like a dog chasing its tail.

Each El Niño has its own personality, building and relaxing according to its own schedule and meting out its own particular set of effects—though each tends to follow a predictable pattern in the tropics at least. Scientists can track it, and they can predict it pretty well, but they still can't say exactly what "causes" it.

The search for a cause to such a global phenomenon may lead to some very long excursions. Some experts, including Lonnie Thompson, believe that the amount of winter snow cover on the Tibetan Plateau may play a role in El Niño's genesis. Even Gilbert Walker knew that a year of heavy snow on the plateau tended to weaken the monsoon. If the plateau is covered with snow, the sun's energy will go toward melting it rather than

*Rossby postulated the existence of these waves while he was working at the University of Chicago in the forties. They are induced by the rotation of the earth, and they are so subtle—sometimes thousands of miles between wave tops that are less than four inches high—that it took fifty years for their existence to be verified: with a high-tech satellite monitoring a grid of ocean-based sensors.

heating the sand, and the air above the plateau will remain cold. Cold air means high pressure, which prevents warm, moist monsoon air from crossing the Himalaya. In a weak snow year, the air will heat up, the pressure will drop, and the wet air over the Bay of Bengal will be sucked north across the mountains.

Yes, Tibet's monsoon winds do come from the southeast, but its snows are carried in on west winds that can be traced back across the Hindu Kush, through the Middle East, across the Mediterranean, to the Atlantic Ocean. In other words, El Niño's indistinct trail leads most of the way around the world. . . .

ONE SUSPECTS THAT THE MOCHE PRIESTS HAD LITTLE IDEA JUST HOW POWERFUL A FORCE they were dealing with.

By the early sixth century, the priests in the Moche River Valley, who lived and "worked" in and on the huacas to the sun and moon, appear to have gained control of at least eight of the neighboring watersheds, mainly through military conquest. (War had a dual purpose: sacrifice victims tended to be prisoners of war.)

These priests would have been more than usually concerned about the weather during a drought that settled in about halfway through the sixth century. Seasonally—perhaps monthly or weekly, judging by the alacrity with which they seem to have performed their civic duty—they would march in procession to the top of one or the other of their huacas and offer up the magical essence to the gods. At the height of their political and, to their minds most likely, magical powers, an unusually strong El Niño backed into the coast of Peru.

They were probably happy at first, for their rituals had produced the first rain in decades. But the gods of the atmosphere gradually bared their teeth. It rained furiously for weeks, perhaps months. The fields around the city were flooded, and the crops were buried in mud.

One day, one of the lords in his luxurious living quarters near the top of the Huaca de la Luna might have cast an eye toward the magnificent mountains to which he directed his magic and perceived an odd brown swell, an apparent rising of the earth, in the distant V-shaped canyon at the valley's neck—as a flash flood from the high Andes surged onto the plain and began making slow and steady progress across the seven intervening miles. He and the rest of his royal clan probably stood dumbstruck, above it all, as they watched the god-given waters wash their city away.

Moche lay near the Pacific delta of the Moche River, on its southern bank. The great flood of the mid-sixth century scooped ten to fifteen feet of topsoil from the fields in the Moche River Valley—which were more extensive then than they had ever been before or have ever been since—and dumped it in vast quantities, together with large sections of the city, into the sea. The cane and masonry huts of the peasants were swept from their lowly positions on the valley floor, as were many corpses of both the citizens and their livestock. Even the friezes at the bases of the huacas were etched away.

Then, over the next few decades, the sea gave the topsoil back. Southwest winds created dunes at the shoreline and proceeded to blow them across the coastal plain, burying the damaged city and most of the valley south of the river in sand. By century's end, the splendid capital of Moche had been abandoned.

IZUMI SHIMADA AND HIS FELLOW ARCHEOLOGISTS PIECED TOGETHER THE GENERAL OUTlines of this story in their usual ways, using stratigraphic evidence from the city itself and a refined but inexact dating method based on ceramic sequencing and radiocarbon dating: the track of materials, vessel shapes, fabrication methods, and iconography as they passed from one South American civilization to the next. Archeologists weren't sure of the precise timing of the events, however, and they were missing the crucial climatic details until the late 1980s, when Shimada and his student Crystal Schaaf joined Lonnie and Ellen Thompson in a close look at what Quelccaya might have to say about the sixth century.

They found that a drought "remarkable relative to the entire 1,500 year glacial data set in its severity, abrupt onset and relentless persistence for over thirty years" had hit Peru from A.D. 562 to 594. Annual precipitation dropped 30 percent, and the telltale signs of El Niño appear in A.D. 546, 576, and 600.

So it seems that when El Niño induced a flood that washed away their fields, obliterated their irrigation channels, and left a large fraction of the population either dead or destitute, the people of Moche were already enduring or were about to endure a drought that would have limited their agricultural output by about a third even in the best of times—and for more than a generation. The El Niño that devastated the city of Moche came either sixteen years before the drought or right in the middle of it. Either way, by the time of the third El Niño, in A.D. 600, the capital city was a ghost town.

Unlike the Tiwanaku, however, the Moche left enough evidence behind to permit scientists to paint a reasonably convincing picture of the way climate change caused this particular society to lurch and finally fall.

• • •

THE DOWNSIDE OF THE PRIEST'S SHAMANIC STATUS WAS THAT IF THE GODS TURNED wrathful he might be held accountable and replaced—or, in the direst circumstances, the belief system supporting the whole civilization might fail.

There are signs that the particular cast of characters who probably sat high in their pyramid as the flood inundated the commoners below grasped desperately at keeping control. Sometime near the end of the sixth century they held a horrific sacrifice, probably aimed both at preventing another flood and at instilling a new level of fear in their subjects.

In 1995, archeologists discovered a mass grave in a burial chamber inside the Huaca de la Luna, containing nearly forty victims who had been executed with more than the usual brutality. Fingers and toes had been sliced off and inserted in cavities carved into other bodies. Fingers had been crushed. Feet had been pierced with copper lances. Some corpses were missing heads. Those that weren't had signs of cutting on their necks, as if they had been bled to death. Some bodies were hacked up and rearranged in ritual positions. Heads were placed between legs. Bodies were placed one on top of another. And all were buried in a matrix of mud comprised of the adobe that had been eaten from the base of the Huaca de la Luna by the great El Niño flood.

Since the bases of both huacas were refaced after the devastating El Niño, it seems that the "holy men" who performed this butchery held on for a few years at least. There are indications, however, that the ideology that gave them power had begun to erode. There may even have been a rebellion. A new theme crops up in the Moche myth cycle after the flood: the "Revolt of the Objects," in which strong, heavily armed priests in ritual costume fight off smaller, humble, and helpless-looking commoners. And in a sign that the lords had to search for new ways of manipulating belief, they imported foreign symbols for the first time. Tellingly, one came from Tiwanaku, which rose to empire status on the Altiplano at about that time. A figure reminiscent of Viracocha, the Lord of the Atmosphere, was painted on a mural in the Huaca de la Luna just before it was abandoned.

A similar erosion of belief seems to have taken place among the Anasazi six centuries later, when chaotic weather forced them to abandon their cliff dwellings. Linda Cordell points out that "the post abandonment era . . . witnessed a remarkable flourishing of novel religious forms," among them the katsina religion, which is still practiced by the Hopi and Zuni today. "When [an irrigation] network fails, the gods have failed," she adds. "Ritual and ceremony fail. Former allies may become enemies. Trusted elders may become witches, and the world must not seem a very safe place."

AFTER MOCHE'S FAILED REVOLT, THERE WAS PROBABLY A SUCCESSFUL ONE—OR MAYBE A coup by military officers or some subset of the priest clan. What was left of the population moved—or was forced to move—seven miles inland, to the north side of the Moche River, at the valley neck where the river issues from the mountains. The soil was good there; but more important, it was the most strategic and reliable place in the valley to tap the river for a new system of canals.

They built a new town, named Galindo by anthropologists, above flood level on the western slope of the Andean foothills. Galindo's fields descended to the plain, but they were limited to the north side of the river, beyond the reach of the blowing sand.

In comparison to the former capital, the new town has a claustrophobic, paranoid flavor, all compartmentalized and enclosed. Unlike the open city surrounding the huacas, its layout testifies to secular control. What huacas have been found are tiny and lie at the edge of the city, behind high walls. It would seem that the sacred rituals that had so long maintained power were now suspect.

The common man lived in closer urban quarters than he had on the plain. Walls restricted his access to the spacious residential zones of the elite, to the centers of power, and, significantly, to the river that was his source of water. Galindo was also the first Moche township to build large-scale storage chambers for beans and maize kernels—against the onslaught of another drought perhaps. The chambers were also guarded by walls.

But Galindo never became more than a threatened frontier town. Some archeologists believe it survived only as long as its first autocratic ruler and that a second popular revolt led to its final collapse. The wall that hid the hillside homes of the elite from the view of the lowly commoners was used at least once as a fortification. In any event, by A.D. 700, the people had left the city and reverted to a simple country life. They didn't need the priests to grow food for themselves, after all.

THIS CHRONICLE OF FLOOD AND DROUGHT FOLLOWED BY DRIFTING SAND AND HUMAN dislocation was repeated to varying degrees in all twelve watersheds of the Moche empire. It was worse in the south than in the north, generally speaking, so there was an overall retrenchment northward to escape sand, and eastward toward water. The priests at the Huaca de la Luna lost control of the other watersheds, of course.

The largest of the twelve valleys is the northernmost Lambayeque, about a hundred miles north of the former capital. After the cataclysmic events of the sixth century, the inhabitants of Lambayeque also moved to the valley neck and built a new city, again on the north shore of the river. However, their priests seem to have ridden the crisis out; for they managed somehow, probably with the help of some physical coercion, to mobilize the populace into building the last great monument of Moche civilization, Huaca Fortaleza, which stands at the center of a planned city, now called Pampa Grande, which may have been the last Moche capital.

The layout of this city suggests that the people still believed just enough in the myths that empowered the priests. Thirty or more smaller adobe mounds and terraces radiate from the central huaca to the city's perimeter, and the only restricted areas are the chambers for large-scale food storage. Nevertheless, Pampa Grande has a more controlling, urban quality than any of the older cities that had previously stood out on the plain. Socially, it stands about halfway between Moche and Galindo. And it also looks as though the extraordinary measures its rulers took to keep control through extraordinary times planted the seed of their own demise.

Izumi Shimada estimates that it would have taken a workforce of a thousand men several years simply to place the bricks in the Huaca Fortaleza—ignoring the making of the bricks, the supporting cast required to feed and manage the workers, the effort to house them, and so on; yet it went up in less than a generation. Signs point to massive administrative systems, such as special "cafeterias" and housing for the workers and new construction techniques requiring labor gangs. A large secular bureaucracy was probably interposed between priest and commoner. Even the craftsmen who made the thousands of ceramics, murals, and textiles for the public monuments and priestly precincts lived and worked in compounds.

But Quelccaya shows that the first three decades of the seventh century saw more than the usual amount of rain. In this era of plenty and with the new city built, the weighty bureaucracy and what may have become habitual policing of daily behavior were unnecessary for the survival of the state—or of its citizens, at least. Yet the priests continued to live in splendor above the city, removed from the everyday affairs of their people, dressing up in outrageous costumes every once in a while, climbing to the top of the giant huaca, taking psychedelic drugs, and cutting people's heads off.

The common folk tired of these spectacles somewhere near the turn of the century. In about A.D. 700, they torched Huaca Fortaleza and most of the other adobe mounds and terraces in Pampa Grande—virtually every structure that had required a labor gang to build—and promptly abandoned the city. The most spectacular fire burned at the summit of the

Huaca Fortaleza itself, where the heat was so intense that it baked the adobe walls and ceilings to the hardness of ceramic, an inch deep. The priests may very well have been burned alive.

NOW, JUST AS THESE TRAGEDIES WERE UNFOLDING IN THE LOW DESERT, THE TIWANAKU were beginning to flourish in the high one. In fact, the sixth-century drought coincides with the first of the two dust events at Quelccaya, which the Thompsons attribute to the building of aqueducts and raised-bed fields by the shores of Lake Titicaca.

The lake sits at the southern end of a north–south El Niño seesaw, similar to the one that reaches west across the Pacific. While El Niño brings rain to the coastal desert, it brings drought to the southern Altiplano. The way the Thompsons identify El Niño events at Quelccaya is to look for unusually thin annual ice layers, which indicate dry years. The severe El Niño of 1982 and 1983, for example, brought devastating floods to the land of the Moche, dealing a new round of blows to the bases of the huacas, but drought to the land of the Tiwanaku. Most of the fields around Lake Titicaca were devastated, but Alan Kolata's experimental raised beds soldiered through: the level of the lake didn't drop enough in that short a time span to drain the irrigation ditches completely.

Similarly, although the sixth-century drought was more severe than the one that would bring Tiwanaku down five hundred years later, it was not long enough to lower the lake disastrously as the empire was being built. The Tiwanaku seem to have responded by constructing more raised-bed fields.

Simultaneous to the Tiwanaku, the Huari rose to empire status in Peru's northern highlands. They soon extended some influence down into the northern coastal valleys, but it wasn't until A.D. 1000, as Tiwanaku and Huari simultaneously collapsed, that the next real lowland empire, Chimor, flourished in the former land of the Moche.

From the few Chimú who survived into the sixteenth century, a Mercedarian friar named Miguel Cabello de Balboa learned the story of the Sican, a people who lived, built monuments, and produced their own quota of sacrifice victims in an area limited to the Lambayeque watershed during the centuries between the fall of Moche and the rise of Chimor:

> The people of Lambayeque say that in times so very ancient that they do not know how to express them, a man of much valor and quality came to that valley on a fleet of balsa rafts. His name was Naymlap. With him he brought many concubines and a chief wife named Ceterni. He also brought many people who followed him as their captain and leader. Among these people

were forty officials, including Pita Zofi, Blower of the Shell Trumpet; Ninacola, Master of the Litter and Throne; Ninagintue, Royal Cellarer (he was in charge of the drink of that lord); Fonga Sigde, Preparer of the Way (he scattered seashell dust where his lord was about to walk); Occhocalo, Royal Cook; Xam Muchec, Steward of the Face Paint; Ollopcopoc, Master of the Bath; and Llapchillulli, Purveyor of Feathercloth Garments. With this retinue, and with an infinite number of other officials and men of importance, Naymlap established a settlement and built his palace at Chot.

Naymlap also brought with him a green stone idol named Yampellec. This idol represented him, was named for him, and gave its name to the valley of Lambayeque. [Note the similar sounds of the names.]

Naymlap and his people lived for many years and had many children. Eventually he knew that the time of his death had arrived. In order that his vassals should not learn that death had jurisdiction over him, his immediate attendants buried him secretly in the same room where he had lived. They then proclaimed it throughout the land that he had taken wings and flown away.

The empire and power of Naymlap were left to his oldest son, Cium, who married a maiden named Zolzdoñi. By her and other concubines he had twelve sons, each of whom was father of a large family. After ruling many years, Cium placed himself in a subterranean vault and allowed himself to die so that posterity might regard him as immortal and divine.

Subsequently there were nine rulers in succession, followed by Fempellec, the last and most unfortunate member of the dynasty. He decided to move the idol that Naymlap had placed at Chot. After several unsuccessful attempts to do this, the devil appeared to him in the form of a beautiful woman.* He slept with her and as soon as the union had been consummated the rains began to fall, a thing which had never been seen upon these plains. These floods lasted for thirty days, after which followed a year of much sterility and famine. Because the priests knew that their lord had committed this grave crime, they understood that it was punishment for his fault that his people were suffering with hunger, rain, and want. In order to take vengeance upon him, forgetful of the fidelity that is owed by vassals, they took him prisoner and, tying his feet and hands, threw him into the deep sea. With his death was ended the lineage of the native lords of the valley of Lambayeque, and the country surrounding remained without patron or native lord during many days.

Then a certain powerful tyrant called Chimu Capac came with an invincible

*The friar seems to have been employing some devout editorial license here. The Sican had no notion of a devil particularly. They saw the attractive seductress as an evil sorceress.

> army and possessed himself of these valleys, placing garrisons in them. In Lambayeque he placed a lord called Pongmassa, a native of Chimu . . .

So the Sican arrived by water and died by water, in a theme that recurs throughout the Andes. Furthermore, they went down in exactly the same way as the Moche. Fempellec's end, involving flood and famine, recalls the effects of El Niño and La Niña; and the Quelccaya ice cores show evidence of a drought in about A.D. 1020 that was almost as strong as the sixth-century drought and was interrupted, similarly, by a major El Niño. Furthermore, in a remarkable example of history repeating itself, the Sican civic monuments were set afire and abandoned at about that time.

It is dangerous to take these oral traditions too seriously, of course, but one version of Chimor's dynastic myth tells of the great man Taycanamo journeying south from Lambayeque to Moche to found the Chimor empire—based, as usual, on large-scale irrigation—soon after a flood had destroyed his homeland to the north: Sican perhaps. The Chimú then expanded back to the north, conquering what was left of Sican in about A.D. 1100.

Finally, according to Balboa, the lowland Chimú fell to the last great highland empire in the Andes, the Inca, in about 1470, just at the end of Alan Kolata's four-hundred-year drought. (By the time the friar met them, of course, both the Inca and the Chimú had been conquered by his fellow Spaniards.) The Inca took up the old cosmology of the Tiwanaku, whom they saw as their ancestors. Their myth held that Viracocha, Lord of the Atmosphere, had created the sun and moon at Lake Titicaca, and that the first man and woman, born of the sun god Inti, had emerged from the lake to reign at the city by its shores that was ancient even in their day.

THERE IS A PATTERN HERE. AS EARLY AS 1976, THE ARCHEOLOGIST ALLISON PAULSEN pointed out that highland and lowland civilizations had ridden a sort of cultural seesaw, which rocked with about a four-hundred-year beat for around two thousand years. From about 500 B.C. to A.D. 1532 (the year the first Spaniard, Francisco Pizarro, landed on the Peruvian coast), as highland civilizations rose, coastal civilizations fell, and vice versa.

Paulsen's own field studies had taken place on Ecuador's Santa Elena Peninsula, which is just as dry as the Moche heartland and juts into the Pacific a few hundred miles to the north. The prehistoric peoples on the peninsula used walk-in wells, or catch basins, to store water for what Paulsen believed was some method of "intensified agriculture" like that of the Tiwanaku. Reading the dates of occupation and abandonment from the ceramic sequence of pottery sherds buried in the walls of the wells "during regular prehistoric cleaning and maintenance," Paulsen found that the wells had been occupied and abandoned twice in the twenty centuries of the archeological record. The first phase of scattered occupation began in about 500 B.C. and ended in about A.D. 600—just as Moche fell and Tiwanaku emerged. The Guangala, who lived on the peninsula at the time of the collapse, abandoned the wells and the settlements around them and moved to the edge of the sea. A few survivors managed to eke out an impoverished existence for about two hundred years, then abandoned the peninsula entirely. It was essentially uninhabited for another two hundred years; then in about A.D. 1000, just as Tiwanaku fell and Chimor rose, the peninsula was reoccupied by the Libertad people—who disappeared in 1400, as the Inca emerged in the highlands.

Farther south on the coast, at least two more cultures gave last gasps in chorus with the Moche. The Lima culture abandoned a major ceremonial and civic center, and the Nazca—famous for the Nazca Lines, believed to be the largest works of art on the planet—abandoned the ingenious *pukios* they used to tap the underground water that was all they had for agriculture in their particular watershed.

In his book *Catastrophe: An Investigation into the Origins of the Modern World,* amateur archeologist David Keys argues that the mid-sixth-century climate change was global in extent and caused not only these simultaneous tragedies in South America but also the Dark Ages in Europe and severe dislocations in China, Southeast Asia, and essentially everywhere else. He also argues, unconvincingly to me, that adaptations to this catastrophe gave birth to most modern societies. Since the book is cast in the mold of a detective story—it was made into a two-part TV series on PBS—it requires a satisfying denouement; so, at the end, Keys makes the dramatic discovery that the catastrophe was caused by an enormous undersea volcanic eruption in the strait between Java and Sumatra.

The best evidence he can muster is an old document from China that alludes to some loud bangs coming from the southwest. He attempts to buttress this, shall we say, anecdotal evidence by citing modern ice core records from Greenland and Antarctica. It is true that the polar ice cores provide an excellent record of all the major eruptions known to history, including Vesuvius, which buried Pompeii about five hundred years earlier than Keys's putative blast, as well as many that have escaped historical notice. His Indonesian event would have had to have been a hundred or a thousand times more powerful than Vesuvius; however, unfortunately, when I scoured the polar ice core papers cited in his book, I found no evidence for a massive eruption in the mid-sixth century. None of Lonnie's ice cores has picked it up either. Keys interviewed Lonnie, who told me, "He found what he wanted to find."

There *has* been one climatic catastrophe that has changed civilizations around the world, but it took place in about 2000 B.C. Lonnie would find its fingerprints in the snows of Kilimanjaro.

WHEN HE, ELLEN, AND MARY DAVIS FOUND THE TIME TO STEP BACK FROM THE DETAILS of Tiwanaku and Moche and view the situation panoramically, they realized that the epochal alternation between lowland and highland dominance—first the Moche, then the Tiwanaku and Huari, next the Chimú, finally the Inca—bore an extraordinary resemblance to long-term oscillations in the Quelccaya climate record.

What's more, the cultural seesaw followed an El Niño–like pattern. When the Moche fell, mainly because of drought, the Tiwanaku emerged with the aid of more rain. This is exactly the way La Niña behaves in the places these peoples lived. Both the highland and lowland civilizations subsisted at the margins of agriculture, so it is not surprising that slight changes in rainfall might have caused them to rise and fall; but the two-to-ten-year El Niño cycle beats too quickly, of course, to pace the advances and retreats of civilization.

However, the Thompsons also found longer, gentler oscillations in the Quelccaya record, with periods of two to four hundred years, which dovetail quite well with Paulsen's archeological timing. Quelccaya's story goes back only 1,500 years—not as long as the archeological record; but from the time of its oldest layers, which date from A.D. 500 until about 750, conditions were drier than average on the Altiplano. They were wet from A.D. 750 to 1050 or so, dry from A.D. 1050 to about 1500 (Kolata's drought), quite wet from A.D. 1500 to shortly after 1700, dry again until about 1900, and they have been wet ever since. The wet periods correspond reasonably well with the flowering of highland cultures, and the dry periods with the flowering of coastal cultures. (The match would be exellent if the boundary differentiating "wet" from "dry" were shifted slightly toward the wet. There is no reason to expect the *average* amount of precipitation at Quelccaya to correspond to an agricultural balance point.) The Thompsons believe an El Niño dynamic may explain this: over the long term, slightly drier periods at Quelccaya may have coincided with slightly wetter periods at the coast, and vice versa.

The archeological record certainly supports this idea, but it is difficult to obtain good climate records on the coastal plain. Indeed, one of Alan Kolata's present projects is to retrieve sediment cores from an ephemeral lake bed, or playa, in the land of the Moche and Chimú.

There is supporting evidence, however, from other parts of the world. At Methuselah Walk, California, in the rain shadow of the Sierra Nevada, where El Niño is felt very strongly, the tree rings of bristlecone pines provide records of annual precipitation reaching back five thousand years. They show that rainfall at Methuselah Walk oscillates in phase with Quelccaya's. And Lonnie's ice cores from Dunde show the same thing, half a world away.

"There is strong evidence in our documented history that monsoons and El Niño are connected," Lonnie says. "With an El Niño, you generally have a failure of the monsoon. Well, if they're connected in this short time frame, then you might expect to see the same long, low-frequency signal on the center of the largest and highest plateau on earth [the Tibetan Plateau] that you do on the second largest and second highest plateau on earth [the Altiplano]. Both lie in the belt of the planet that's impacted by the east–west oscillation across the Pacific Ocean [the Walker Circulation]."

This is teleconnection. Since the Altiplano, Methuselah Walk, and the Tibetan Plateau all feel El Niño, since El Niño has a similar effect on rainfall in each, and since their rainfall seems to oscillate together over the course of centuries, it seems reasonable to suggest that the same circulation of air and moisture that underlies El Niño might also oscillate at a slow frequency, affecting moisture patterns in Peru's lowlands in the opposite way. Then rainfall in the low desert would oscillate out of phase with that in the highlands, and cultures might rise and fall accordingly.

However, like the Mercer Problem, this was a question that would take decades to answer. The problem with the Dunde record is that it doesn't have the resolution of Quelccaya's. Few do. Visual layer counting goes back only four hundred years, not much longer than the historical record, less than one long-term rainfall cycle ago. It would take a few complete cycles from China, as at Quelccaya, to validate the transpacific teleconnection beyond reasonable doubt.

That evidence would emerge a few years down the road.

PART V

MORE PIECES FOR THE PUZZLE

STUDY NATURE AND NOT BOOKS.

—LOUIS AGASSIZ

12

CASTING ABOUT

From an ice coring standpoint, Quelccaya and Dunde were both ancient history by the end of the eighties; and in fact, less than six months after returning from the latter, Lonnie went off exploring again. Over the "summers"—that is, Novembers to Februaries—of 1988–89 and 1989–90, he drilled on the Antarctic Peninsula.

The peninsula nearly qualifies as South America, because it is actually the half-submerged southern extension of the Andes. Some sort of tectonic wave seems to have swept the range on its side for a stretch of about six hundred miles south of Tierra del Fuego. It then reemerges, winds south for another twelve hundred miles, and connects to West Antarctica.

The place has a reputation for bad weather. About a century ago, Ernest Shackleton lost his ship, *Endurance*, in the Weddell Sea, to the east of the peninsula's curving arm, and went on to endure the astounding two-year survival epic that has been the subject of just as astounding a number of books, museum exhibits, movies, and television shows recently.

Although Ellen—with the assistance of Keith Mountain and Bruce Koci, among others—had recovered a 302-meter core at Siple Station, near the base of the peninsula in 1985, and a team from the British Antarctic Survey (BAS), led by scientists David Peel and Robert Mulvaney, had recovered a few shallow cores on the peninsula itself, no deep cores had been retrieved there, and the community realized that Lonnie's was the only group in the world that stood a reasonable chance of doing so. Peel and Mulvaney got in touch with him, and the three decided to collaborate. It was a "Have drill, will travel" arrangement. Lonnie's lightweight equipment and ability to work without station support were perfect for the place. The BAS maintained a fleet of Twin Otter aircraft, which could not possibly carry the weight of the standard polar drilling rig; and the British field camps were notoriously spartan, so Lonnie's formative experiences with John Mercer were also a boon.

"It's a tough place, and you also have to contend with the Brits," Lonnie confides wryly. "They have this mentality about being independent—that you shouldn't be comfortable on a field trip; it's a man-against-the-environment type of thing. It was amazing. At the end of the twentieth century they still used these man-boxes, with food allocation for eleven days, that were prepared in the days of Scott tents.* The base commander was running this experiment, trying to find out why it was that when they sent people out into the field they always came back twenty, thirty pounds lighter. It didn't take long to figure that one out."

Although this polar location lay a bit to the side of Lonnie's main life's work, it did fit a plan he and Ellen had conceived to produce a north–south transect of the Americas—the only connected landmass on which a complete pole-to-pole transect is possible. In the mid-seventies Ellen had studied a core retrieved by other scientists at the South Pole for her doctoral dissertation. She had then produced the next data point herself a decade later, at Siple. Now came the peninsula. Lonnie still has plans for the next site to the north, on one or the other of the Patagonian ice sheets. Sajama lies north of there; north again lie Quelccaya and Huascarán. He also has plans for Ecuador: either 19,000-foot Cayambe, whose summit is the highest point on the equator; 18,700-foot Antisana, twin-summited and, at least for now, abundantly glaciated; or 21,000-foot Chimborazo, the highest mountain in the country. (For many years "Chimbo" was thought to be the highest mountain in the world, and its summit is still believed to be the farthest spot on the surface from the center of the earth, owing to the equatorial bulge.) Surprisingly, the first site north of the equator is Alaska's Bona-Churchill glacier, which Lonnie drilled in the spring of 2002.

So far, the gathering of these individual data points has taken almost thirty years. Perhaps he and Ellen will complete the task by the end of this decade.

*Pyramidal canvas tents derived from the design used by the bumbling British aristocrat Robert Falcon Scott, who died on the Ross Ice Shelf with four companions in 1912, having lost the race to the South Pole by one month to Roald Amundsen, a Norwegian. As it happened, 1912 was an El Niño year, and El Niño brings cold temperatures to that part of the continent. The temperatures Scott recorded in his logbooks show that it was one of the coldest seasons on record, which was an important factor in the tragedy. After man-hauling sleds for 1,600 grueling miles, south to the pole and most of the way back, he and his men died just 170 miles from their base camp and only 11 miles from a food and fuel depot.

• • •

THE PENINSULA JUTS FROM THE DEEP-FREEZE OF ANTARCTICA INTO THE SO-CALLED CIRcumpolar trough, disrupting the great sea and air currents that flow in a circle around the frozen continent and form a barrier between its unique conditions and those to the north. Prevailing winds are easterly, from the Weddell Sea, the coldest spot on the Antarctic coast (as Shackleton discovered), so there is a stark difference between the weather on either side of the mountain range that forms the peninsula's spine. To the east, it's as cold as in the Antarctic interior; to the west, it's more maritime—almost eleven degrees Fahrenheit warmer on average, and much snowier.

From an expedition standpoint, the only remarkable aspect of the first season—in which they worked to the west of the divide—is that the weather was appalling enough to merit notice in Lonnie's written work for the first and only time: "A 4-by-13 meter snow pit was excavated by hand to 4 meters and capped with plywood. Fluorescent lights were installed in the ceiling, and all the drilling was done in this trench. This proved absolutely essential because it permitted drilling despite poor weather conditions."

That year the group recovered two cores, about hundred meters each—nowhere near bedrock, since radar measurements put the thickness of the ice at more than a kilometer—and the next year, with Bruce Koci in the company, they recovered two more cores on the east side, about 235 meters each. These reached about two-thirds the distance to bedrock, since the ice is thinner on that windward side.

It was four years before the Thompsons, Mulvaney, and Peel managed to write up the results from the peninsula, and even then they covered only the cores from the east side, which, as it turned out, reached back five hundred years. While this remains the definitive study of *recent* climate change in the region, Lonnie sees it as preliminary. He plans to return to drill to bedrock west of the divide, where he believes the ice may be as thick as two kilometers and the record may go back past the Last Glacial Maximum, twenty thousand years ago. Since the heavy snow on that side makes for thick annual layers, it also holds the promise of a dating precision to match Quelccaya's.

Timing has always been a problem for Antarctic ice cores, because the interior receives very little snow—it actually qualifies as a desert—so a well-dated core from the peninsula might help determine whether epochal events such as the retreat of the Wisconsinan glacial stage began earlier or later in Antarctica than they did in Greenland or tropical South America, for instance. This type of question has risen to the fore over the last decade, as various researchers have tended to stake a claim for the region they happen to be studying as being the place where all great climate changes begin—and, by implication, for their work being especially crucial and, incidentally, worthy of public funding.

WHILE YOUR AVERAGE PERSON MIGHT LOOK BACK ON TWO LONG TRIPS TO ANTARCTICA as the experience of a lifetime, Lonnie's were little more than a footnote. And as his attention returned quickly to the center of the planet, geopolitics came to his aid. Mikhail Gorbachev's perestroika and glasnost programs had led to an opening of the Soviet Union at the end of the eighties, and one of the first tangible benefits was an opening of scientific channels, just as in China. Thus, early in 1990, fresh from the Antarctic Peninsula, Lonnie met in Washington with "four Russians who did ice," and they decided to explore together what was then Soviet Central Asia. Lonnie traveled for seven months that year.

In July and August, he and Mary Davis undertook a joint expedition with a group led by Mark Dyurgerov from the Moscow Institute of Geography, to the Pamir, in what is now Tajikistan, and the Tien Shan, in what is now Kyrgyzstan. One lasting benefit of that trip was the emergence of Vladimir Mikhalenko, an accomplished mountaineer and glaciologist who has participated in all of Lonnie's major expeditions since then—including Sajama, where I never heard the taciturn Vladimir utter a word.

From Kyrgyzstan, Lonnie and Mary went north to Almaty, Kazakhstan, whereupon she flew west for home and he went east, across the Chinese border to Ürümqi, where he rendezvoused with Keith Mountain and nine scientists from the Lanzhou Institute, including Yao Tandong. (Keith had traveled west via Beijing.) They proceeded to drive south for six hundred miles to the Kunlun Shan, a range that emerges from the confluence of the Pamir, the Hindu Kush, and the Karakoram at the cultural crossroads where western Tibet meets Tajikistan, Afghanistan, and Kashmir, and runs east along the northern edge of the Tibetan Plateau for more than a thousand miles. On the Guliya glacier in the western Kunlun, they hand-drilled three cores, including one from the twenty-two-thousand-foot summit that was the highest ever retrieved at the time.

Afterward, Lonnie stopped briefly in Ohio, then left for Peru's Cordillera Blanca to look for a site that might complement Quelccaya. He would have preferred to visit the Blanca in summer, which is the dry season and the optimal time for climbing in Peru; but in this and subsequent years his summer trips to China made that impossible. He points out that "it really got touchy for a while," because the wet season brings frequent thunder and lightning to that spectacularly high cordillera. (On the other hand, he managed to squeeze some lemonade from that lemon by experiencing a flash of scientific insight one day, while staring up at huge wet-season thunderheads from the 19,800-foot Garganta Col on Huascarán. More on that later.)

And the excitement was not limited to the mountains, for this was also the heyday of Sendero Luminoso, the Shining Path. In 1990, with the election of Alberto Fujimori to the presidency, Sendero shifted its attention (meaning car bombings and assassinations) from the rural southern zone around Quelccaya to urban centers in the north—especially the capital, Lima, which Lonnie passed through on every trip. He didn't run into many recreational climbers on these forays.

THAT SAME YEAR, HE, DAVE CHADWELL, AND A FEW PERUVIAN COLLEAGUES, SUPPORTED by Quelccaya veteran Benjamín Vicencio and his fellow guides from the Casa de Guias in the town of Huaraz, gateway to the Blanca, drilled short cores—all higher than 16,500 feet—on three separate mountains: Caullaraju, Qopap, and Pukajirka. It didn't go well for Chadwell, however. On Caullaraju he had an odd reaction to the altitude.

"I maintained consciousness, but I lost my vision," he recalls. "You've heard people describe tunnel vision? It was tunnel-length vision that popped on and off. I think we timed it at one point. It would come on for fifteen minutes or so, then be gone, and then come back for a few minutes."

They put him in a Gamow bag* until he stabilized, then escorted him down the mountain, where his symptoms disappeared. Now, Dave had worked for months at altitudes higher than the site of this incident—on both Quelccaya and Dunde—so this demonstrates just how capricious altitude sickness can be. It comes in a variety of forms, and there's no predicting whom or when it will hit. Hoping this was a one-time event, Dave joined Lonnie in the Blanca the following year, but the symptoms recurred, so he was forced to give up high-altitude work altogether. He eventually moved to Scripps and turned his interest to the other end of the altitude spectrum: he now uses remote sensing techniques to produce highly accurate maps of the ocean floor.

In 1991, despite Dave's second bout of tunnel vision, the group succeeded in drilling on the last two possibilities in the Blanca: 19,193-foot Hualcán and 22,205-foot Huascarán, the highest mountain in the range. (This was Lonnie's second visit to Huascarán; he and Keith Mountain had hand-drilled in the Garganta Col on Keith's first trip to Peru, in 1980.)

*The Gamow bag—invented just that year by Igor Gamow, a professor of chemical engineering at the University of Colorado, in Boulder—looks something like a huge duffel bag with a foot pump attached. The victim of altitude sickness lies inside, the bag is zipped and made airtight, and the pump is used to raise the air pressure inside. The device has saved a number of lives. Lonnie takes one on all his trips.

They also placed the world's two highest satellite weather stations on Hualcán and 19,836-foot Pukajirka and took a side trip south to visit their old friend Quelccaya, where they retrieved yet another short core and photographed the Qori Kalis outlet glacier in order to measure its retreat since the "cleanup" visit seven years earlier.

AN INTERNATIONAL GROUP THAT INCLUDED MARK DYURGEROV FROM RUSSIA, YAO Tandong from China, Ping-Nan Lin* from Taiwan, and Lonnie, Ellen, and Mary Davis from the United States presented the results of the quick strikes in Central Asia in a 1993 paper that was the first to reflect what was now becoming the global reach of the Thompsons' fieldwork. The paper synthesized results (going north to south) from Camp Century, Greenland; the Gregoriev glacier, in the Kyrgyz Tien Shan; Dunde and Guliya, in China; Quelccaya, in Peru; and Siple and the South Pole, in Antarctica. While the polar cores showed little sign of change in the twentieth century, every core from the tropics and subtropics demonstrated twentieth-century warming. At Dunde, the authors were now prepared to say, "the last 60 years . . . have been the warmest in the entire record"—that is, at least forty thousand years.

When the results from the east side of the Antarctic Peninsula were finally presented the following year, they bucked the polar trend. The peninsula had warmed at least as much as the tropics (owing, perhaps, to its maritime location, which isolates it from the inertia of the vast subzero ice sheets that blanket the Antarctic interior). The peninsula had started warming in the 1930s, and the seventies and eighties were among the warmest decades in the five-hundred-year record.

The Russian-American team had done no drilling on their reconnaissance to the Firn Plateau in the Pamir, but they had retrieved two short cores at fifteen thousand feet on the Tien Shan's Gregoriev glacier; the Chinese-American team had retrieved the two from Guliya. Although none of these reached back more then a few decades, they all evinced a gradual enrichment of the heavy oxygen isotope, indicative of a warming, in the periods they did cover. On the Gregoriev, the team had also dropped a temperature probe into the drill hole to measure ice temperature as a function of depth. Comparing their numbers with a similar profile obtained by a Soviet team twenty-eight years earlier, they found that the temperature of the ice had risen at every depth. Twenty meters down, the shift was four degrees Fahrenheit.

*Ping-Nan, a.k.a. Ahnan, had joined the Thompsons' group as an analytical chemist in late 1989, the second permanent addition after Mary. He was later base camp manager (and guardian of the alcohol) at Sajama.

The short core from the 1991 side trip to Quelccaya bore particularly sobering news: the ice cap was dying. Meltwater had begun to bleed down from the surface into the firn and to homogenize the chemical signals in the upper layers. All signs of annual variation had been obliterated. "If the survey to assess the quality of the record preserved in this ice cap had been conducted in 1991 [rather than 1976]," they wrote, "Quelccaya would have been eliminated as a possible site for acquisition of a long ice core."

Photos of Qori Kalis were also presented. The images from 1991 showed that the glacial tongue had retreated more than five hundred feet from the terminal moraine it had occupied when they first photographed it in 1978.

This international team concluded that the tropics and subtropics had experienced a significant and widespread warming since the start of the industrial era, and they suggested that the center of the planet might be more sensitive to the early stages of greenhouse warming than the polar regions, especially at high elevations.* They also pointed out, as the Thompsons had in the Dunde paper, that computer models had predicted this: Jim Hansen had suggested that Central Asia, so far from the mitigating influence of the oceans, might be one of the first places to demonstrate unambiguous greenhouse warming, and other simulations had shown that the maximum temperature change should occur at twenty-six thousand feet (or eight thousand meters), not much higher than Lonnie's field sites.

The paper ends with something like a plea. "These data also make it very clear that some of these unique archives of climate and environmental history are in imminent danger of being lost forever if the current warming trend persists."

IT SEEMS WORTHWHILE TO NOTE THAT THESE SOVIET AND AMERICAN SCIENTISTS, WHOSE opportunity to collaborate resulted from a thaw in superpower relations, put one of the most terrifying features of the Cold War to scientific use: fallout from atmospheric thermonuclear bomb tests conducted by their respective nations in the fifties and sixties enabled them to assign dates to specific layers in the Gregoriev ice cores.

Twelve meters down they found a pronounced spike in beta-radioactivity from the largest aboveground nuclear blast of all time, which had been set off by the Soviets on the Novaya Zemlya Archipelago in the high Russian Arctic in October 1961. The explosive yield of Novaya Zemlya was equivalent to fifty-eight million tons (megatons) of TNT—2,600 times that of the bomb that destroyed Hiroshima and about twice that of its closest competitor.

*This confirmed a prediction Guy Callendar had made in 1949.

It takes a thermonuclear, that is, hydrogen, bomb with at least a megaton yield to leave a global fingerprint, because only then will the infamous mushroom cloud penetrate the stratosphere, where high-speed winds will broadcast the fallout around the globe and eventually mix it with the entire atmosphere. And the polar ice coring community was on top of the scientific possibilities even before what was, it is hoped, the historical era of atmospheric testing came to an end. Tracking the fallout from the biggest blasts as it came down, the polar scientists discovered that the cloud from an individual blast usually wanders around the earth for a year or two while it slowly dissipates via snow and rain (". . . and it's a hard rain's a-gonna fall," sang Bob Dylan in 1963). It took two years for the huge cloud from Novaya Zemlya to reach Central Asia, for example, so when the Russian-American collaboration found it in the Gregoriev core, they knew that particular layer had been laid down in 1963.

Ice cores are routinely examined for two "nuclear horizons": the spikes from the biggest atmospheric bomb blast, Novaya Zemlya, and the first, 10.4-megaton Ivy Mike, which rang in the thermonuclear age from the Eniwetok Atoll at 7:15 A.M. on November 1, 1952, having been set off by the United States. (The "Mike horizon" would be deeper in an ice core, of course.) According to Richard Rhodes, "Mike's fireball alone would have engulfed Manhattan; its blast would have obliterated all New York City's five boroughs."

The history of atmospheric thermonuclear testing is written in glacial ice around the world. It began with Mike, stopped briefly during a unilateral moratorium observed by the United States from 1959 to 1961, resumed dramatically with the Novaya Zemlya blast, and stopped for a few years in 1963, when the United States and the Soviet Union signed the Limited Test Ban Treaty, prohibiting underwater, atmospheric, and—God forbid—extraterrestrial bomb tests. The ice cores show a rise in radioactivity again from 1968 through 1975, owing to tests conducted by China and France in their scramble to join the thermonuclear club before signing the treaty.

The earliest blasts from the U.S. testing program, which occurred over the ocean, produced tremendous quantities of radioactive chlorine-36 from the salt (sodium chloride) in the seawater vaporized by the blasts. A Swiss group has measured the level of chlorine-36 in Lonnie's cores from the Antarctica Peninsula, Guliya in western Tibet, Huascarán in the Andes, and a core they themselves retrieved on the Fiescherhorn in the Swiss Alps. In every one of these widely separated locations, the level of chlorine-36 rises sharply in the early fifties, reaches a peak at one thousand times its natural level in 1960 (which indicates that while the United

States may not have set off the biggest bomb, its extensive testing before fallout was recognized as a hazard sent the greatest amount of radioactivity into the air), drops slowly after 1960, and finally reaches background again in about 1980, a few years after the French and Chinese completed their testing.

Let us hope this is not the way some ice coring expedition from another planet eventually guesses how our civilization came to an end.

THIS INTERNATIONAL WORK ALSO FORMED THE BASIS FOR TESTIMONY LONNIE DELIVERED in February 1992 to the Senate Committee on Commerce, Science, and Transportation, at the invitation of Al Gore.

The political campaign that would see Gore installed as vice president had just begun, and Gore—who was still seeking the presidency at that point—had just published a bestselling book, *Earth in the Balance,* about global warming. The official title of the hearing was "Global Change Research: Indicators of Global Warming and Solar Variability," and it seems to have been a kind of sanity check on the first "assessment" by the Intergovernmental Panel on Climate Change, issued in mid-1990, which had pointed to "clear evidence" that the global mean temperature had risen by about one degree in the previous century, but had not gone so far as to attribute the rise to human activity.

In his opening remarks, Gore demonstrated a superb understanding of climate science (as he had in his book) and even came out with a prediction of his own that later proved true: "If the models are right, it is likely that the 1990s will replace the 1980s as the hottest decade on record."

In this conversational setting, the most dramatic aspect of Lonnie's testimony proved to be his vivid evidence for the worldwide retreat of tropical glaciers, Qori Kalis being the best example. From the transcript of the hearing:

[A slide was shown.]

DR. THOMPSON. This is a margin of this ice cap showing a boulder about the size of the middle of this circle here, being pushed by the glacier in 1977.

[A slide was shown.]

DR. THOMPSON. This is the same boulder in 1978. You can see the wall pulling back from the ice cap.

[A slide was shown.]

DR. THOMPSON. This is in 1979. This wall is about 15—

SENATOR GORE. Is that the same boulder?

DR. THOMPSON. That's the same boulder. This wall was about 15 to 20 meters high.

[A slide was shown.]

DR. THOMPSON. This is 1983.

[A slide was shown.]

DR. THOMPSON. And this is 1991. It is very difficult to get the boulder and the ice wall into the same frame.

[A slide was shown.]

DR. THOMPSON. And if you look at the wall, if you go down to where the wall is, this is the boulder back here. You can see it is only about 2 feet thick under present conditions.

[A slide was shown.]

DR. THOMPSON. Now we have surveyed the Qori Kalis glacier. This is the largest outlet glacier coming off of this ice cap. This is what it looked like in 1983. It is a convex glacier. There is no lake in the front of it. In 1991, 5 months ago, you can see that it has retreated. There is a vertical wall here. There is now a lake in front of this glacier, and it is now a concave glacier.

SENATOR GORE. Now, wait a minute. That's just in 8 years?

DR. THOMPSON. That's 8 years, and if we look at the actual record—

SENATOR GORE. It's not complicated to interpret that.

DR. THOMPSON. It is very obvious the change that is taking place on this ice cap. This was the position of the glacier in 1963, based upon aerial photographs. This is the position in 1978. We set up a survey line, took terrestrial photographs so that we could actually map that position. This was the position in 1983. This was the position in 1991, and you can see this lake has formed.

If you look at the rate of change, we see that in this period from 1963 to 1978, it is 5 meters per year. From 1978 to 1983, it is 8 meters per year and from 1983 to 1991, it is 14 meters per year, almost triple what it was in the earlier part.

Lonnie also mentioned Stefan Hastenrath's work in Africa, which showed that the glacier on Mount Kenya was disappearing, and studies in Uganda's Ruwenzori Range, which showed that the rate of retreat of the Speke glacier had increased by more than a factor of three in the period from 1977 to 1990 as compared to the period from 1958 to 1977. Lonnie told the senators that he knew "of no tropical glacier that is advancing under present day climatic conditions."

• • •

THIS WAS A TIME OF HEIGHTENED INTEREST IN GLOBAL WARMING, NOT ONLY BECAUSE of Gore's evangelizing but also because the warmth that had come to the world's attention during the summer of 1988 had continued to build. In a lecture to a group of climatologists in 1990, Jim Hansen had offered to bet anyone in the audience one hundred dollars that one of the next three years would be the warmest on record. There was just one taker, and he lost that very year. The years 1990 and 1991 each set new records as the warmest year since the start of instrumental measurement. (Hansen later dropped this way of making his point, because his wife felt he was setting a poor example for young people by gambling.)

The contrarians had been hammering away at Hansen's global temperature studies for years—claiming that methods used to measure temperature had changed over the decades, that the growth of the cities where many of his weather stations were based had led to a "heat-island" effect, which biased his results toward a warming, and so on. In fact, Hansen had corrected for these factors, but the political damage had been done.

Perhaps this is why Gore chose to focus the Senate hearing on indicators that are immune to the tinkering of mankind: ice caps, which, as Lonnie points out, "have no known political agenda"; bore hole measurements (used to determine past temperatures by looking at the temperature profile in a deep hole drilled into the earth's crust); and tree rings—all three of which happened to be agreeing with Hansen. As Lonnie also testified, "The evidence is very clear that warming is taking place."

SO, BY THE EARLY NINETIES, THE BIG PICTURE WAS COMING SHARPLY INTO FOCUS: WHILE mainstream scientists remained cautious, detailed agreement between the models and the data was beginning to emerge. The one computer modeler to testify at the hearing, Jerry Mahlman, director of the Geophysical Fluid Dynamics Laboratory, in Princeton,* said that although they were still "confounded by the natural variability in the system" (because the instrumental record was less than two centuries long, and proxy data such as Lonnie's were coming in from isolated spots that might not represent the earth as a whole), "many" climatologists believed the measured warming was "due to the added greenhouse gases." Owing to "the normal variation of the system," Mahlman demurred, "it is very difficult to point and say, 'Yes, this is the smoking gun, and this is unambiguously due to the greenhouse effect.' We think this is likely so, but the case is still being assembled."

As for Lonnie, although he was quite positive the planet was warming

*Home to GCM expert Syukuro Manabe.

and all but positive humans were causing it, these were side issues in his research. As he testified to the committee, "These programs were not undertaken to look for evidence for global warming. They were undertaken to create high-resolution records of our past in these low-latitude areas to compare with the records we are creating for the polar regions."

While this lends credibility to his statements about global warming, it also underlines the fact that serious researchers had begun to lose interest in the scientific side of this small subject area more than ten years ago—believing the main question to have been largely answered. And while Lonnie has continued to produce what many see as the most compelling evidence there is that the warming continues to build, his more interesting work has had little to do with global warming, per se. And his *most* interesting work was still to come.

·13·

FROM TIBET . . .

Not long before Gore's senate hearing, Lonnie and Ellen submitted two proposals for deep drilling in Central Asia: one for the Firn Plateau and the Gregoriev, the other for Guliya.

The funding agencies, balking at the idea of supporting two large simultaneous efforts on the other side of the world, asked the Thompsons to choose one or the other. Russia being even more unstable than usual, the Thompsons chose China.

My guess is that Lonnie also found Guliya more inspiring. It is very remote, very high—the highest point in the western Kunlun—and he believed at the time that it was the largest ice cap outside the polar regions. (As it turned out, when he and Yao Tandong went to drill on the Purogangri glacier in central Tibet at the end of the nineties, they discovered that Purogangri was larger.) Guliya is seventy-seven square miles in area and lies in the middle of a larger three-thousand-square-mile ice sheet as big as the state of Rhode Island. Its name means "Roof of the World."

THE KUNLUN SHAN ARE ALMOST AS HIGH AS THE HIMALAYA, LESS WELL EXPLORED, AND a place of legend even in China and Tibet. One Chinese myth holds that the Immortals live in a jeweled palace in the range. In the sixth or fourth century B.C., the Taoist philosopher Lao Tzu is said to have left China on a water buffalo, out of disgust at the country's corruption, and taken up residence in the palace, where he lives to this day. The idyllic land of Shangri-La in James Hilton's novel *Lost Horizon* is also located in a hidden valley in the Kunlun. Hilton drew his inspiration from the myth of Shambhala, an extraordinary enlightened kingdom, ringed by mountains, believed by many Tibetan Buddhists, including the Dalai Lama, actually to exist on the material plane. Some Buddhist guidebooks to Shambhala place it in the Kunlun—but it's a difficult place to reach: you have to get lost in order to find it, and everyone who leaves forgets how to get back.

"WHAT WERE EXCITING FIELD EXPERIENCES FOR YOU?" I ASK BRUCE KOCI. "WHAT'S THE first thing that springs to mind?"

"Well . . . one of the reasons Guliya was special was that we did live up on top, and the air was very dry so the starry sky at night was unbelievable. I always slept in the drill dome, which has a five-foot hole in the top, so I could watch the stars. It was cold, but I had lots of sleeping bags. (None of my sleeping bags are any good, but I make it up in sheer numbers.) And of course I had to take a leak about every twenty minutes and go out and look at the stars, because it was just—it was unreal. I've never seen a sky like that in my life, even in Fairbanks in the wintertime. It was very wild; there were no lights anywhere on the horizon. You could see a hundred miles in any direction and not a light in sight, so none of this light pollution that we have here. It was just a long ways from anywhere—one of those places where, you know . . . you could die? . . . and nobody would know? . . . So that makes it interesting.

"It's a little like, you know, you shut the trucks off and everything, and it gets real quiet? It's the same thing as having a plane drop you off somewhere, on a canoe trip, and there's only one way out. It gets real quiet when the plane leaves [chuckle]. You kind of stand there and say, '. . . Aahh, do we really want to be doing this?'

"I've been places in Antarctica that feel like that. At the Pole of Relative Inaccessibility with Ellen Thompson we were a thousand miles from anywhere. It was really cold and steady twenty knot winds and we were literally worn out. You couldn't go out in it, because the wind chill was below a hundred below out there all the time. It was just bedlam; and, you know, we just went there and did the job and got out.

"I hate to say it, but, I mean, that's the way it works. By Guliya, it was almost becoming routine. . . . But I think we also have a little of the Amundsen* philosophy: 'Only fools have adventures.' "

ON THE FIRST RECONNAISSANCE, IN 1990, THE DRIVE FROM ÜRÜMQI HAD TAKEN SEVENTEEN days, the last two on roadless terrain; yet the team had covered

*Roald Amundsen, who beat Scott to the South Pole and lived to tell the tale, mainly because he had thought everything through and prepared well.

only six hundred miles, which means they averaged only thirty-five miles a day. All they had to lead them to the ice cap were a few satellite photos showing a large white splotch in the western half of the Kunlun Shan. Lonnie says the "whole idea" was simply to "find the ice cap."

"We knew it was out there; people had been in the area maybe five years earlier. We went in from the north, out of the Tirim Basin region, crossed the western Kunlun to get up onto the [Tibetan] plateau, went down by K2 [the second-highest mountain in the world, on the China-Pakistan border], and turned back to the east, then left that road and took a two-day journey so that we approached from the south, because on the satellite images you can tell where the dome is located. It's on the northwestern edge of the plateau."

This was the last trip Keith Mountain would take with Lonnie until I would meet him on the Kilimanjaro reconnaissance in 1999. "This place was really out in the boonies," Keith writes, "and one of the things we did really well was to get these six-wheel-drive trucks buried up to the axles. Unloading and reloading was the most likely outcome for the day. There were absolutely no roads, and how these folks had any real idea of just where they were was beyond me. Still, they seemed to get us there, and I viewed this as no small accomplishment."

By the time of their second reconnaissance, in 1991, the idea had evolved, according to Lonnie, to, "Okay, let's bring out frozen core from this site; see where the problems are going to be." And there *were* problems, but they were no preparation for the problems they would encounter the following year.

Mary Davis says, "Some of these expeditions, I look back on them and I don't understand how we did it. Lonnie doesn't mind building up his hopes and then seeing he has to change his perspective—or, as I see it, building up his hopes and seeing them dashed. I would just as soon guess on the conservative side and be pleasantly surprised. I guess that's how he manages to succeed. I mean, you have to be convinced that you *are* going to succeed, otherwise you'll never start.

"Guliya was the toughest. I was there in '91 and '92, and '92 was the deep-drilling expedition. That was a tough one from a logistics standpoint—just the sheer scope of what we were trying to do."

Even base camp, at eighteen thousand feet, was six hundred feet higher than the summit of Dunde, while Guliya's summit was four thousand feet above that, not to mention ten or eleven up-and-down miles (mostly up) from the glacier's edge. That makes Guliya a thousand feet higher than the highest mountain in North America, Alaska's Denali. The only local inhabitants were a few scattered Tibetan nomads and the occasional wild yak, so there was no equivalent to the Mongols at Dunde with their horses, to guide and porter on the ice cap.

Even Lonnie was willing to admit that it would be impossible to man-haul six tons of equipment to the summit, so they decided to use snow-mobiles. In 1990, they purchased a Japanese model in China and left it at base camp. In 1991 they hauled in another all the way from Ohio.

In 1999, I spoke with Zhongqiu Li. A research associate in the Thompsons' laboratories at the time, he had participated, as a student at the Lanzhou Institute, in both the 1991 reconnaissance and the 1992 deep-drilling expedition: "I've been on a lot of expeditions to Tibet and the Tien Shan," he said. "I think Lonnie's group works the hardest and achieves the most, because they have clear goals. They moved faster than all the other groups that came through, even the Germans. The Russians were the slowest; the Germans came next. The Japanese climbers practiced all the time, but when they got to the mountains they were exhausted."

In 1991, the team drilled three short cores, including a second from the summit, and went through the usual "drill" of surveying, sounding, and planting accumulation stakes. They followed Lonnie's plan of bringing one core back frozen, and it proved a logistical challenge, as usual, to keep it frozen all the way. The nearest freezers were in Kashgar, which lies to the north of the range, across a high pass; and the one- or two-day drive from the pass to Kashgar traversed the Tirim Basin, which is one of the hottest places on earth—it can reach 120°F, according to Bruce. It was then another three or four days across the basin to Ürümqi.

"We actually got off the plateau and back to Yecheng, then to the freezers in Kashgar," recalls Lonnie, "and halfway across the desert, going from Kashgar to Ürümqi—I think it was near the town of Aksu—a driveshaft fell off the big six-wheel-drive truck. So there we were in the middle of the desert, no place to store ice core. We went into a little shop in Aksu and bought all the ice cream out of their freezer and gave it away—distributed it to the local kids—and took the cores out of the boxes and put them in their ice cream freezer until we could get the truck fixed. Then we loaded the stuff up and took it on to Ürümqi."

They resolved to take extra backup trucks the following year and to fly the cores from Kashgar to Ürümqi, but the Kunlun Shan had more tricks up their sleeves.

BRUCE WAS PROBABLY AT THE HEIGHT OF HIS CAREER AT THIS TIME. HE HAD RECENTLY moved with PICO from Nebraska to Fairbanks, so he was shuttling back and forth seasonally from the Arctic to the Antarctic, following the sun.

He was involved in both the GISP2 project at Greenland's summit and the first exploratory work for the gargantuan AMANDA drill at the South Pole—more comfortable a place than Plateau Remote thanks to South Pole Station, but equally incredible weather-wise, to judge from Bruce's stories. Since he missed the second Guliya reconnaissance for the surprising reason that he "actually took the summer off," he did not get a chance even to see the glacier before deciding on the technology he would use to drill there.

"Lonnie called up about six months before this thing and said, 'I think the ice cap is three hundred meters thick.' With a two-hundred-meter winch, that's not good. And I knew we couldn't drill it electromechanically all the way, because there would be hole-closure problems, so we took along an old Russian thermal alcohol drill that had been used on the Ross Ice Shelf project. We always take a suite of drills, because you never know what the conditions will be."

On this trip the group began to assume its present international complexion: the roster included Vladimir Mikhalenko, whom Lonnie had met two years earlier in the Tajik Pamir; William Tamayo Alegre, a civil engineer from Huaraz, who also did the surveying on the Andean probes of this period; and an assistant driller named Bill Barber. Yao Tandong joined them in Lanzhou and flew with them to Kashgar, where they met the rest of the Chinese contingent, who had driven the equipment and trucks from Lanzhou—about 1,500 miles as the crow flies, and we can be sure they did not fly as crows.

"This was a long, drawn-out trip," Bruce continues, "because we flew from Beijing to Ürümqi and then down to Kashgar—of course, flying on a [Russian] Tupolev 154S is an adventure in itself, because it's a plane that climbs, flies, and stalls at the same speed. In Kashgar, I remember, when we went down to the market, it was like being back in biblical times; there were mostly mules and carts.

"Guliya was one that, again, people said we couldn't do; and there were some reasonable reasons for that [the only time I've heard him voice any doubt on that score]. It was way the heck out nowhere. That was back when the Chinese had said you couldn't go through that way, because the highway had fallen off; and a lot of people were saying, 'Aw, they're just making that up.' Well, we went there, and this thing was gone! Miles of it were gone. I guess it was a minor earthquake—a minor earthquake and the mountain disappears—like the Huascarán thing [of which, more later].

"You went to a couple of cities, got to Pakistan, took a left, went up the road a piece—which I later found out is rated the worst road in the world, the road between Pakistan and Lhasa—and went about one-third of the way out that thing, and sort of took a left and headed up into the mountains."

The only accommodations to be found between Kashgar and the ice cap were three utterly remote army bases.

Mary remembers that "they had rooms for tourists, and as you went higher onto the plateau the rooms had these big oxygen tanks and masks next to the beds. Down off the plateau they're not too bad—you're still in green valleys, near towns—but the last army base up on the plateau is . . . If I were to die and go to hell, that would be my idea of hell. And there are people who are stationed out there for three years! I don't see how they keep their sanity! It's out in the middle of nowhere. The landscape is just blasted. It's nothing but sand and rocks, and it's always gray and brown. There are no colors, or at least there were no colors that we could see. The weather was always overcast.

"At that point, acute mountain sickness was setting in for me, from sitting in a truck day after day and climbing in altitude. I had shortness of breath, Cheyne-Stokes breathing, nausea, headache, insomnia, lethargy, lack of appetite—the works. The army base food doesn't help your appetite any either. What also didn't help at one particular base was a pack of dogs roaming around. These were not nice dogs; they weren't house pets. One thing that characterizes these army bases, too, is that no one seems to use the outhouses inside the walls. So it's just in this perimeter around the base where everyone tends to relieve themselves. And of course, strangers, when they go out there to walk or do anything else, have to keep one eye out for the dog pack. Despite my mountain sickness, I remember being chased back inside the wall, into the base, by these dogs.

"I've got pictures of the place. As a matter of fact, I've got an eight-by-ten blowup of that base. I look at it every once in a while when I get tired of Columbus, to remind myself that there are worse places."

Zhongqiu Li says the only payment the base commanders required for these accommodations were letters of recommendation from Lonnie and Yao that might improve their chances for a transfer. And their men often feigned sickness in order to take breaks from the bleakness of the bases in a hospital located in a slightly less bleak town.

BASE CAMP WAS HIGHER THAN THE FINAL ARMY BASE, OF COURSE. THUS, INAUSPICIOUSLY, one of the Chinese scientists fell unconscious in the truck before they reached it. But Bruce recalls an auspicious omen on their arrival:

"We got there towards dusk and were immediately charged by a wild yak. He wasn't mean really; it was just that we were in his territory. Magnificent animal!"

One advantage at Guliya was that they could back the trucks right to the edge of the ice. This advantage was offset, however, by the near verticality of the ice cap's wall. (This "polar" quality was part of the reason Lonnie had chosen Guliya in the first place: there were obvious annual layers on the wall, and the ice was evidently frozen to its bed.) They situated base camp near a spot where the slope was gentler—forty or fifty degrees—but even there it was difficult to drag the equipment up to the flatter top surface. They planned to use generators rather than solar power, owing to relentlessly overcast skies and the promise of temperatures warm enough to require drilling at night, as on Dunde, so they had taken tons, literally, of fuel.

After digging out the snowmobiles from the previous year, their first disappointment was to discover that the vehicles weren't powerful enough to climb the initial slope unloaded, much less dragging sleds of cargo; and they had brought only three or four porters from Lanzhou.

According to Bruce, "The porters looked at all our stuff and were about ready to die. So I said, 'Well, I think we can pull it up, but you may owe us something later.' Fortunately, we had these strong winches for drilling, so we wound up pulling all the fuel, all the food—everything for the camp—up top with winches. They had this type of gear reducer called a harmonic that was invented by a guy from United Shoe Manufacturing Company. Interesting gear reducer design. NASA uses them on their spacecraft because they're really light. We could pull two fifty-five-gallon drums up a fifty-degree incline at the same time."

They moved up in four or five stages, determined by the length of the cable, employing a sort of a slingshot arrangement: the winch and generator sat below, and the cable ran up through a pulley held by a stake at the top of each pitch. When they got all the gear to the top of the first pitch, however, they were faced with the problem of getting the winch and generator up as well.

Lonnie relates that they "sat down as a group, you know, 'How are we going to do this?' It was just amazing how creative we could be when we put our heads together. We ended up putting the winch in a sled and running the drill cable up through the pulley system on the ice and back down to a sled in which sat the generator. So you have the generator running in one sled and a power cord going across to the winch; they're in sleds tied to each other, side by side; and when the winch starts pulling cable, they pull themselves up; they both go up at the same time."

They planned to drill duplicate cores in three locations, if possible: the first about two miles from the glacial margin; the second at the top of a dome two miles farther in, where sounding had shown that the ice was about three hundred meters thick (almost a thousand feet); and the third and most audacious at the distant twenty-two-thousand-foot summit, where the ice had the one saving grace of being only about a hundred meters thick.

Mary, who remembers riding to the summit in a snowmobile with Lonnie one day near the beginning of the trip, describes it as "constantly windy, incredibly cold; any exposed skin is just sandblasted by the wind; it's a miserable place."

The snowmobiles managed to get them to site one, where they proceeded to drill for about a week and a half. Only Lonnie's group slept at the drill site; the Chinese scientists walked up and down from base camp every day. Water samples from this core were split with the Chinese, as on Dunde, and Mary did the melting and bottling down at base camp with "Smiley," the agreeable grad student who'd assisted in the same task five years earlier. Then she joined the others at the drilling site, as she also had on Dunde.

But as Bruce points out, "The first hole was sort of a bust. It wasn't a place I would have chosen to put a hole. It was too close to a moraine, so the surface was sloping up, and we were having a heck of a time with it. The ice was really strained. You could see elongated bubbles."

Bubble deformation of that sort indicates pent-up strain in the ice, engendered by years—in this case, centuries—of horizontal shear. Not only did this make it difficult to drop the drill into the hole and pull it out again, but it also resulted in core segments that tended to shatter.* To make matters worse, at about ninety-four meters (later dated at about A.D. 1400), they hit a discontinuity.

"The dirt layers are supposed to be horizontal," says Mary. "Well, suddenly we hit a strange horizon where the layer was tilted, indicating that there may have been a gap in the timing and a flow problem, so we knew we couldn't trust that record. We pulled up stakes and moved to site two. Unfortunately, at about this time the snowmobiles died—died forever. They just burned out, one within a couple of days of the other, so everything had to be man-hauled to site two."

Site two seemed more promising, perched as it was on the summit of a wide, flat dome at twenty thousand feet; but without snowmobiles there was now no chance of drilling at the summit.

"ANYWAY, WHEN THE DRILLING BEGINS AT SITE TWO, THEY IMMEDIATELY START RUNning into the same problems they had at site one," continues Mary. "They can't drill during the day, because it's too warm; water runs into the hole and freezes and the drill gets stuck. So, after getting the drill stuck and

*See source note about elongated air bubbles on page 399.

released a couple of times—it's a terrible feeling when it gets stuck—we started drilling at night. And this is where you get into the part where you wish you hadn't come along, because it gets very, very uncomfortable.

"Drilling at night was quite an experience. It was very cold; the wind would start blowing at night. We were in the dome, but still—one thing I remember was my feet freezing. I'm in Sorrels;* and, of course, when you're processing cores you're standing all the time, taking notes, waiting for the core to be drilled; and my feet were bothering me badly. I had already frostbitten them once when I was a grad student, so my feet were ultrasensitive. I remember, Bruce took off a little poncho he was wearing and gave it to me to stand on, so I could have some insulation for my feet.

"So, long about midnight or one in the morning, we would start opening up cans of anchovies. The Chinese would send up these cans of Chinese Spam—Spam!—and fish. It was just—I hate fish! We would open up the can and pass it around, and everybody would kind of stick their finger in or their Swiss army knife and pull out a chunk of Spam—cold, frozen Spam—and chew on it and chase it down with hot water. We had an electric kettle inside the dome, which we used for heating water. And that would be one of our meals. . . .

"It's just that . . . I look back on it now, and it seems so depressing! Everybody sitting around chewing on this cold fish and cold congealed Spam. What a glamorous life. I try to tell graduate students who come to me when they're applying to go on an expedition—I sit them down, because I've seen this happen over and over. They're so excited to be going, because they think it's going to be the great adventure—and they're miserable! So I sit them down and tell them they're going to be miserable; I tell them they're going to hate it; I tell them they're going to be homesick. They don't believe me until it happens."

She is relating this in the comfort of her office, as the group is in the planning stages of a deep drilling expedition to the Purogangri glacier in central Tibet, which promises to be even more of a challenge than Guliya. I observe that she seems excited about Purogangri and ask if she thinks one forgets the bad stuff.

"Sure you forget the bad stuff. If you remembered the bad stuff, you'd never go back."

The same thing happens on mountaineering expeditions. You might endure weeks of boredom and discomfort in pursuit of a few days of good climbing, and when you get home all you'll remember is the climbing.

*Very warm winter boots with thick felt liners, leather uppers, and thick rubber shoes and soles.

Speaking of Guliya a few minutes earlier, Mary had said, "It's a shame we couldn't man-haul that stuff to the summit. I think we could have gotten a lot of interesting results from up there."

IN ORDER TO AVOID THE SNORING OF THE MEN IN THE MAIN THREE-ROOM TENT (as well as what she describes as a "ripe" smell therein), Mary pitched a small, one-person tent next door.

"Now, Bruce also slept alone," she adds. "He slept in the dome. I have a picture of him emerging from his ten sleeping bags one morning, and all you can see is his top, up to here [mid-chest], and he's got a Snickers bar in his hand and it's kind of unwrapped, and he's got this look on his face, this stunned look that perfectly captures my perception of Bruce. It's kind of, 'Here I am, a basic organism with my breakfast in my hand, waiting for the day to start.' "

Bruce was having a great time as we know, not only sneaking out to stare at the stars as he peed at night, but also watching the thunderstorms march in:

"We could see the mountain ranges stacking up, getting bigger and bigger until they reached the Himalaya. You could really see the effect of the [tectonic] collision. The storms would come in off the plateau. As soon as you started to get graupel* you knew there was going to be lightning, and everybody's hair would stand on end. I had put some sharp things on top of the tower, to make a sort of lightning rod to dissipate the static, because what you don't want is to have the drill down the hole making a conductive path to bedrock, because then you have a true ground at the end and you know the lightning's gonna find it: 'The good news is we drilled to the bottom; the bad news is we blew the cable to bits; the drill is a melted mess at the bottom.' So, anyway, we had to haul the drill out as the storms approached, and then you could hear the tower start to buzz—it sounded like a million bees. It was fun, because you could go out and, if you stood at more than a forty-five-degree angle, put your fingers up and feel the electricity discharging—feel the power. Your fingers would tingle. It was like a massage. You were out of the range where the static could do anything to you. The tower made a kind of shadow where it was dissipating the charge.

"The ice is a good insulator. We found that out at Dunde, where we used to walk away from the tower and actually hide in a tent—lie there hoping nothing would happen. We experienced these things on Quelccaya, too, to

*Small pellets that seem to be a cross between snow and hail. For some reason, graupel comes down more frequently in the mountains.

some extent. On Dunde, though, you could see the leaders [main lightning bolts] come down and go off to the side, horizontal. They wouldn't strike the ice cap at all, because ice is such a good insulator. Basically, we just didn't want to make a conductive path through the insulator. We were in a good place."

They drilled in these alternately uncomfortable, alternately hair-raising conditions for three weeks; so between there and site one, Lonnie, Bruce, Vladimir, and the other "core" members of the drilling team spent more than a month on the ice. They managed to get a single core to bedrock at site two, but only by the skin of their teeth.

Bruce remembers that "the main hole was three hundred nine meters, as it worked out. We had to go to the absolute limit of the mechanical drill before switching to the Russian thermal alcohol drill, and the thermal drill was pretty slow. It was really heavy, and it's physically hard to drill with that thing, because you have to try to keep it just a little bit off the bottom of the hole and keep bobbing it around. At three hundred meters, you're really losing track of the cable. Even with Kevlar, there's a lot of stretch."

They had fitted their three generators with special carburetors to allow them to run at high elevation, but they ran hot nevertheless, and two of the three burned out by the time they reached about 308 meters—not having a firm idea, at that point, how much farther they still had to go. The third and final generator burned out roughly half an hour after the team pulled up the last core segment, but Lonnie points out that "it could just as easily have been half an hour beforehand, too."

On the other hand, there was exhilaration amidst the anxiety. As Bruce says, "We knew as we were drilling that it was a good record—no question about it. When we got down near the bottom, every core came up looking different. That doesn't necessarily mean anything scientifically, but at the time it was fun, because it was very interesting visually.

"Anyway, we finished drilling the hole, and two days later it filled with water. That would have been the end of it. We wouldn't have gotten to the bottom. So, talk about life on the edge! When you think about it, drilling is a lot like flying: moments of terror, hours of boredom."

But the job was only half, or, perhaps, in view of subsequent events, less than half done. They still had to get the ice off the glacier and halfway around the world. Not only that, they were running very low on food; and with no generators, they could not employ their ingenious winching method to lower the core boxes down the steep final slope. A single box weighed up to 150 pounds, and they would put several on a single sled, so the act of lowering a sled bore some resemblance to group wrestling with a large game fish—at high altitude, while walking on ice. The sleds sometimes broke free and went sailing down the hill.

"A couple of groups had the bright idea of letting the sleds go on their own," sighs Mary. "This was definitely the low point of the expedition. They thought they could ride the sleds. They would climb up on top, let them go from the top of the hill, and of course they lost control.

"There we were, struggling with our own sled, and the next thing I know this thing goes shooting past me. I look up and there are four core boxes in it and one of the drivers sitting on top. We could see he was in trouble as he went shooting past. He understands no English, so I'm yelling at the Chinese next to him to tell him to jump off, because I was afraid he was going to get killed. So they start yelling at him in Chinese; and just as he jumps off, the whole sled goes—it must have hit a rise or a rock—it goes flying up in the air, turns over a hundred and eighty degrees and comes slamming back down on the ice. Nobody was injured, luckily, but Lonnie was not happy, to put it mildly. It was the angriest I've ever seen him. Yao was very angry also; they were both yelling. It was understandable; I think I was yelling, too."

"After all that effort and all that pulling and man-hauling," agrees Lonnie, "to lose it because of some stupid thing. Yes, I was mad. But we've learned, Yao and I. We get mad at each other in the field, but we remain friends.

"There were some fractures in the ice, but it wasn't as bad as it might have been, because the sled got going down some lateral moraines where the rocks slowed it down. It did damage some of the ice, though. We can always tell by checking the logbook from the drilling site."

Then they discovered that a pit near site one, where they had stashed most of the core boxes, had filled with water, and the water had then refrozen. It took a day and a half to chop the boxes out with ice axes.

Mary believes "the hardest thing we ever did as a group was to get those ice cores off that ice cap, down the hill, and into the trucks." Yet even then, the task was far from over. As they were wrestling the boxes down the ice cap, they were also keeping an eye on the calendar, for the nearest refrigerator was in Kashgar, four or five days away, across a pass that was open only on certain days of the month. They left base camp the morning after they got the cores down.

"We pulled down camp in the middle of the night and left at four the next morning," Mary continues, "racing not for the army base, but for Kashgar. We had to make Kashgar."

Lonnie picks up the story:

"Between '91 when we did our dry run with the two core boxes, and '92, the Red Army decided to fix the road going over the western Kunlun. While we were up drilling, they closed that road. You could only go north, I think it was, three times a month, on something like the first, the eleventh, and the twenty-second; and we were four days out on the plateau away from that pass.

"And when we got to that pass—the day before it opened—there were forty-five trucks ahead of us. Now the pass opened at midnight, and a lot of people didn't want to get up; the whole road was occupied. So our group went from truck to truck, getting the people up—they were sleeping underneath the trucks—and getting them out and getting them in their trucks and then actually getting out, pushing and pulling and pushing, because even though *we* had two new trucks, we were *behind* forty-five of the worst vehicles in the world—just barely running! We worked all night to get through that pass."

"It was surreal," says Mary, shaking her head. "We got up before dawn to wake all the truck drivers ahead of us, under the stars with the mountains all around. At first I thought someone might get angry and punch one of us out, but it turned out they were all pretty accommodating. If they had been American truck drivers it would have been a different story."

"I remember the next morning getting down to Yecheng," Lonnie continues. "It's in the desert, and the temperature is going up, and then we had to make a run from Yecheng to Kashgar; it's all across desert, and there are no freezers in Yecheng, so you're running. We knew we had five days, and we were on our fifth day of traveling to get to those freezers. We made it—just!—but we made it."

In Kashgar they filled three of Bruce's beloved Tupolev passenger planes with ice core, and Lonnie and Mary accompanied the cores to Ürümqi, where it took ten days to split the 309-meter core from site two with the Chinese.

"We had taken along this freestanding band saw," says Mary, "and we set it up in a meat freezer next to this huge pile of pigs' heads—we've got video of this; it's quite interesting. The reason it took so long was that it was hard to get the voltage and amperage right for the band saw; it needed American current. And I think we went through about fifteen saw blades, but eventually got it done. It was just amazing, trying to get those cores, intact, all the way from the ice cap, through all this, to Columbus."

BRUCE AND A FEW OTHERS HAD REMAINED BEHIND WITH A COUPLE OF 4×4S, TO CLEAN up base camp—a task he always tends to do—so he enjoyed a more leisurely ride home:

"Yeah, I always wind up being about the last person off the ice. You know, it's kind of, 'Well, it's your drill now; you gotta pack it up.' So I'm usually the last dog that leaves the summit, but I don't mind in the least. It gets nice and quiet. I enjoy that after all this other stuff. It's nice to have that pressure off.

"The morning we left, you couldn't tell where the road was; there was snow. So we just kept on. It was a beautiful journey back down the valley, slowly, slowly into green; and we met some interesting people living on the land—a Buddhist family living in a tent at seventeen thousand feet, with yaks and horses. They were curious about us, and we were curious about them, too—neat people—and continued down to the military camp and headed back.

"We stayed in Kashgar for a couple extra days, and at Yecheng, which was south of there. Closed to tourists. Wonderful place. Went to the market there. Excellent lunch for a dollar and a half. Shish kebab, tea, biscuits for two people. Bill Barber and I were there together. So we went on to Ürümqi and met the other folks. Actually, got brave and took one of the buses. We even had, on that expedition, folks who got really brave and took the Beijing subway.

"Anyway, we flew back from Ürümqi, caught another great Russian airliner that had about as much power as two ants—I've been on freight trains that accelerated faster than that. We scraped out of there, got back to Beijing, and then headed out. That was a satisfying expedition, because things at GISP were not going particularly well that year; and I sent out a fax from Beijing that started out 'Greetings Earthlings,' and, boy, that thing was pasted up all over the PICO office in Alaska."

·14·

. . . TO PERU

Meanwhile, three years of "scrambling amongst the cordillera" had resulted in a decision to drill on Huascarán the very next summer.

Having scoured all the possibilities in the Cordillera Blanca, Lonnie had found that he had no choice but to drill on its highest mountain—for the simple reason that Quelccaya's was not the only glacier that was dying. The team had found no sign of an annual signal in any of the cores they had drilled in the previous few years, on Qopap, Caullaraju, Pukajirka, or Hualcán.

Furthermore, the cores they had obtained six and more years earlier showed this to be a new development. In 1984—on their return from the "cleanup" expedition on which they had first noticed the drastic retreat of Qori Kalis—Lonnie, Keith Mountain, and Dave Chadwell had retrieved a shallow core at 16,800 feet on Pukajirka that *did* exhibit seasonality, whereas a second they drilled in the same spot in 1990 did not. This meant, basically, that snowlines were rising: the altitude at which the average temperature remained cold enough to keep snow frozen year-round had risen above the summits of all but Huascarán; and while the others still held year-round snow only because it would take years or decades for their sizable glaciers to melt, the snows were slowly climbing the mountainsides and diminishing in size. At the drilling sites near the summits of a few of the lower peaks, in fact, the team had discovered deep layers of meltwater, which had percolated from the surface down into the firn. Rising temperatures were forcing Lonnie to go to higher and more dangerous locations.

• • •

"IT'S A TREMENDOUS SIGHT," HE SAYS. "HAVE YOU BEEN THERE? IT'S A FANTASTIC PLACE, South America. It's like the Alps three hundred years ago—just beautiful."

Huascarán is the highest mountain not only in the Cordillera Blanca but in all Peru and, in fact, all the tropics. Its twin summits are separated by the 19,800-foot Garganta Col, and the southern and higher of the two, at 22,205 feet, towers more than two and a half vertical miles above the valley of the Río Santa, which runs north–south only nine horizontal miles to the west, parallel to the crest of the Andes.

The Garganta is about five hundred feet lower than the highest site at which they had previously drilled to bedrock—Guliya's site two—and the Cordillera Blanca are much less remote than the Kunlun Shan; but, as Bruce points out, Huascarán "was really the first serious mountain we did." The route to the col involves technical climbing.

"They are all different in their own way," Lonnie reflects. "Guliya had its moments, but in some ways it was not the challenge of Huascarán for other reasons. I think it's very hard to rank them as to which is the more difficult.

"I saw Huascarán as the challenge of a lifetime in that we had over ten thousand pounds of equipment and twenty-three people in the expedition, plus porters. We were going to take this stuff up to twenty thousand feet, and on Huascarán that's going almost straight up from the valley. Even today, whether you can get into the col depends on the year, because there are huge crevasses in the Garganta glacier, which are sometimes too wide to get across; it varies from year to year. So even if you've checked a place out, as we had the previous year, you don't know exactly what you're going to face when you do the actual program."

On the positive side, Peru was enjoying a temporary measure of stability that summer, owing to the recent capture of Abimael Guzmán, the leader of Sendero Luminoso. This had not ended the conflict, however, and scars were still evident in Lima. Mary remembers a strong military presence around the U.S. embassy, which had recently been bombed, and a number of "bombed-out police stations."

On the other hand, there was some ironic value in the continued threat, for, as Lonnie points out, the mountain guides at the Casa de Guias, in Huaraz "had very few customers, so we had good rates. They were lining up for work. We could also set up the whole lower floor of the Casa de Guias as a laboratory for processing." The Casa being Quelccaya veteran Benjamín Vicencio's home base, he led the group of guides whom Lonnie's team has come affectionately to call "the Peruvian mountaineers." At that point there were only two besides Benjamín: Maximo Hinostroza Z. and Magno Camones G.; a fourth stalwart, Jorge

Albino S., would join a year or two later. I would meet them all on Sajama in 1997.

Also helping out were the national weather bureau, SENAMHI,* and the national power company, Electroperú. Lonnie points out that the air force colonel who headed up SENAMHI managed to ease their passage into the country by obtaining permits allowing them to label most of their gear EQUIPMENT OF WAR, NOT SUBJECT TO CUSTOM INSPECTION. Then, he says, "because of the Shining Path activities, they removed the military labels from their trucks, so we appeared to be a simple climbing expedition. There were a lot of uncertainties. . . .

"A fellow from the national fishery cold room in Huaraz—they brought fish up from the coast—also took a special interest in our project. When I told him we were going to be sending down frozen ice core and we needed a place to store it, he said, 'You've come to the right place. I think this is a very important thing to do.' So as the ice cores came down, they would be put on trucks and taken down to his freezer. We actually had ice there for over a month. He even got his people to help us unload the stuff and put it in the freezer and, when we were getting ready to go, load up the truck and ship the stuff back to Lima. We had supper at his house and met his family and all, and at the end when I went to ask for the bill, all he wanted was a letter from our university stating that he had helped with the project. A remarkable guy. It was just amazing how much help we got from the local people. We do a lot of things in these places, but if it weren't for the local people and the contacts, we just couldn't do it."

The last stop for the covert vehicles was the town of Musho—two miles high, but still two vertical miles below the Garganta Col. They hired forty-five burros and thirty porters from Huaraz to ferry their six tons of equipment from there to the edge of the ice.

HUASCARÁN IS ONE OF THE MOST IMPOSING MOUNTAINS ON THE PLANET. "IT'S REALLY steep," says Bruce straightforwardly, "and as you approach it, it keeps getting bigger and bigger. Also, the closer you get, the more crevasses you see. But it was beautiful. It makes you realize why people climb mountains. They really are big chunks of rock.

"It was also a humbling reminder just to see that north face, where the hundred million cubic meters of stuff had come off and buried the town of Yungay in 1970. There were a few palm trees around, but just the tops. They should have been fifty feet tall, but they were about three feet tall. It was amazing to think of, having that much stuff free-fall ten thousand

*Servicio Nacional de Meterología e Hidrología.

feet and hit the town doing one-forty or whatever avalanches tend to do. It must have first kind of blown the town away and then buried it. You keep looking over your shoulder to the left as you're going up the mountain, and you see this bare wall of rock where the avalanche came down."

The west face of the north peak, Huascarán Norte, which is steeper and rockier than its slightly higher sister, Huascarán Sur, has avalanched twice in recent years. The first event, in 1962, wiped out several small villages on the lower flanks of the mountain and killed more than three thousand people in Yungay. Afterward, the survivors decided to move the town a few miles down the valley—out of harm's way, so they thought. Unfortunately, the first avalanche was just a warning.

On May 31, 1970, a magnitude-eight earthquake rattled through the Cordillera Blanca and shook loose a large section of the high rock band that protects the summit ice fields on the west side of the north peak. The initial cascade was about three thousand feet wide and a mile long, made up mostly of rock from the band and ice from the hanging seracs that sat on top of it. By the time the cascade reached the Río Santa Valley, nine miles away, it comprised one hundred million cubic meters of rock, ice, and mud, just as Bruce recalls. That would be one square mile of debris, 130 feet deep. It took about four minutes for the cascade to reach the valley, which works out to an average speed of one hundred thirty-five miles an hour, so Bruce was dead on there, too. I've noticed that he has a good head for numbers.

The debris went airborne at least twice: once at the top of its run, where the slope is nearly vertical, and again about two-thirds of the way down, where a smaller stream split off from the main flow and made a lateral jump across a six-hundred-fifty-foot ridge into the valley above the new, relocated town of Yungay. This side flow, estimated at only 10 percent of the total, was nevertheless sufficient to kill all but a hundred of Yungay's twenty thousand inhabitants—all but instantly—burying most of them alive.

Part of the neighboring village of Ranrahirca was also destroyed, and the flash flood induced by the avalanche flowed all the way to the Pacific, sixty miles away. The earthquake killed another fifty thousand people in other parts of Peru, which brought the death toll to seventy thousand all told. It was the worst natural disaster ever to have hit the southern hemisphere.

OVER THE WEEK THAT IT TOOK FOR THE PORTERS AND PACK ANIMALS TO CARRY THE equipment straight uphill from Musho to the 14,000-foot base camp and up again to 15,600-foot Camp I, at the edge of the ice, the scientific crew had plenty of time to fret not only about avalanches but also about the crevasse field in the glacier they would soon be climbing. Bruce was fretting about both: "If there's an earthquake, you'll probably die. You'll either fall into a crevasse or a few will band together.* You'd be like a bug on the windshield of life."

And one of these crevasses is remembered by everyone who has recently climbed the Garganta Route, which is the easiest and, therefore, the most popular, or "standard," route on this very popular mountain. This particular dark chasm in the ice (which the scientists came to call the "dreaded crevasse") was eight or nine feet wide on their way up and had widened to twenty-five feet or more by the time of their descent, two months later. It is usually crossed technically: a pair of climbers will set up a belay on one side and lower the leader into it; he will swing across to the opposite wall, ice-climb out, and set up a belay; then his partner will do the same thing in reverse. Even a quick pair of climbers with just the packs on their backs could easily spend an hour doing this.

It would have been difficult if not impossible to get six tons of gear up the mountain this way—not to mention that plus an additional four tons of ice back down—so the Peruvians positioned a ladder across the crevasse. At first, while it was relatively narrow, they used a roll-up, aluminum unit, which sagged like a hammock between anchor points on either side. By that point on the ascent the team was moving in single file, roped together, with one guide at the front and another at the back, everyone wearing a climbing harness and crampons. But they crossed the crevasse one at a time—one guide belaying from the front, another belaying from behind. If anyone had fallen, he would have found himself suspended in midair like a kid straddling a swing.

THEY REACHED THE COL ON JULY 2 AND COMMENCED DRILLING ON THE FOURTH, BUT an undergraduate named Paul Kinder had altitude problems from the outset. He developed a crippling headache, and Lonnie remembers him "breaking a number of thermometers when we were trying to do a temperature profile in a snow pit." They sent him home within a few days.

Bruce had decided to take two thermal drills and one electromechanical drill this time; and of the thermal units, one was the original—one might even say historic—drill from Quelccaya, while the other was the Russian drill they had used on Guliya. Theoretically, the electromechanical drill would have been the fastest, but when Bruce saw what they were up against he realized it didn't make sense even to take it out of its box. Since the ice was composed of alternating soft and hard layers, the antitorque skates—metal flanges mounted longitudinally on the top half of

*Meaning a large section of the glacier will collapse as a piece.

an electromechanical drill and designed to hold that half in place and "skate" down the hole as the bottom half spins and cuts out the ice core—would simply have spun whenever it was located in a soft layer.

Since they were using solar power, they could drill only when the sun was up; and to make matters slower, they were forced to rely mostly on the old Russian drill because it permitted the use of an alcohol solvent, even though, as Bruce recalls, it "could only make twelve, maybe fourteen meters a day." Although the Quelccaya drill was faster and, in fact, succeeded in coring twenty-eight meters in one day—in the firn at the top of the second hole—Bruce points out that they "couldn't get alcohol down the hole with it. It's just a straight thermal drill—all you have is water—so if it's below freezing slightly, which it was, you can't use it."

As the core segments were brought to the surface, the guides and porters carried them down—in the case of the first core, to Ahnan Lin at base camp, who melted and bottled them, and in the case of the second, to a truck that ferried them to the fishery cold room. On the climb back up, the guides and porters would take food and other sundries for camping or drilling, and by all reports Benjamín did an excellent job with the logistics: there was plenty of nourishing food—and the drillers definitely needed it.

The occasional climbing party also wandered through the col, and Bruce recalls that "it was kind of hard on some people's egos," in the middle of what was to many the epic struggle of their lives, to find a comfortable, casual encampment at almost twenty thousand feet.

"We'd offer them a cup of tea; we helped some out with food and water. We had an electric hot plate, which was a good way of making water both for drilling and for tea, because it didn't involve flame. We needed a resistive load anyway, to keep the panels happy, and we had to have water in an alcohol solution to get the right mix for the hole."

But the air was not entirely free of anxiety, as a few snowstorms blew through early on, blocking the sun and the solar power in consequence. And during the long intervals in which the drill pursued its slow task down in the ice, the team had plenty of time to scheme and worry—both sleeplessness and anxiety being well-known symptoms of acute mountain sickness. It took them three weeks to reach bedrock on the first 160.4-meter core: they averaged only 7 meters a day.

Since Bruce was especially concerned about avalanches, it is no surprise that he took a special interest in the quality of the ice as the layer count told them they were crossing 1970, the year the big avalanche might have crashed through the col: "You have thoughts like that. Most of us were claustrophobic. Oh, God! To be buried in an avalanche just sends shivers up my spine. And we would have seen it; but, luckily, there was no evidence at all."

Meanwhile, Ahnan was sitting directly in the path of any and all prospective avalanches, processing samples. The plans called for Mary to replace him near the end of July, so she completed the bottling of the first core after he went home, then decided to move up to the col to help with the drilling of the second. She climbed with Benjamín and "a Peruvian air force major named Evaristo" (who was arrested a few years later, incidentally, for using air force planes to smuggle cocaine out of the country).

By the time Mary made her ascent, the dreaded crevasse had outgrown the original flexible ladder, and the guides had replaced it with a long wooden ladder. "I went scampering across as fast as I could," she says. "I don't enjoy being up on ladders even around my house. I'm a little acrophobic, not horribly acrophobic—I mean, if I were horribly acrophobic I never would have gotten on the ladder to begin with—but nobody enjoyed it.

"Even near the bottom, I remember, Evaristo and I were exhausted—the altitude was catching up with us—and the ascent becomes rather dramatic when you get up toward the col, of course. We had to cross an avalanche zone. There was no way to get around it. I remember Benjamín trotting across and calling back, *'Más rápido por favor,'* trying to speed us up, but warning us not to make loud, sudden noises. He was afraid it might trigger an avalanche.

"I don't remember much more about the climbing, except actually reaching the col and taking a look at everybody. They had been up there for about thirty days. I remember noticing how absolutely horrible everybody looked. They were dirty and scruffy and already pretty ripe looking. As far as their health went, they looked pretty good; they just looked really *scrungy*.

"I had AMS [acute mountain sickness] for the first couple of days. It didn't kick in at base camp, but it sure kicked in up there. So I was feeling the usual loss of appetite, headache, the usual routine."

Then, just as Mary's maladies were beginning to subside, the event they *all* remember from Huascarán occurred: the most violent storm any of them have ever experienced in their combined *years* altogether of living at high altitude. It was evidently a freak diversion of the jet stream. The wind rose abruptly to hurricane force, while the sky remained unnervingly clear. When Lonnie happened to peruse some meteorological charts for that week on a visit to Yale sometime after the expedition, he discovered that an unusually large air pressure differential had set up between the

Amazon Basin and the Pacific Ocean and pulled the jet stream low enough somehow to couple itself to the cordillera—only in the immediate vicinity of Huascarán. Southerly winds roared along the spine of the range, non-stop, for three days.

Bruce says they "kept drilling anyway. Well, there were a couple days when it was so bad we couldn't drill, but we still had to go out and tie things down. And it would make you mad, because half a mile to the east you could see puffy white clouds just sitting there over the Amazon. You could tell it was absolutely still. Talk about something that would piss off the Easter Bunny. You knew that if you were half a mile away, life would be good."

Lonnie remembers "solar panels flying like kites, five of them hooked together—the only thing that's got them tethered is the power cord. Now, for the wind to make those things fly, that's a lot of wind. We had them staked down, but the wind eroded the surface, undermined the stakes, and sent them airborne. We lost panels; the glass covers would crack; we had to rewire them afterwards. Amazing stuff! We've got it on video. You can see the kitchen tent ripping; the wind is just ripping the place apart. You couldn't even go out and prepare a meal.

"And at the time, see, you don't know what the full strength of this thing is going to be. There's a difference when you talk about it afterwards: you've been through it, you know what it was, but when it's actually happening it's very real and unknown. You don't know where you're headed."

Speaking of not knowing where you're headed: in the middle of the second night Lonnie's tent collapsed and he found himself trapped inside it, sliding along the snow.

"I yelled in the night, and the only person who showed up was Vladimir, because everyone else was going through the same type of thing. We actually put an ice axe through the floor to stop my tent from sliding along the surface. So I spent the night in this collapsed tent. Again, while it's happening, you don't know the full range of possibilities. It might even get worse. . . . I mean, it was definitely moving, and there's a big drop-off at the edge of the col; so it gives you pause for thought."

Pause for thought?

THE STABILITY AND QUIET STRENGTH LONNIE HAD NOTICED WHEN HE HAD FIRST MET Vladimir in the Pamir certainly showed its value that night. Lonnie has taken him on every major drilling operation since they met, paying his salary and travel expenses. He points out that for Vladimir this has been a "godsend, because when everything fell apart in Russia, they were going for months without pay."

One of Lonnie and Ellen's purposes has been to help preserve the science infrastructure of Russia, which was once one of the greatest in history. With the aim of forestalling the emigration of top Russian scientists to the West, the Thompsons have written proposals to NSF for joint projects in which the Russians do their portion of the work in their home country. Since the best scientists are usually the first to receive offers from abroad, they are generally the first to leave when funding dries up.

Having traveled with Vladimir twice myself, I tend to think of him as the quintessential member of the field team—not only a hard worker, a good scientist, and the only mountaineer among them, but an authentic team player. He's quite tough physically and mentally, but he gets along well with people, and he's not a prima donna. He will often be Lonnie's sole companion on the initial reconnaissance to an especially remote new area—Purogangri in central Tibet being a good example.

But he's not the only one. Lonnie says, "It would be very hard to duplicate this group of people. When things go to hell, you've got to have people who've been there, are dedicated to getting the job done, and on whom you can count. There have been a number of cases over the years when things have gone wrong in the field, storms or whatever, and Vladimir has been there. He's not off hiding in his tent; he's there helping. That type of thing is very important for morale in these remote parts of the world. You need to know that if something goes wrong with any one of you, there is going to be a group of people looking out for you—and he's that type of person.

"He's very much interested in mass balance* and densification processes, so we've sent him data from our core sites, and he's working it up. We'll publish it jointly. He'll be the lead author. So he contributes on the science side and he is also a fantastic field person. That's often the case in the European community. And I've found that in our group they like to go out in the field. If you had to stay in the lab all the time, it would drive you crazy. But if you can go out and see where the samples are coming from and participate on that end—well, heh, heh, it can make the lab look like a beautiful place."

EVERY ONE OF THE TWO-PERSON DOME TENTS MADE BY A CERTAIN WELL-KNOWN MANufacturer was destroyed by the storm (while the non-domes made by a certain other manufacturer were not), as was the large three-room kitchen

*Mass balance is the study of the net growth or decline in the mass of a glacier, the tally of accumulation through snow or rain, and loss through melting and sublimation. A mass balance calculation might predict, for example, whether a glacier will advance or retreat.

tent they had also used on Guliya. Below, at the edge of the glacier, says Mary, "Camp One was utterly destroyed. The winds got that far down—all the tents disappeared—but at base camp they never felt a breeze. That's how sharp the line of the storm was."

With the cook tent destroyed, the porters fashioned an elaborate snow cave using chain saws (which, in retrospect, would have been just the haven in which to endure the storm). Mary points out that "they even carved benches and a table out of snow. That's how we took our meals for the last week. It worked out very well." Thus an innovation was added to their bag of high-altitude tricks.

TWO ITALIAN CLIMBERS HAPPENED TO BE VERY HIGH ON THE MOUNTAIN'S CASAROTTO Route—named for another Italian, Renato Casarotto, who had made its first ascent—just as the storm hit. Even in the best of conditions (and they could never be said to be "good"), this is not only the most difficult climb on Huascarán but one of the most difficult and dangerous climbs in the world. As a British climbing journalist wrote in 1999, "In a remarkable *tour de force* characteristic of the man, [Casarotto] spent twenty days alone on the North Face of Huascarán Norte, creating a direct line . . . on both good and bad granite, interspersed with tricky mixed ground. Since the first ascent, all attempts to repeat this [extremely difficult] directissima have failed and at least three climbers have died in the attempt."*

The drilling team learned by radio that Bonali and Ducoli were missing, a day or two after the storm. The Casa de Guias leads most rescues on Huascarán, and Lonnie's guides were in a good position to help, so they immediately joined in the search, communicating by radio with the hotel and with a helicopter that flew around the mountain—it even zoomed through the Garganta Col at one point—looking for the climbers. A few weeks later, after Mary's time on the mountain had ended, she emerged from her room at the Casa de Guias one morning to find the climbers' wives and families "standing in the courtyard crying," having just received word that the bodies had been found.

"The storm had actually blown them off the north peak," she says. "They'd fallen about one thousand meters and landed on rock. Benjamín, who was in the party that found them, said they were messed up bad; they had hit hard."

*Mixed terrain combines both rock and ice, and though mixed climbing can be breathtakingly beautiful in a gothic way, and extremely elegant, it is sometimes horrifically dangerous. A directissima is a route that goes straight up a face. It is said that the truest directissima follows the line that "a single drop of water" would take if it were dropped from a summit.

. . .

THE DRILLING SLOWED TO AN EXCRUCIATING PACE AT THE VERY END, FOR FOG BEGAN rising from the Amazon Basin nearly every day to shade the solar array by midafternoon. Since Bruce left a few days early, owing to what he calls "PICO politics," it was only Mary, Lonnie, and Vladimir in the end—and, since Mary had come up halfway through, only Lonnie and Vladimir who had lived in the col from beginning to end: a stunning total of fifty-three days. Lonnie brushes off this feat of high-altitude endurance by pointing out that "Benjamín was moving back and forth, transporting and taking care of logistics and things—a much harder job."

But the central drama, of course, was the drilling.

Bruce says, "There's always the pressure of getting the cores, the usual feeling of depression when you first get started: 'Are we really going to do this?' But we certainly had one hundred percent success. The record was good. Once you start getting into it, and you start reading the ice, you forget all that stuff; you're looking at the ice and trying to get things done.

"We could see that there were a lot of changes in the ice. You know you have to get back in the lab to get confirmation, but that's what's really exciting. Every core—especially when you get near the bottom—gets interesting. The thrill of the chase is always there, even though sometimes you have no idea what you're chasing. And that's the interesting part: you don't know what you're going to find. Then this light goes on and you're from Minnesota and you go, 'Ooohh, yeeaah. . . .'"

THE FIRST OBSTACLE LONNIE, MARY, AND VLADIMIR ENCOUNTERED ON THEIR DESCENT —the day after they finished drilling—was a wide swath of debris sweeping across the "death zone" through which Benjamín had rushed Mary on the way up, from an avalanche triggered by the ferocious windstorm.

Then Mary caught a crampon in the flexible aluminum ladder that was now draped along the wooden ladder, across the dreaded crevasse—and the fact that the crevasse had widened by then to "twenty-five feet with a twenty-four-foot ladder" did not help.

"The foot was jammed; I couldn't move it. I was stuck there for about ten minutes, and they didn't want anyone to climb out and help, because they were afraid of the thing cracking and breaking. Lonnie and Vladimir were up at the top of the ladder waiting to come down, and a bunch of climbers were down at the bottom. Everybody's yelling at me to get moving, and I'm trying to tell them I'm stuck. Finally I had to actually take my hands off the ladder, brace myself with a shoulder, and work my foot free. And I did it, but my gosh, I think it exacerbated my acrophobia.

They say the way to get over phobias is to face them, but I don't think that's always true.

"Got through the camps; Camp One was missing. Got down to base camp and on down the mountain, just kind of went down the path toward Musho, where some trucks were waiting for us. You know how you use different sets of muscles to climb up and climb down? Well, my climbing-down muscles had atrophied, so by the time we got down to base camp the fronts of my thighs hurt so badly I could barely walk. In Huaraz, I had to walk backwards down stairs for about three days."

Lonnie experienced similar discomfort—perhaps worse, since he'd been high for about twice as long as Mary—but a celebratory emotion shone through: "It's always amazing when you descend from the ice into the forest with its plants, particularly in the tropics. There is a variety of odors that you don't normally notice. I guess a lot of different emotions come together. The success of the project is always thrilling, especially if you're doing something that's never been done before and you don't know whether it's going to succeed—plus the fact that you're going from a low-oxygen environment into a high-oxygen environment. Along with the vegetation and the green and the flowers and all that—yes, it's an exhilarating feeling."

·15·

ALTERING THE COURSE OF THINKING

The main reason for drilling on Huascarán was to obtain a record from the same climatic zone as Moche, in the hope of shedding more light on the trajectory of that bloody civilization, and of those of the other evanescent civilizations that had risen and fallen on Peru's northern coastal plain in pre-Incan times; the archeological implications of Quelccaya were making an impact by this time. But Quelccaya stands a thousand miles from the Moche heartland, on the southern end of the north–south El Niño seesaw, whereas Huascarán stands only a hundred miles to the southeast and experiences the same increase in precipitation that Moche does during an El Niño.

The Thompsons had figured the Garganta glacier would have roughly the same flow characteristics as the Quelccaya ice cap. Based on the thickness of the firn layers and the total depth of the ice in the col, they had suggested in their grant proposal that the record might go back 1,500 or 3,000 years, maybe 4,000—not much farther than Quelccaya's. Lonnie says they "had hoped to have annual layers back to the demise of the Moche, around 600 A.D., but, unfortunately, you have this tradeoff with glaciers. Either you have high resolution and a short record or you have low resolution and a very long record."

They got the latter, and it was a blessing in disguise. Although Huascarán did not have the resolution to say anything about the Moche, it reached far enough back in time to answer questions that lay at the heart of climatology in the mid-1990s—some of which had first been raised by John Mercer after his pathbreaking explorations in Patagonia almost half a century earlier. The new ice cores from Peru propelled Lonnie to the center of his field's scientific and political maelstrom.

And except for a strange effect that he still can't explain, the cores might very well have had something to say about the Moche.

IN THE DAZZLING SUNLIGHT AT TWENTY THOUSAND FEET, THE DRILLING TEAM HAD counted annual layers by eye. As always, Lonnie would set a blank page of his special meter-long spiral pad on the stand next to each core segment as it was extracted from the drill, draw a simple schematic of it, marking obvious visual features—among which the dark lines separating annual snow seasons were probably the most important—jot a note or two, assign the core a number, bag it, and place it in a similarly numbered tube, which would then be buried in a snow pit for the Peruvian mountaineers to carry down later.

They quickly passed 1970, allaying Bruce's fears of an avalanche, and as the days trudged by, about seven meters at a time, they kept a running tally of the lines, which grew closer and closer together the deeper they drilled. It took 120 meters to reach the time of the American Revolution, two hundred years ago, yet only forty more to reach the time of Egypt's Old Kingdom, four thousand years ago. Three meters lower than that, at 163 meters, they reached the end of the Younger Dryas, about eleven thousand years ago. When the spacing became so fine that they could no longer distinguish individual lines, they had counted almost twenty thousand of them!

They seemed to be mining ice age ice.

Indeed, the basal three meters, down by bedrock, were very different from the 163 meters above them. At about the time, according to the layer count, that the polar ice caps would have receded for the last time (or, since the team was drilling backward in time, as the ice caps began to advance), Huascarán's ice turned dark and dirty. This hearkened to Lonnie's earliest work on the Byrd and Camp Century cores: in the tropics as at the poles, the air seems to have turned more dusty in glacial times.

This was not *proof* that they had obtained the first glacial-stage ice from the tropics, but as Lonnie says, "We certainly expected it, and we were excited about it. When you think about what could cause layers like that in the tropics, it's hard to come up with anything outside of seasonality."

But a funny thing happened on the way to Ohio.

"WE WERE COUNTING THESE LAYERS," LONNIE CONTINUES. "NOW, I'VE GOT TO CLARIFY that. I was counting these layers as we were drilling these cores, and I have dates. I have the field books. For each core I have a meter-long page on which these layers are marked and counted. I have dates that go back seventeen thousand or eighteen thousand years—counted in the sun on the col of Huascarán.

"But when we brought the cores back we couldn't find them. We couldn't see the layers on the light table. So I thought it was the light up there or the fact that we were looking at the cores the moment they came out of the drill barrel: our light table has only a certain wavelength range. So we bought some full-spectrum lights [which mimic sunlight]. Still couldn't see them. Well, we could see more of them. And you know what's even more striking? They're in the film. You can see them in the video I took as the cores came out, and they're beautiful. They're stacked like layer cake.

"Now, things happen to ice when it's taken out of its environment. They've seen it in polar cores, below one thousand meters. When the cores come up, they're beautifully clear cylinders of ice. Two days later, they're full of bubbles, because the gas comes out of solution when the pressure is released. . . . And again, the lighting's quite different at the top of the earth's atmosphere, up at the drill site where you've got over half the atmosphere below you, tropical sun beating down.

"I've had tremendous discussions with Keith Henderson, one of my Ph.D. students. He said, 'You didn't see those layers; there *were* no layers. You were seeing things.' Well, look at the video."

("It's funny," muses Bruce, "I didn't notice them smoking anything funny up there either.")

"I thought we'd be able to put this core on a light table," continues Lonnie, "use our tree-ring system and just count the layers—measure the thickness and reconstruct accumulation rates from ten thousand years ago. But in the end we never used those layers, because we couldn't reproduce them on the light table; we couldn't validate what was done in the field. We had to go with other techniques for dating the core. It's amazing, though: they're almost identical.

"The record is twenty thousand years old. We knew when we were drilling that it was a long record, and we knew that the lower three meters were different from everything above, because you could see it. I couldn't think of any other explanation for what was at the bottom of the core. But when you come back and do the laboratory analysis and get a confirmation, that's fantastic!"

CONFIRMATION DID NOT COME EASILY, HOWEVER. WITHOUT VISUAL STRATIGRAPHY, they had to rely on chemistry and dust to count back the years. The oxygen isotope ratio and the levels of dust and nitrogen-based aerosols ("nitrates") oscillated annually and in unison—all peaking in the dry season, May to August; and the peaks lined up with the visual layer count from the logbooks. But the chemical methods required almost an inch and a half of ice core per data point; these methods were not capable of resolving layers thinner than that, so chemical layer counting took them back only to A.D.1730.

It must have been tantalizing to know the dates from the logbooks but not be able to use them. Lonnie found himself in a familiar, somewhat desperate situation—a situation scientists tend to seek out, actually: he needed a revelation, and there is no recipe for those things. The true scientist knows from painful—perhaps deliciously painful—experience that it is patently ridiculous to put one's faith in some scripted "scientific method" that can be drawn from one's shirt pocket to cut through the confusion in each bewildering new encounter with the unknown.

"It must have been two in the morning," he recalls. "I was down here* looking through some old journals, and I came across this figure from a marine core off the coast of Portugal. It was Ed Bard's core, SU81-18, a record of sea surface temperatures based on isotopes in forams. I looked at that curve, and I said, 'I've seen that curve before.' It was the isotope curve from Huascarán."

"Forams," more formally known as "foraminifera," are a kind of tiny planktonic shellfish that has proven extremely useful to the scientists studying the North Atlantic. The plankton live at the surface of the ocean, but their shells eventually end up in the sediment on the bottom, and the oxygen isotope ratio in the calcium of the shells serves as a measure of ocean surface temperature at the time the plankton died. Edouard Bard, an excellent, younger climatologist from France, had blazed new ground in the study of seabed sediments, a sort of muddy version of an ice core, as a postdoc at Columbia's Lamont-Doherty Earth Observatory, and his uniquely precise radiocarbon work has made SU81-18 one of the most well dated marine cores in existence.

Portugal is rather far from Huascarán, admittedly. It is not even tropical; it's subtropical, at about the same latitude as New York; and it lies not only north of the equator but on the eastern edge of the Atlantic Ocean. During Huascarán's wet season, however, when the mountain receives 80 to 90 percent of its snow, some portion of its winds do originate in the general neighborhood of Portugal. Air from that vicinity circulates west,

*It may be significant that the Thompsons' offices, hard by their labs, are literally in the basement of the Byrd Polar Research Center. They are as close, dingy, and confining as their field sites are spacious and spectacular.

across the Atlantic, and some will flow into the Amazon Basin and up the eastern slopes of the Andes.

Now, aside from the distinct possibility that Huascarán's temperatures may track those off Portugal on geological timescales anyway, the logic here is that some of the Atlantic surface water by the Portuguese coast—the same water whose temperature is recorded by the forams in SU81-18—will evaporate into the air and eventually make its way to the mountain. By the time it gets there, its oxygen isotope ratio will be depleted somewhat, but it will still reflect its initial value—and the amount of depletion should also depend on temperature. Therefore, in considering variations in the isotopes from both places, averaged over centuries or millennia, it is not unreasonable to expect those at Huascarán to mimic those in the subtropical eastern Atlantic.

"We made the assumption that it was the same record," explains Lonnie, "and, therefore, we could use the carbon-14 dates from SU81-18 to date the lower part of our core. What is this, really? It's curve matching. The curves matched beautifully, but curve matching is not independent data from Huascarán. Some people have raised issues with that. Then along came Sajama. . . ."

This argument may seem far-fetched, but it seems less so when one observes the remarkable similarities in both the large and small features of the Huascarán and SU81-18 isotope curves. Moreover, the larger features in both are also seen in the curves from the Greenland, Vostok (Antarctica), and Dunde ice cores. There is a gigantic dip in the isotope ratio at about the time of the Younger Dryas and again at the onset of the last glacial. Put another way, Huascarán's oxygen isotope curve, when laid out on the timescale derived from SU81-18, is remarkably similar to those from ice cores retrieved over a geographical range spanning the Arctic, the northern subtropics, and the Antarctic. And although Lonnie has never permitted himself to say so in a scientific journal, nor even casually to his colleagues as far as I can tell, the curve-matched timing sequence lines up very well with the dates in his logbook, estimated by visual layer counting in the Garganta Col.

WHEN MARY COMPLETED HER ANALYSIS OF THE DUST, SHE TAGGED A NUMBER TO ANother hunch they'd had in the col: the air above Huascarán was about two hundred times dustier in glacial times than it is now. This helped confirm the timing, for in every ice core containing glacial-stage ice that had been recovered at the time, dust content had risen and the isotope ratio had simultaneously dropped at the transition to the glacial.

The Thompsons believe nitrate levels at Huascarán measure the size of the Amazon rain forest, which lies to the east of the Cordillera Blanca, upwind during the wet season. Forests spew all sorts of nitrogen compounds into the air through biological activity, and when the isotope ratio drops at Huascarán, the nitrate concentration generally follows course—with a lag of about two thousand years: roughly the time it might take a rain forest to reach steady-state conditions in response to a climate change. During the glacial, nitrates drop by a factor of two or more, and the pollen count in the ice drops as well; thus the rain forest seems to have shrunk significantly.

Together, these various indicators paint a coherent picture. It makes sense that the rain forest would have shrunk during the cold glacial era, when the enormous growth of the ice caps lowered sea levels by as much as four hundred feet. Land that is now forested might have turned to savannah, something like modern-day Oklahoma, in which case it makes sense that more dust would have been whipped into the air by random gusts of the wind.* These signs also suggest that the air was drier; and here a most telling correspondence emerges: since water vapor is the most powerful greenhouse gas, drier air should lead to lower temperatures.

This last observation goes some way toward answering a question Lonnie and Ellen had first posed in 1981, in Ellen's first publication in *Science,* written by just the two of them. After demonstrating that dust levels rose and isotope ratios simultaneously dropped at the transition to the Wisconsinan in all three polar ice cores that had been recovered at the time—Byrd, Camp Century, and Dome C—the Thompsons stated that "the variations in these two parameters must be satisfactorily resolved in any successful hypothesis that addresses the causes of climate change." A significant change in tropical water vapor would certainly have had the power to turn both tricks.

Huascarán also resolved one aspect of the Mercer Problem: Mercer, remember, had believed either that there was *no* Younger Dryas in South America or that whatever cold reversal may have occurred down there must have preceded the Younger Dryas in Greenland. Huascarán was now saying that there *had* been a cold event in South America, and at about the right time. Its timing, however, based as it was on a record from the

*An interesting potential consequence of these changes was proposed in 1969 by Jürgen Haffer, a petroleum geologist who had worked in South America for a few years, for the Mobil Corporation. As an amateur bird watcher, Haffer was amazed at the diversity of Amazonian fauna. He explained it by suggesting that the drying out of the river basin during glacial episodes may have caused the rain forest to retreat into small islands amid the savannah. In these isolated "refugia" different groups from the same species may have evolved separately and spawned new, divergent species (just as Darwin had observed on the different islands of the Galápagos). Thus climate change seems to have acted as a "species pump."

northern hemisphere, was neither precise nor independent enough to say anything about precedence.

NOW, THE NOTION THAT THE TROPICS MIGHT COOL ALONG WITH THE REST OF THE world during an ice age might seem obvious to a nonscientist, but in the early 1990s, believe it or not, it came as a great shock to the leading lights in climatology. The most recent study of ice age surface temperatures, conducted in the mid-1970s by an influential consortium known as CLIMAP,* had concluded that nothing much had changed in the tropics as the ice sheets crept down from the poles. The central belt of the planet had remained balmy; temperatures at low elevations had dropped four degrees Fahrenheit at most. But Huascarán was now saying they had dropped between fourteen and twenty-two degrees at twenty thousand feet and, by deduction, between nine and eleven degrees at sea level—as much as they had in the North Atlantic.

Even the polar mafia took notice of that.

CLIMATOLOGY HAPPENED TO BE EXPERIENCING A RENAISSANCE AT THIS PARTICULAR time—but for ironic reasons: the strategy taken by former oil executive George H. W. Bush upon assuming the presidency had been to sow confusion in the global warming debate by throwing extra money at research. He instituted the greatest funding increase of all time for basic climatology as he worked at the same time to discredit the solid science behind the greenhouse effect—not to mention a few leading climatologists, such as Jim Hansen. Bush's idea seems to have been to give the scientists enough money to keep them arguing among themselves for a while, so that he could avoid doing anything on the policy front—and it worked quite well for the four years he was in office.

The North Atlantic School was also enjoying a heyday, for scientific reasons. After the pettiness engendered by the Camp Century and Dye 3 efforts, the funding agencies on either side of the Atlantic had finally reached a compromise that resulted in separate European and U.S. teams drilling simultaneously seventeen miles apart, near the summit of the Greenland ice cap. The Europeans had produced the 3,028-meter GRIP (Greenland Ice Core Project) core in 1992 and the Americans the 3,053-meter GISP2 core in 1993—both just shy of two miles long. (Dye 3 is sometimes referred to as GISP1.)

*Climate: Long-Range Investigation Mapping and Prediction. The list of CLIMAP's principal investigators, who hailed from a variety of institutions, reads like a who's who of the North Atlantic School.

The first results from these summit cores appeared while Lonnie was pursuing his field programs on Guliya and Huascarán, and they told of wild times in the North Atlantic. During the coldest stretches of the Wisconsinan glacial stage, central Greenland was more than forty degrees colder than it is today; however, temperatures oscillated wildly and in a sort of saw blade pattern: every 1,500 years or so they shot up by twenty or more degrees in as short a time as two or three years, then, over the course of centuries, decayed to roughly their former lows. And during the abrupt warmings, climate sometimes "flickered" between warm and cold episodes, two or three years in duration, before finally settling into the new warm mode.

Willi Dansgaard had found the first indications of these flickers and abrupt warmings in the deepest layers of the Camp Century core in the late sixties. His early findings didn't cause much of a stir, however, for they were unexpected and there seemed a high probability that summer melting and horizontal ice flow had distorted the record at Camp Century, since it lies close to the edge of the ice cap. But when Dansgaard and his Swiss colleague Hans Oeschger found similar indications near the base of Dye 3, others began to take notice: it was unlikely that records so far apart would have such similar and unusual features as a result of a random effect. GRIP and GISP2 finally established them as credible evidence for real climate change. Dozens of what came to be called Dansgaard-Oeschger events stood out clearly in the undisturbed median layers of the two summit cores.

Climatology being a science of seesaws and oscillations, it then turned out that even Dansgaard-Oeschger events came in cycles. Successive warmings tended to diminish in strength until, every three to five events, a massive warming set in to "re-prime the pump."

A German graduate student named Hartmut Heinrich had found a clue to the nature of this longer 4,500- to 6,000-year cycle in the early eighties, when he had discovered a series of striking and unusual layers in a suite of seabed cores from the floor of the North Atlantic. These "Heinrich Layers" consist of shards, pebbles, and other small fragments of rock that chemically match the feldspar and other constituents of the Canadian Shield in the Arctic province of Churchill as well as the limestone underlying Hudson Bay and the Hudson Strait, which border the shield. The Heinrich Layers are thicker in the western North Atlantic than in the east: they are almost ten feet thick in the Labrador Sea, but only an inch or so off the British Isles.

The thinking there is that vast fleets of icebergs must have periodically surged from the landmasses of northern Canada onto the surface of the sea, carrying with them the debris they had scraped from the rocks beneath them. Upon melting, they would have released the debris, which would then have sunk to the ocean floor.

Comparing the timing of the Dansgaard-Oeschger events in GISP2 and GRIP with that of the Heinrich Layers in the seabed cores, Lamont's Gerard Bond showed in 1995 that the layers had been deposited in the cold periods just preceding not every Dansgaard-Oeschger event but only the strong "re-primings" that occurred every third or fifth time.

Doug MacAyeal from the University of Chicago then provided an explanation for the pump priming by suggesting that the Wisconsinan glaciers, flowing ponderously north across the surface of Canada and into Hudson Bay, must have grown over the course of several Dansgaard-Oeschger cycles until they became so thick that their deepest layers melted from geothermal heat, which the thickening ice would have contained like a blanket. With water beneath them, they periodically lost their grip on the land and skated catastrophically into the sea.

The final piece to the puzzle seems to have been provided in 1997, when Gerard Bond explained the 1,500-year spacing between Dansgaard-Oeschger events with two elegant studies: the first showing that the cycle has persisted (in less dramatic form) all through the present ice cap–free Holocene epoch; the second that it is paced by a 1,500-year oscillation in solar output.

Now, according to the Heinrich Layers, these huge icy armadas invaded the North Atlantic at least eight times during the late phases of the Wisconsinan, most recently about fifteen thousand years ago, in the midst of the glacial's violent death throes; and after that final invasion, temperatures in Greenland soared to their highest levels in more than one hundred thousand years.

The Last Glacial Maximum—that is, the termination of the ice caps' last great advance—occurred about eighteen thousand years ago, just before the last Heinrich event. In other words, by the time of the last surge of ice onto the sea, the sheet covering North America had been receding for a few thousand years. On the surface of the ice, above the depression on the land made by its great weight, lay Lake Agassiz. Seven hundred miles long and two hundred miles wide—larger than all of today's Great Lakes combined—it covered vast stretches of what are now the northern Great Plains: the states of North Dakota and Minnesota and the provinces of Manitoba, Ontario, and Saskatchewan. At first the lake drained down the Mississippi watershed into the Gulf of Mexico.

The ice melted for another two thousand years, and then something new occurred: the Younger Dryas, a sharp drop in temperatures that caused the glaciers to re-advance for about a thousand years and came to a halt with a Dansgaard-Oeschger warming as strong as the Heinrich re-priming that had preceded it—this time, however, unheralded by icebergs. There is no rock layer in North Atlantic sediments at the end of the Younger Dryas, and only a thin layer exists on the floor of Hudson Bay.

ONE OF THE MORE REMARKABLE ASPECTS OF THE GRIP AND GISP2 ICE CORES IS THAT IT has been possible to count their annual snow layers back about forty thousand years. With a nod to the errors of a few percent inherent even to this method, the isotope ratio at GISP2 tells us that the Younger Dryas began in Greenland precisely 12,880 years ago, lasted for 1,240 years, and ended precisely 11,640 years ago, at which point the present Holocene epoch began.*

So, how is this thousand-year cold spell, which ended with no help from ice-rafting, reconciled with the eight previous Heinrich/Dansgaard-Oeschger events, which produced warmings of similar magnitude?

There is evidence that two channels formed in the walls of ice cupping Lake Agassiz at about the time the Younger Dryas began, the larger channel near the present St. Lawrence River Valley on Quebec's Atlantic coast, the smaller near what is now the Nelson River, in northern Manitoba. The lake seems to have drained, primarily into the Labrador Sea and to a lesser extent Hudson Bay, for about a thousand years.

Like the eight great surges of ice evidenced by Heinrich Layers, these more or less simultaneous dam bursts would have sent a deluge of freshwater into the North Atlantic Ocean. Based on this common feature, the scientists of the North Atlantic School took a stab at explaining this particular form of climate change: as these pulses of cold freshwater flooded the North Atlantic, so the thinking goes, they may have interrupted an enormous ocean current called the thermohaline circulation (*thermo* for hot; *haline* for salty), or great ocean conveyor belt, which normally carries warm water along the surface of the Atlantic Ocean from the equator to the vicinity of Greenland and Iceland. The best known of the many currents that make up this conveyor is probably the Gulf Stream.

The dominant contributor to this line of thinking, the foremost apologist for the great conveyor, the intellectual leader of the North Atlantic School, and probably the world's most well known climatologist (which may not be saying a lot) is Wallace Broecker, from Lamont. In 1997, he described the conveyor's beginnings with characteristic self-regard.

*For convenience, these numbers are given relative to 1950, a standard date for carbon-14 calibration, since it predates atmospheric bomb testing.

> The basic idea came to me in 1984, while I was listening to a lecture given by Hans Oeschger at the University of Bern in Switzerland. He pointed out that the Greenland ice core record suggests that Earth's climate was jumping back and forth from one state of operation to another. . . . I began to ponder what these states might be. It soon dawned on me that they could be related to a change in a major feature of the ocean's thermohaline circulation system, which I subsequently termed its conveyor belt. People now refer to it as Broecker's conveyor belt, but I have a colleague, Arnold Gordon, who thinks it's his conveyor belt rather than mine. . . .
>
> My idea can be summarized as follows . . . One of the most prominent features of today's ocean circulation is the strong northward movement of upper waters in the Atlantic. When these waters reach the vicinity of Iceland, they are cooled by the cold winter air that streams off Canada and Greenland. . . . The Atlantic is a particularly salty ocean, so this cooling increases the density of the surface waters to the point where they can sink all the way to the bottom. The majority of this water flows southward, and much of it rounds Africa, joining the Southern Ocean's circumpolar current.
>
> The importance of this current to climate is the enormous amount of heat it carries. The conveyor's flow is equal to that of 100 Amazon Rivers! . . . The amount of heat carried by the conveyor's northward-flowing upper limb and released to the atmosphere is equal to about 25% of the solar energy reaching the surface of the Atlantic north of the Straits of Gibraltar.

Broecker guessed that pulses of freshwater, whether they came from Lake Agassiz or from melting fleets of icebergs, may have diluted the surface water in the North Atlantic to the point that it no longer sank, and thereby shut the conveyor down. At his prodding, a few computer modelers simulated a shutdown and showed that it could lead to a drop of nine to eighteen degrees Fahrenheit in the average air temperature over the North Atlantic basin. The strongest effects would be felt in Newfoundland, Greenland, and northern Europe.

So far so good. The existence of a massive warm current flowing north at the surface of the Atlantic has been pretty well verified. There is also good evidence that it sinks in two primary locations: around Iceland in the Norwegian and Greenland seas, and between Greenland and Newfoundland in the Labrador Sea. There is further evidence that at least part of the same current then flows south along the bottom of the Atlantic to about sixty degrees south latitude, where a portion of it wells to the surface as it merges with the Antarctic Circumpolar Current, the largest in all the oceans. There is also proof that deep-water formation in the North Atlantic has switched on and off from time to time, during Heinrich's ice-rafting events in particular.

Most climatologists have no trouble with the idea that a significant change in the flow of North Atlantic surface water might affect temperatures in the local area. Those of the North Atlantic School, however, with Wally Broecker in the lead, have taken the concept many steps further than that. They have frequently referred to the two sites of sinking water near Iceland and Newfoundland as "master switches," capable of toggling the entire Earth back and forth between ice ages and warm spells; and Broecker, ever the wordsmith (he named both the Dansgaard-Oeschger and the Heinrich events), has gone so far as to call "his" conveyor belt the "Achilles' Heel of Our Climate System."

He is fond of referring to a cartoon that shows the conveyor, still cold and at the bottom of the ocean, diverging from the Antarctic Circumpolar Current southeast of Australia, flowing north into the Pacific, welling to the surface north of the equator, gaining back solar heat as it flows back west through the Indonesian archipelago and the Indian Ocean, rounding the Cape of Good Hope, entering the Atlantic again, and, finally, flowing north, ready for another cycle. This goes somewhat beyond the geographical range in which the conveyor has actually been observed, but it does provide it with enough breadth to permit Wally to lay at its feet virtually every climate change that has taken place anywhere on the planet for the past two million years. All through the nineties, he and his enthusiastic followers in the North Atlantic School—who have made many discoveries of undeniable importance and come up with undeniably creative and credible explanations for the tumultuous ice age events in the far north—have touted their region as the source of all climate change.

The nineties were an exciting time for them. Dansgaard-Oeschger events, Heinrich Layers, and the draining of Lake Agassiz are the sort of evidence that slaps a scientist in the face, demanding explanation. In their feverish race to explain it—not to mention the significant public relations job required to justify the mobilization of small armies and the expenditure of tens of millions of dollars to keep their drilling projects alive—these fine scientists may have succumbed to a well-known course hazard in scientific work: they may have "fallen in love" with their subject.

Most climatologists prefer to believe that the region they are studying holds the keys to climate change; the scientists of the North Atlantic are not the first to have blown the horn of their own research. Science, like many human endeavors, runs on money, and excitement attracts money. What's more, it sometimes takes the concern and commitment of a "loving mother" to nurture a controversial idea into acceptance, and public relations is part of that game. But just because you love an idea does not mean it is true. It may take decades, even centuries, for a concept as intangible as the great conveyor, much less its global effects, to prove out. Witness Arrhenius's carbon dioxide theory.

IN ANY EVENT, CLIMATOLOGY WAS VERY MUCH CAUGHT IN THE THRALL OF THIS DRAMA as Lonnie Thompson was quietly carrying Huascarán's ice down from the Garganta Col. Most climatologists hadn't paid much attention to him since he'd struck off for Quelccaya twenty years earlier. And, as it turns out, he and Ellen were additionally estranged from the North Atlantic School just then, for political reasons. During the delicate transatlantic negotiations that had led to the GRIP and GISP2 compromise, Ellen had unknowingly incurred Wally Broecker's wrath.

This is not a hard thing to do. Wally has hung up on me both times I've spoken to him, and when I mentioned this fact to James McCarthy, a biological oceanographer at Harvard who was a lead author on the most recent IPCC assessment, he told me I was "in good company."

In a 1998 *New York Times* profile entitled "Climatology Guru Is Part Curmudgeon, Part Imp," William K. Stevens noted that Wally is possessed of an "extroverted personality as volatile as the climate system he studies. He is known for his blunt honesty, a self-confessed volcanic temper and a reluctance to suffer sloppy thinkers gladly, balanced by a sense of fun that has made him an incorrigible practical joker." These qualities make him an entertaining conversationalist, and most people seem to like him in spite of his foibles. I know I do.

In the more successful of the two conversations I have had with him, I never found a chance to ask any of the questions I had prepared, because nothing could keep him from describing the simultaneous births of GRIP and GISP2.

On the earlier Camp Century and Dye 3 projects, he pointed out, "The National Science foundation felt that the United States had put in most of the money, and the Europeans had reaped most of the scientific glory. It's true, but it wasn't malicious cheating or anything. It's just that the way it was set up, the American scientists involved, mainly Chet Langway from Buffalo, weren't doing any of the really interesting things. The interesting measurements were made mainly in Denmark [where Willi Dansgaard's group carried out the oxygen isotope analyses] and Switzerland [where Hans Oeschger pioneered methods for measuring carbon dioxide in trapped air bubbles]. So there was a stalemate, and there was a real asshole at the National Science Foundation, who had no imagination, heading Polar Programs at the time. He controlled an enormous amount of money—because polar science is ninety percent logistics and five or ten percent science.

"NSF was stalled, because they felt the same thing was going to happen—and without the U.S. nothing could have happened, because we had the airplanes* and so forth. So I got the idea of getting the people together to see if I couldn't break this logjam—an outsider coming in. I proposed that the principals from both sides of the Atlantic meet at the Bostonian Hotel. I think it started on a Super Bowl Sunday. I remember, because the New York Giants were playing Denver. . . .

"So we met there for two days, and we came to an agreement—something suggested by Dansgaard—that there be two holes drilled. . . . The Europeans would pay for theirs; they would control the ice from their hole. We would pay for ours; we would control the ice from our hole. This was signed by the Danes; it was signed by the French; it was signed by the Swiss (so they're the big players); and it was signed by us.

"Now, it had no official, heh, heh, status, and this guy at the National Science Foundation didn't want any part of it. He put out this Request For Proposal that infuriated us, because it clearly stated that he wanted a consortium of universities to run this thing—from the political point of view, not according to the individual investigators' strengths. We got that RFP killed before it went out, and I don't remember the details, but it would have led to Ohio State and a couple other universities becoming the principal centers—a very bureaucratic way of doing it.

"The next step was to get the young people in the U.S. involved. I suggested that to beat down the National Science Foundation, we get everybody in the country to write a joint proposal that they couldn't work around with political tricks. This is where the Thompsons come in. . . . We got everybody together, and only one obvious person did not get involved: Ellen Thompson. I never knew for sure, but I think Ohio State really wanted to take over the Greenland ice core—have it run out of Ohio State. *I think*. Anyway, she and Lonnie were the only two young, active scientists in the country interested in ice coring who didn't get involved.

"So, our proposal was written; it was rammed through the National Science Foundation, and it led to one of the more successful programs ever run . . . and it wasn't bureaucratic at all; it was grassroots. Instead of having

*The United States owns the world's only fleet of LC-130 Hercules cargo planes. This is the polar version of the more familiar C-130, equipped with skis for landing and other less visible improvements that allow it to function in extreme cold. (Ellen and Bruce used LC-130s to get back and forth from Plateau Remote.) Not only does this technology give Americans a unique ability to undertake large projects in the polar regions, it also demonstrates yet again the boost that Cold War paranoia has given to advanced science.

our federal government in all of its bureaucratic idiocy deciding who did what and how, we beat 'em down. . . . And, you know, I wasn't a member of the team; I was just trying to get things moving. I'm not an ice scientist. I've never touched an ice core, and I never will.

"Now, because Ellen would not get involved—and we never had a long talk about it or anything—she was invited, but she chose not to participate, that sort of soured me on both the Thompsons, heh, heh, because, you know, I figured they're a team."

Had Wally bestirred himself to speak to Ellen he might have been surprised (although he might still have found an excuse to exercise his volcanic temper). Ellen says that she "was not aware that Polar Programs had a plan to situate anything at Ohio State. Wally travels in higher circles than I do. But, in fact, if I had known, I would have opposed it. At that time it was just Lonnie, myself, and Mary in the lab. How could we have done it?"

Having chaired the committee that had developed GISP2's overall scientific plan, Ellen understood its scientific importance more than most. The main reason she gives for refusing to sign Wally's proposal is that she and her husband objected on point of principle to the way the science would be managed. They had been troubled long before by Chet Langway's monopoly on scientific decision making during Dye 3—remember his threatening conversation with Lonnie at Jay Zwally's dinner party and Ellen's statement that he had cut the Thompsons out of Dye 3. They saw the same spirit at work here.

"All the science proposals went through a blue-ribbon panel," Ellen points out. "We saw that as counter–peer review. This is an important issue. Later on, in fact, once they actually did get up and running, we submitted a proposal through [the normal peer-review process at] NSF. It was rejected, of course."

She agrees with Wally that GISP2 was a successful project; however, she expresses "some doubt as to whether it needed to cost as much as it did." Rumor has it that GISP2's logistics alone cost twenty-five million dollars. That's drilling, mostly; it does not include chemical analysis or any other science. The Thompsons' projects generally run between half a million and seven hundred thousand dollars for everything.

BUT WALLY WAS RIGHT IN ASSUMING THAT ELLEN AND HER HUSBAND WERE ACTING AS A team. They batted the idea back and forth, and if they had signed on to GISP2, they would both have been involved. Another factor was their own ambitious research program. They employed the same drills and drilling crews; Lonnie was in the midst of what has thus far, arguably, been the most productive period of his life; and Ellen was not only co-managing their group, co-writing all of his grant applications and nearly all of his scientific papers, and supporting his fieldwork, but also pursuing her own projects in such logistically challenging locations as Plateau Remote. The Thompsons have a small, self-contained, highly functional group and a strong aversion to bureaucracy—be its seat in the National Science Foundation, their own university, or anywhere else. They've built their careers by avoiding bureaucratic entanglements, and they are straight, exceedingly polite people. Wally's rude, big-city manner and his propensity for livid outbursts offended them greatly.

With the perspective of years, Ellen now says that she "gets along with Wally just fine. It's amazing with Wally. You can be the greatest one minute and a complete jackass twenty-four hours later. If he weren't such a good—I'm searching for the right words here—if he weren't one of the leading thinkers in the field he wouldn't be able to get away with that behavior. But you get used to that with Wally. Nowadays whenever he yells at me, I just figure he's having a bad hair day and let it go."

BUT THE SENSE OF ESTRANGEMENT WAS VERY MUCH ALIVE WHEN LONNIE REVEALED THE first Huascarán results at a CLIVAR (Climate Variability and Predictability) meeting in Venice in the fall of 1994. Since CLIVAR's emphasis is the recent past, Lonnie's prepared talk covered only the last two thousand years.

"At the end," he recalls, "I asked the chairman if I could have five extra minutes to present the bottom of the core. I showed the isotope curve, the nitrates, the dust—in both the frozen core, which is the best record, and the adjacent [melted and bottled] core, which shows the same thing. Wally Broecker was there, Nick Shackleton. They really got excited about having an ice core from the tropics, above the Amazon rain forest, that went back into the last glacial cycle. Wally came up to me after the talk and invited me up to Lamont for a week to give a series of lectures so we could discuss this record."

That visit took place in the spring of 1995. In July, the Thompsons published the Huascarán results in *Science,* and a few weeks later Wally published the following comments in *Nature*:

> . . . before discussing its scientific aspects, we should note that this story has another important dimension. It reveals how the tenacity of a lone scientist moving against the grain of conventional wisdom can alter the course of thinking. While most scientists in his field were obsessed by the record kept in ice cores from Greenland and Antarctica, Lonnie Thompson

pushed to extend these studies to small glaciers capping the Earth's highest mountains. For years, he fought not only the cold condition of his field sites but also the lukewarm reception by many of those in our field (including me). In 1993, with the help of a small army of strong bodies recruited in Peru, he succeeded in hauling 6 tons of equipment to a col located at 6,050 metres elevation on Huascarán in the Andes (9° S). Power derived from solar panels allowed him to drill two 160-metre long cores to bedrock. Ten tons of equipment and ice had to be lugged back down. This accomplishment alone places Lonnie Thompson in the ranks of our great explorers.

These words touched a nerve. Accolades poured in from all quarters of the climate community. Colin Bull, who ran across Wally's comments in the library at the University of Seattle on a trip across Puget Sound from his retirement home on Bainbridge Island, sent Lonnie a letter in which he pointed out that this was especially high praise coming from Wally, who rarely compliments anyone outside the circle of his own protégés and Lamont colleagues. Lonnie's answer read in part:

Dear Colin:
Thank you for your very kind letter of August 5. I have just this week returned from Peru where I have spent the last month in the splendid high Andes.* I must say I was surprised with Wally Broecker's article on ice cores in *Nature* given what has been a stormy relationship due to his attitude toward Ellen over the GISP2 project. In fact about three years ago I refused to shake his hand at an Ocean conference because of the way he treated Ellen. . . . However time changes many things. This spring I spent a week at Lamont as a visiting scientist at the invitation of Wally and Rick Fairbanks. . . .

I believe that we are all the products of our interaction with many people and we can never forget whatever we achieve in life it is a collective achievement. I was very fortunate to have had you as an advisor and to have gone through the department when there were a number of other outstanding faculty members. I truly love what I do and thus consider myself to be a very fortunate individual. . . . Thank you for all your efforts 357 years ago when I needed it. . . .

*Not missing a beat, Lonnie had returned to the Andes in the summer just following Huascarán to drill another shallow core on the summit of Quelccaya, take more photographs of Qori Kalis's retreat, and begin the search for a third Andean drill site, which would finally converge on Sajama.

Willi Dansgaard, who had last weighed in on Lonnie's errant mountain forays a decade earlier with a skeptical review of the grant proposal for deep drilling on Quelccaya, sent Lonnie a fax in which he stated his regret at not having co-authored the Huascarán paper and cast his apology in the form of a poem:

We must confess
that we did less
but don't rephrase
a single line.
Infested with
conceitedness
most minds like praise
and so does mine.

Thus, twenty years after he broke from the polar fold and went to the tropical mountaintop, Lonnie first caught the attention—and gained the respect—of his peers. Make no mistake, however: their accolades came not for his physical tenacity but for the "course-altering" science it produced.

"Lonnie was going on and doing the mountain stuff," Wally continues, "but we weren't paying much attention to him—none of us. We thought Greenland was the place to be. Nobody thought there was glacial-age ice in the mountains. That's because we were ignorant. People had done a lot of work in the Alps and never found ice more than a few thousand years old, you know. Finally, they got the 'bronze age man,'* or whatever, and that's what, five thousand? You're only about halfway back to the glacial. So when Lonnie's Huascarán data came out, we just were gaga-eyed about it.

"We had known since the fifties that the snowlines in tropical mountains had dropped about a kilometer during glacial times. CLIMAP had a *big* problem with that. They tried to pass it off as a difference in precipitation or something, which was utter nonsense. The clear answer is that it was colder in the tropics at those elevations. And the drop in snowlines was roughly the same all over the world, so it wasn't as if the poles got colder and the tropics stayed the same. But as far as a detailed record that showed something about the pattern of tropical temperature or precipitation change with time, the records were few and far between—and kind of

*Wally is referring to the unfortunate prehistoric hiker who was discovered by two modern hikers in the Tyrol in 1991. "Ötzi," as he was nicknamed, since he was found poking out of the ice in the Ötztaler Alps, served as a message about global warming as well: why did he choose to appear just now, after hiding in the ice for five thousand years?

ratty, poorly dated. . . . So Huascarán just wowed people, because it gave a detailed record that showed that the oxygen isotope changes were far bigger than anyone would have guessed, even from snowline lowering."

Wally said these words in June 2000, just a few months after Lonnie had returned his favor and invited him to lecture at Ohio State.

"I got to know Lonnie better, and you know what I realize? . . . Well, it's clear that he and Ellen together run an amazing operation. They have well-operated labs to do the oxygen isotope analysis and the chemistry, and it's amazing to see how many dimensions this group has, how well they work together, and how high the morale is. You realize that Lonnie is just a *demon* to keep all this stuff together. He's fair and he's honest and he really keeps the group humming. It's an amazing thing, because not many people in the world are able to, single-handedly, be the brains *and* the brawn—and, you know, it's clear that it *is* him. He isn't sitting at a desk having other people do this; he's involved in every detail."

ANOTHER HANDSOME TRIBUTE—AND A TESTAMENT TO A MIND UNHINDERED BY THE hobgoblin of consistency. Wally's attacks stoop all too often to the personal level, but the victims usually realize, once they have recovered from their wounds, that he is motivated by a near-religious zeal for clear thinking, a pristine belief in science as a noble calling. (He was raised in a fundamentalist Christian family, and though he now disclaims that sort of religion, his present fervor for the great conveyor has a similar ring.) He thrives, more than most scientists, on the "ah-hah!" experience, walking the boundaries of present knowledge to seek contradictions that will upset the applecart and lead through catastrophe to a new understanding. Lonnie points out that Wally "will often publish a few lead papers on a topic and, by the time the rest of the community comes along, be on to something new." He has written or contributed to an astounding number of seminal papers, and his seminal thinking is cited in an astounding number more. For more than a decade, from the late eighties through the end of the century, he more than anyone else set the agenda for climate research. As Jim McCarthy puts it, "When Wally speaks, the earth shakes."

So it was natural that Huascarán would excite him. Unfortunately, however, it started a revolution that may have upset the applecart too violently even for Wally to absorb complacently. Near the end of his comment in *Nature,* he used a phrase that may have portrayed a certain lack of clear thinking on his own part:

> . . . For those of us in search of a mechanism capable of propagating the effects of reorganization of the oceans' thermohaline circulation across the

> globe, Lonnie Thompson's discovery points us to the water vapour pumping capacity of tropical convection systems. His findings should also send a strong warning to those who choose to discount the potential of the ongoing greenhouse buildup. The paleoclimate record shouts out to us that, far from being self-stabilizing, the Earth's climate system is an ornery beast which overreacts even to small nudges.

Fine prose. The last line has been repeated many times. It speaks eloquently to the concern for the future that is shared by virtually all serious climatologists. But might Wally have been demonstrating a little too much motherly love for his pet theory by "searching" for a way to extend its influence all the way to the tropics? His colleagues would certainly agree that changes in tropical moisture have the power to change climate drastically via the greenhouse; however, those such as Lonnie who have not been so closely involved in the exciting work in Greenland see no reason to connect changes in tropical water vapor to the flipping of a switch in the obscure North Atlantic. Wally's protestations notwithstanding, he was no outsider to Greenland. His ideas provided the intellectual context for GRIP and GISP2, not to mention the main justification for spending so much money on those gargantuan projects.

Although it attracted far less interest than the goings-on in Greenland—primarily because fewer, less vocal scientists happened to be involved—the recognition that the tropics probably weren't as dull and unchanging a place as most had believed was starting a quiet revolution just then. In the years since, the line of investigation opened by that recognition has come to undermine—even topple—the notion of North Atlantic primacy. Ironically, but not surprisingly, Wally himself played an early role.

In 1993, at the annual December meeting of the American Geophysical Union, in San Francisco—almost a year before Lonnie revealed the Huascarán results at CLIVAR—a group led by Rick Fairbanks of Lamont had presented evidence from Barbados corals indicating that surface waters in the tropical Atlantic had cooled by about nine degrees Fahrenheit in glacial times. (This was part of the reason Fairbanks later joined Wally in hosting Lonnie at Lamont.) At the same meeting the following year—only a month after CLIVAR—another Lamont group, led by Martin Stute and including none other than Wally Broecker, inferred a drop of about ten degrees in a study of carbon-dated groundwater in lowland Brazil. The year after that, a group led by Dan Schrag (who helped propose the Snowball Earth hypothesis a few years later) published an isotope analysis of water trapped in the pores of seabed sediments, which indicated that *deep* tropical ocean water was about seven degrees colder during the glacial than it is today. These studies dovetailed very well with Lonnie's inference from Huascarán that temperatures had dropped between nine and eleven degrees at sea level.

From what I can tell after speaking with a number of climatologists who were involved and interested at the time, Huascarán had the greatest impact among these four more or less concurrent studies—perhaps, as Wally points out, because ice cores archive climate in such fine detail.

HOWEVER, AS THE SAYING GOES, "HIGH TREES CATCH THE WIND." WITH TROPICAL ICE cores having entered the limelight, intense intellectual (and emotional) interest quickly focused on the conundrum Lonnie had uncovered in the very first ice he had retrieved on Quelccaya, some twenty years earlier. Just after his first, ill-starred trip with John Mercer, you will recall, Lonnie and Willi Dansgaard had revealed the presence of a puzzling "seasonal reversal" in Quelccaya's ice. The oxygen isotope ratio oscillated as expected over the course of a year, but it went the wrong way: it *peaked* in winter, when it is definitely colder than summer. The same thing happens at Huascarán. On the timescale of centuries to millennia, on the other hand, Huascarán's isotopes behave sensibly: the ratio rises and falls in a manner consistent with records from around the world—from Greenland through the tropics to the South Pole, at all of which places the ratio was long ago shown to measure temperature directly—and not only in ice cores, the handiest example being Ed Bard's Portuguese seabed core, SU81-18.

Dansgaard had achieved legendary status in 1964 by using snow samples from a variety of locations to demonstrate that the ratio was simply proportional to average annual temperature. When graphed against temperature, the ratios from his different locations—which included Greenland and Antarctica—fell on a straight line: the higher the temperature, the higher the ratio. Basically, if you were to send him a sample of snow from your backyard, he could tell you how cold it gets (in winter, at least) where you live. The present ratio and average yearly temperature at Huascarán, by the way, fall directly on Dansgaard's line.

For reasons that are hard to fathom, however, Huascarán's isotopes quickly evolved into a sort of Rorschach inkblot in which theorists demonstrated a remarkable ability to find whatever they wished to find.

As indicated by his *Nature* comment, Wally Broecker's overriding interest was to show that changes in tropical water vapor might lend global compass to his great conveyor. He quickly developed a primitive mathematical model of ^{18}O depletion on Huascarán, which led him to the surprising conclusion that in the Andes, uniquely, the isotope ratio measures atmospheric humidity alone; it has nothing to do with temperature! He explained away the mountain's contemporary agreement with Dansgaard's temperature line as being possibly "fortuitous."

From the dip in the isotopes during the glacial, Wally found, not surprisingly, that the calculated dip in humidity would have had quite enough of a greenhouse effect to drag temperatures down all over the world. He admitted, however, that a certain contradiction in his model left his "interpretation . . . open to many challenges": in order to simplify the mathematics, he had assumed that there was no change to the greenhouse at all.

The next to register an opinion was another otherwise fine theorist named Raymond Pierrehumbert, from the University of Chicago, who agreed with Wally that Huascarán's isotopes had nothing to do with temperature, but presented a more plausible and erudite, but, as we shall see, ultimately confused argument. Pierrehumbert found that the ratio reflects mainly the amount of rain that a typical parcel of airborne water vapor drops as it crosses Amazonia on its way from the Atlantic to the Garganta Col. This thinking probably comes closest to what has become the prevailing argument *against* linking isotope changes at Huascarán with changes in air temperature. It has been called the "amount effect."

Dansgaard's straight line reflects a tacit assumption that the "life cycle" of that typical parcel of vapor does not change very much: the vapor evaporates from the same part of the same body of water, sitting at an unchanging temperature; it takes the same average path to the spot where it finally drops to earth as snow or rain; and it encounters the same conditions, with the exception of variations in temperature, along the way. But the fact is that every time a bank of vapor condenses to form rain or snow, it loses some of the heavier isotope. Thus, during rainier or snowier seasons—or epochs—the vapor that lands as snow on Huascarán might be depleted simply because it has dropped more rain in its travels.

Since the mountain receives more snow in (the austral) summer than in winter, the amount effect theory would seem to explain Huascarán's seasonal reversal, but it still has a problem with the long-term changes that track the entire Earth's temperature so well. It also leads to the surprising conclusion that it must have rained *more* during the cold, dry glacial period—when even the Amazon cloud forest shriveled for lack of rain—than it does now.

By the way, similar arguments can and have been made about isotope depletion in other regions. For instance, a parcel of water vapor must travel farther to reach the summit of Greenland in winter, when the surrounding seas are covered with ice, than in summer; so a wintertime drop in the ratio may simply reflect the fact that the parcel had many more chances to condense on its longer journey. Yes, the extent of the ice probably reflects temperature, but the point is still made: many factors affect the isotope ratio.

NOW, IN CONTRAST TO THE TWO THEORISTS SITTING IN THEIR OFFICES IN NORTH America—Wally Broecker scratching away with pencil and paper, Ray Pierrehumbert hunched mainly in front of his computer—Lonnie has been meditating on the Andean isotope conundrum for going on thirty years. He has also lived literally for years on the high glaciers of the Andes, so close to the "wondrous factory" that is the atmosphere that he has not only seen but been forced to experience its magnificent and manifold displays. In fact, he had solved the conundrum to his own satisfaction a few years before he drilled on Huascarán—in a flash of insight of the sort he claims often to experience in the mountains.

"Mountains are fabulous places," he once said. "I'm intrigued by the sacred value the Tibetans see in them. It makes sense to me that Moses received the Ten Commandments at the top of a mountain. Some of my best ideas come to me when I'm up there. . . . I always wonder about these things. Where does vision come from? How do you suddenly realize what the story is and what's going on? I think mountain settings, quietness, help you to observe what's going on in the real world, to see and to pay attention to what you're seeing and what it might mean for the records you're obtaining—it's remarkable how that works."

In the exploratory phase of the Guliya and Huascarán programs, when his summer visits to China forced him to visit the Andes in the wet season, he once experienced a flash of insight as he stood on the Garganta Col.

"I was looking out over the Amazon, and I realized that there is a big difference in the height of the clouds in the dry season versus the wet season. In the dry season the anvils are about the height of your vision, six thousand meters, but in the wet season you have to look up almost vertically to see the tops of the clouds. It's obvious that there's a big difference in the height and hence the temperature where condensation is taking place."

The drop in temperature with altitude is known as the "lapse rate." Any climber knows that he'll get colder the higher he climbs.

There are obvious discrepancies between the models Broecker and Pierrehumbert have dreamed up in their small rooms and a few simple observations they might have made had they spent even a week in the mountains about which they speculate so freely. Pierrehumbert notes loftily that his model "implies a precipitation altitude of 6000 meters during the wet season and 6600 meters in the dry season"—higher in the dry than in the wet. Not only is this exactly the opposite of the seasonal cycle Lonnie has seen with his own eyes, Pierrehumbert also seems to believe that rain falls *up*: his wet season "precipitation altitude" is lower than the Garganta Col! In mathematical physics, glaring discrepancies such as these cause a model to collapse like a house of cards.

Lonnie's insight on the mountaintop prompted him to review other scientists' studies of the temperature structure of the atmosphere and the mean height of droplet condensation.

"Our theory right now," he says, "is that isotopes in the tropics *are* temperature, but it's the temperature at the mean condensation level, the height where most of the moisture is coming from. In the wet season, even though temperatures are a few degrees warmer, the convection cells are almost two kilometers higher. Consequently, it's colder where the moisture's coming from, and the isotopes reflect that. It's truly temperature.

"It has always been a puzzle that Huascarán and Quelccaya have provided the best records of the Little Ice Age of any ice cores on earth. Not in Greenland. Not Antarctica. In the tropics. The decrease in isotope values from 1520 to 1880 A.D. translates into about two degrees of cooling, and that's exactly what you'd calculate from the glacier expansion in the Alps and other parts of the world. Huascarán also has the same decrease during the glacial as the polar regions. So you ask, 'How can you be recording temperature on the long timescale and not on the seasonal timescale?' You have to remember that ninety percent of the snow falls in the wet season in the Andes. I think isotopes are temperature everywhere."

His last point resolves the conundrum: the ratio drops in the warmer wet season because droplets form in the higher, colder portions of the atmosphere at that time of year. But it tracks temperature over the long term, because virtually all Huascarán's snow falls in the wet season; so, if you average over a few years, the wet season is the only season you see—and wet-season temperatures evidently reflect the yearly average over South America. Lonnie, Ellen, and Keith Henderson finally published this thinking in 2000.

I once told Lonnie I wasn't sure I believed him on this. He laughed and said, "That's all right. Nobody does. It's been a long time, and we're still working on that. Fact is, temperature and precipitation are not unrelated in the tropics, so this debate will continue for a few more years."

"It's been good," adds Ellen. "You always have to question yourself, because if you keep questioning yourself and you don't find anything wrong, then you may actually be right."

In the face of much disagreement—even by the man whose words are said to "shake the earth"—Lonnie and Ellen have stuck to their guns on this issue as tenaciously as they have drilled ice in the cold conditions of their field sites. In 2003, they published what seems to be their definitive statement on the subject, a review of isotope results from Quelccaya, Huascarán, and Sajama in South America, and from Dunde, Guliya, and Dasuopu in Tibet, over the past twenty-five thousand years. The Thompsons show from a number of angles that isotope ratios in all six locations reflect temperature—and why they should. Most convincing is the remarkably close match between the isotope records from South America and Tibet, both individually and combined, with the famous "Hockey Stick": the graph of average temperature in the northern hemisphere over the past one thousand years that was the centerpiece of the most recent (2001) IPCC assessment. "I guess some might argue that the IPCC curve is the amount effect," jokes Lonnie.

A more vivid image is provided by the plant uncovered by Quelccaya's retreating Qori Kalis tongue sometime between Lonnie's visits in 2001 and 2002. Its radiocarbon age (à la John Mercer) shows that the last time Qori Kalis retreated to that height was about 5,200 years ago, which is the last time South America was as warm as it is now, according to Huascarán's isotopes.

·16·

PUSHING BACK THE BAR

> The cause of differences between the earth's climates attracted the attention of the thinkers of antiquity. As long ago as that time it was established that there is a close relationship between climatic conditions and the mean altitude of the sun, i.e., the latitude of the terrain. This relationship explains the origin of the term climate (from the Greek verb κλινειν [to tilt], characterizing the angle of the solar rays).
>
> MIKHAIL BUDYKO

Lonnie began rolling out the Guliya results a full four years after the drilling took place, at a NATO Advanced Research Workshop in a luxurious hotel in Tuscany—a stark contrast to his encampment in the Kunlun Shan, to be sure.

The timing sequence had turned out to be a head-scratcher, so they still had not analyzed the complete record; however, they had been able to count annual layers down through the top 132 meters, corresponding to about two thousand years, of the long core they had recovered at site two. This made Guliya the longest annually resolved ice core ever recovered outside the polar regions. When Lonnie compared it to the next longest, Quelccaya, which lies almost exactly half the world away, a correspondence hinted at by his first Chinese ice core, Dunde, now clearly emerged: average yearly precipitation at Guliya rose and fell in lockstep with that of Quelccaya over the full 1,500 years of Quelccaya's shorter memory, and both rose and fell with the telltale four-hundred-year cycle that once paced the rise and fall of Peruvian civilizations. This was further, and now probably sufficient, proof of the long-term El Niño–type teleconnection that Lonnie and Ellen had first proposed, based on Dunde's scant four hundred years of layer counting, at about the time he first set foot in the Kunlun Shan.

The definitive report on Guliya appeared in *Science* on June 20, 1997, exactly one day after I received the phone call that sent me to Sajama and my first meeting with Lonnie. Multitasking as usual, he had been living on top of the highest mountain in Bolivia for four days.

AS ON HUASCARÁN, THE GULIYA TEAM HAD BEEN CONVINCED THEY WERE PULLING UP very old ice even as they worked. They had a little experience reading ice layers by then, and they had seen unmistakable signs that the long core reached even farther back than Dunde's, that they were scratching more than just the surface of the Wisconsinan glacial stage. In fact, Lonnie was quite confident they'd reached back past its beginnings: at least two hundred thousand years. He said as much in a press release shortly after his return from China and before they had done any lab analysis at all. Ever the optimist, he also guessed that it would be "the best climate record we've obtained so far." (Come to think of it, I don't think I've heard him *not* say that yet.)

While the Pleistocene epoch, or Great Ice Age, is thought to have begun almost two million years ago, it did not exert its grip tenaciously all through its long life. The northern ice caps advanced and retreated cyclically, with a primary beat of about one hundred thousand years. Thus the Wisconsinan "glacial cycle" began somewhat more than one hundred thousand years ago, ended about twelve thousand years ago with the Younger Dryas, and has three older siblings, named for the Midwestern states on which they left their greatest "impressions": the Illinoian, the Kansan, and the Nebraskan. In Europe, glacials are named for the tributaries of the Danube River, because their gravel terraces, the remnants of glacial moraines, lie in layers under the larger river's bed. There, the Wisconsinan is known as the Würm, after the tributary that flows through Munich. Within the Wisconsinan/Würm and the other four glacial cycles there were smaller advances, called stadials, and retreats, called interstadials.

As they drilled on Guliya, the team saw not only the telltale sign of the transition to the Wisconsinan twelve thousand years ago—the dramatic darkening of the ice they had seen many times before—but also, below that, bands of lighter ice corresponding to all the known interstadials of that glacial stage. About 180 meters down, the alternating layers of clear and dirty ice gave way to a clear zone extending a good distance farther, 128 meters, almost all the way to bedrock. Figuring the layers were extremely thin at those depths, Lonnie guessed conservatively that this clear zone spanned at least the last "interglacial," the Eemian, the warm interval between the Wisconsinan/Würm and the Illinoian/Riss.* That's why he figured the deepest ice was more than two hundred thousand years old.

The Thompsons eventually obtained a firm estimate for the age of the basal layers with a method similar to carbon dating, which uses chlorine-36 as the marker (the same radioactive by-product, by chance, that signaled the nuclear horizons they had used to date the first shallow cores from Guliya and Gregoriev). Chlorine-36 is generated by cosmic rays in like manner to carbon-14, but it has a half-life of 301,000 years rather than 5,730, so it can be used to date far older material. Its limit is about a million years, as opposed to forty-five thousand, and it does not require organic samples. The lowest sample the Thompsons tested, just above bedrock, yielded a chlorine-36 age of 760,000 years. That would make Guliya's basal ice the oldest ever recovered from an ice core, anywhere (although the EPICA core recovered at Antarctica's Dome C has recently tied this record).

This might seem counterintuitive, since the ice cores from Antarctica and Greenland are in the neighborhood of two miles long and Guliya's is only about a thousand feet, but you can't guess the age of a glacier from its thickness alone. Different accumulation rates on the one hand and different flow patterns on the other lead to different thinning rates with depth, and to different relationships between age and depth as a result. Detailed "flow models" are used to determine age/depth profiles; once layer counting runs out, these models join the arsenal of weapons, along with chlorine-36 and carbon dating, correlation with other records (as between Huascarán and SU81-18), and so on, used to piece a chronology together. Generally, a few different methods are applied to a single core to serve as cross-checks for one another.

Take East Antarctica's Vostok core, for example. The Soviets began deep drilling at their Vostok Station in 1980. Drilling problems forced them to give up on a first hole in 1985 and a second in 1990, and in 1989 they began collaborating with the French and the United States on a third. Nine years later, the collaboration announced that they had reached a depth of 3,623 meters (2.25 miles) and were now recovering what appeared to be

*The Holocene, in which we are now living, has also been called an interglacial, although that may turn out to be wishful thinking. Some climatologists think the anthropogenic carbon dioxide buildup has a strong chance of knocking the planet into an entirely new climatological mode in which the rules of the Pleistocene no longer obtain, and a few have already begun calling it the Anthropocene. (See source note, page 399.)

"accretion ice," refrozen to the bottom of the glacier from an underlying lake, estimated seismically to be about the size of Lake Ontario. The ice above the accretion zone produced a record reaching back 420,000 years.

The collaborators stopped drilling at that point, first, to prevent contamination in the event that "Lake Vostok" contained life-forms that had not seen the light of day for a while and might be worth studying (which has proven to be the case); and second, because if they had broken through, the lake water, which sits under very high pressure, might have gushed out of the hole like an oil well, destroying the drilling rig and endangering lives—something like Bruce Koci's early experience on the Ross Ice Shelf.

The lake is the main reason Vostok reaches about half as far back as Guliya, even though it's about ten times as thick; its thickness works against it. All glaciers tend to heat up as they approach the geothermally heated earth, and the pressure of the overburden, which is fantastically high at Vostok, allows water to exist in liquid form at temperatures below its standard, sea-level freezing point. It is also likely that the land below Lake Vostok is tectonically active, which would add more than the usual amount of geothermal heat. As Lonnie explains, "Time is being removed at the bottom. The same thing is happening on Quelccaya; its base is melting, too. One year leaves the bottom for every year added. But Guliya is frozen to bedrock. It's not losing time at its base. It's an advantage that it's thin."

Also consider Greenland: the cores from there are almost as long as Vostok's—both come in just shy of two miles—but they reach back "only" about a quarter as far: 120,000 years.

Nevertheless, the field is so focused on the big projects at the poles that many people who should know better still believe Vostok has a longer memory than Guliya.

THE THOMPSONS ALSO HAD STRATEGIC REASONS FOR WAITING TO UNVEIL GULIYA: THEY knew the record's length would come as a shock to their colleagues, and they wanted to prepare the ground by rolling out Huascarán first. They had actually been taking a conservative tack ever since Dunde, where, you may remember, they had expected only five thousand years and obtained forty thousand.

Lonnie says that for Dunde—which was the first record to contain glacial-stage ice that had ever been retrieved outside the polar regions—"We published forty, but we believe it's a hundred. You have to wade into these things slowly with the scientific community, because you've got to go through review processes and you want to make sure the important concepts get through. To me it was fabulous that we could get such a long record—see the transition to the glacial—outside the polar regions. I'd rather underpropose than oversell. All you have to do is look at the history of science to see what happens if you try to change the system too much.

"Guliya was drilled in 1992, Huascarán in '93. Huascarán was published in '95, Guliya in '97. It had to do with pushing the time line back. On Guliya we were asking people to believe the record is 760,000 years old, the oldest ice in the world—older than Vostok.* Chlorine-36 has a half-life of three hundred thousand years, and the bottom of this core is chlorine-36-dead. In the paper, we said it was more than five hundred thousand years old, but the date on the bottom sample is seven hundred and sixty. Frankly, I think it's probably older. It could be a million years.

"We had to establish some credibility first. The scientific community has been focused on the polar regions, thinking that's where you're going to find the oldest ice on earth. I don't believe it's true because of the physics of ice, but it's a political thing. If you get labeled as being out in left field early in your career, you can't change things; but if you base it on sound scientific fact, you can. We've slowly pushed the bar back by repeatedly demonstrating that we could find glacial ice in the low latitudes.

"Now on Huascarán there were only three meters of glacial ice, and there are legitimate scientific questions that can be asked about what is happening in the lower three meters of any ice cap. You don't want to oversell or overstate, even though you may know . . . even during the field operation. . . . We knew it was a very old record when we were drilling at Guliya. You could see high dust content in cold periods and clear ice during warm periods: cold, warm, cold, warm . . . but seeing it and proving it in the scientific literature are two different things. Even though as a scientist you may have a concept down three or four years before it actually gets published, you have to line everything up."

This was a few months before we were to leave together for his deep-drilling expedition to Kilimanjaro. I suggested that the actual moment of a discovery often occurs years before its truth can be verified.

"Yes. It's intuition," Lonnie replied. "I'm sure I'll have a good feel for the quality of the record by the time we leave Kilimanjaro. You just get a feel for it."

I pointed out that he and the others had been giggling like kids on the top of Sajama, absolutely sure they were hauling in glacial-stage ice.

*He said this long before EPICA.

"We had no data," he laughed, "no data. That's what makes an expedition so exciting. You're doing something for the first time, and you're all part of it."

I SUPPOSE IT IS INTRIGUING THAT GULIYA SET A LONGEVITY RECORD, BUT AS ELLEN pointed out in a press release on the day they published the results (Lonnie being somewhat hard to reach at 21,500 feet on Sajama), "The age of the ice is almost secondary to the amount of detail the core provides."

Even the title of their report, "Tropical Climate Instability: The Last Glacial Cycle from a Qinghai-Tibetan Ice Core," added fuel to the "tropics versus poles" debate that Huascarán had helped launch. Here was more evidence of abrupt change during the glacial at low latitudes (Guliya lies about ten degrees north of the Tropic of Cancer), another blow to CLIMAP's view of the tropics as a passive, boring belt of inertia. Here again the Thompsons found evidence for swift and drastic changes at the transition to the Wisconsinan. As in the Andes, Greenland, and Antarctica, the Kunlun Shan were colder and drier; and now that it was possible to see farther back not only than Huascarán but also than GISP2 and GRIP, they found that the Tibetan Plateau had probably experienced even greater instability than Greenland had during the Wisconsinan glacial stage. The interstadials (glacial retreats) at Guliya are attended by dramatic, abrupt warmings; while in GRIP, GISP2, and Vostok, temperatures don't change all that much. Again, it appeared, the poles did not have a lock on climatological drama.

BUT GISP2'S TIMING SEQUENCE AGAIN PROVED USEFUL. THE THOMPSONS CONSTRUCTED the chronology for the lower 177 meters—below the lowest countable layer—by matching Guliya's oxygen isotope profile with the atmospheric *methane* profile from the trapped air bubbles in GISP2. Surprisingly, while GISP2's isotope profile bears some resemblance to its own methane profile, Guliya's matches the methane profile in almost every detail, with wild temperature swings at the beginnings and ends of the interstadials.

"The thing I find absolutely fascinating about Guliya," says Lonnie, "is those interstadials, where the methane jumps in the polar cores, but there's no change in the isotopes, while there's a tremendous isotopic shift at Guliya. Stage three [an interstadial that peaked about thirty-five thousand years ago] is as warm as the present."

Methane has twenty-three times the greenhouse potency of carbon dioxide. Before the industrial era, it entered the air mainly through plant fermentation in wetlands, so its level in preindustrial times is generally taken as a measure of wetland extent, which would in turn measure overall humidity. In glacial times, the higher latitudes would have been covered mostly in ice, and wetlands and any wet air that produced them would have been found almost exclusively in the tropics. Hence, the Thompsons suggested, tropical humidity may have amplified the interstadial warmings by encouraging the growth of wetlands, which then produced more methane. So the tropics seem to have amplified the interstadial warmings—and the coolings as well, by the reverse process.

But what caused the warmings in the first place?

BEFORE DISMISSING CLIMAP AS AN ALSO-RAN, PLEASE KNOW THAT IT PRODUCED ONE OF the greatest geological discoveries of the twentieth century: a plausible and nearly complete explanation for the ice ages. This puzzle had been vexing scientists ever since 1837, when Louis Agassiz first showed a few scratched alpine rocks to his colleagues at the Swiss Society for Natural Sciences, in Neuchâtel. Astronomical theories had been brought into play from the outset, and the seed that finally grew into a solution was planted in the late nineteenth century by the Scotsman James Croll (whom Arrhenius went out of his way to pooh-pooh near the end of his own incorrect treatise on the subject—which did include, on the other hand, a superb analysis of the carbon dioxide effect). CLIMAP scientist John Imbrie and his daughter Katherine have told the story of the wandering journey from Agassiz's talk to the clinching discovery in their fine book *Ice Ages: Solving the Mystery.*

ALTHOUGH CROLL EXHIBITED A PHILOSOPHICAL TURN OF MIND FROM AN EARLY AGE, HE gave up formal education before the age of thirteen and started his working life as a millwright. As a result of an elbow injury suffered in that line of work, he was disabled by his thirtieth year and began drifting from job to job, eventually finding safe harbor at the Andersonian College and Museum, in Glasgow, where he worked as a janitor by day and pored through the superb library by night. After bringing himself completely up-to-date on the planetary physics of the day, he set to work on a theory of the ice ages, which he based on the notion that the glacial advances and retreats of what we now call the Pleistocene were caused by periodic changes in the fine features of our orbit around the sun. These changes do not affect the total amount of sunlight (insolation) that our planet receives in a year; they affect the average amount received by different latitudes in different seasons.

Croll is the hero of the story, even though he didn't quite get it right. One of his miscues was to suggest that glaciations must have alternated between the northern and southern hemispheres, mimicking the same sort of alternation in insolation. A recent speculation by Wally Broecker notwithstanding, the evidence shows that major glaciations in the Arctic and Antarctic have been more or less synchronous.

The next big step was taken by an extraordinarily persistent Serbian engineer-*cum*-mathematical physicist named Milutin Milankovitch, who provided the crucial insight that insolation in the northern hemisphere should drive the system because two-thirds of the earth's landmass is located there. This synchronized northern and southern glaciations. Milankovitch believed summer insolation levels at sixty-five degrees north latitude, just south of the Arctic Circle, played a crucial role. (This might be seen as a sort of boundary: if the ice caps manage to survive the summer below that latitude, they have a greater chance of advancing year to year, and we will probably experience an ice age.) *Low* summer insolation (less sunlight) at that boundary, Milankovitch believed, would allow the ice to survive the summer and would favor annual growth. Growth in turn would enhance cooling through the same positive feedback behind Snowball Earth: the ice would reflect incoming sunlight back into space. The reason the northern hemisphere dominates is that there is more land in the far north for ice sheets to grow on.

But to me Milankovitch's most astounding accomplishment was his detailed calculation of insolation curves for a few different northern latitudes ranging back 650,000 years. (He did similar calculations for Mars and Venus.) Counting interruptions for the Balkan Wars and a year of imprisonment by the Turks during World War I (during which he had the foresight to take his briefcase and so worked diligently on his calculations the entire time), this heroic effort took thirty years.

He also describes once experiencing a revelation similar to Lonnie's on the Garganta Col: He happened to be stuck on a conceptual aspect of his theory in 1912 when the First Balkan War broke out and he was assigned to the Danube division of the Serbian army. Early one morning, his division crossed into Turkey to lay siege to an encampment on Starak Mountain, and just as they captured the summit, Milankovitch had the flash of insight that solved his problem. He later wrote that he had "conquered a mountaintop" in his private inner struggle.

Twenty-nine years later, after publishing his full theory in a book titled *Astronomical Methods for Investigating Earth's Historical Climate,* he displayed the type of naïve arrogance that explains why physicists often incur the resentment of scientists in other disciplines. The causes of the ice ages, he wrote, ". . . lie far beyond the vision of the descriptive natural sciences. It is therefore the task of the exact natural sciences to outline this scheme, by means of its laws ruling the universe and by its developed mathematical tools. It is left, however, to the descriptive natural sciences to establish an agreement between this scheme and geological experience."

In fact, theory and experiment almost always proceed hand in hand, and that's how it would work out here. Those in the "descriptive natural sciences" may take comfort in the knowledge that, while Milankovitch's basic idea was close to correct, his exact predictions were undoubtedly wrong. He linked only two of the three major orbital effects to the ice ages, and it turns out that the one he dismissed—concluding quite rationally based on his understanding of the "exact natural sciences" that its effect should be small—is actually the most important. Scientists still don't know why it's so important, which just goes to show how opaque nature, in its muteness, can be.

JOHANNES KEPLER POINTED OUT IN THE SEVENTEENTH CENTURY THAT THE EARTH DOES not revolve in a circle around the sun, but in an ellipse with the sun at one focus.* An ellipse is a sort of flattened circle, and its eccentricity, speaking very loosely, is how flattened it is. The plane in which we orbit the sun is called the ecliptic. The moon orbits the Earth in almost the same plane (which is why we have eclipses), as do most of the other planets.

If the sun and Earth were the only two bodies in the solar system, our elliptical orbit would maintain the same shape year in/year out. Actually, however, the gravitational pull of other planets—mainly Jupiter and Saturn because they're so heavy, and Venus because it comes so close—jostles us very gently, causing the eccentricity of our orbit to vary in a fairly regular way with a period of roughly one hundred thousand years (a familiar number). At its minimum eccentricity our orbit is almost circular, which means, of course, that Earth finds itself at essentially the same distance from the sun year-round. We're not far from a minimum in eccentricity right now: at perihelion, the point on the orbit at which we are closest to the sun, we receive 7 percent more sunlight than we do at aphelion, when we're farthest away. At a maximum in eccentricity that occurred about two hundred thousand years ago the difference was close to 30 percent.

At the risk of inducing flashbacks to kindergarten, I shall remind you that day and night occur because we spin on our axis like a top. Our spin axis is tilted with respect to the ecliptic—about 23.5 degrees off the vertical, on average—and this, not the distance to the sun, leads to the seasons. Summer solstice in the northern hemisphere occurs when the axis tilts

*A circle has one focus, at its center. As it is flattened into an ellipse—and our orbit is flattened ever so slightly—two foci emerge from the center and move out symmetrically along the ellipse's longer axis.

directly toward the sun, and winter solstice when it tilts directly away. These two moments lie at opposite points on our orbit: a line through them would also go through the sun. The vernal and autumnal equinoxes, about March 21 and September 23, take place when we cross the line running perpendicular to the line between the solstice points—at which times our spin axis tilts in planes perpendicular to the sun. At the equinoxes most of the Earth experiences twelve hours of daylight and twelve hours of darkness, while the sun rises for six months of day at one pole and sets for six months of night at the other.

Besides giving rise to the seasons, the tilt also causes our axis to "precess" in a manner entirely analogous to the precession of a top spinning on a floor. In the top's case, once it begins to tilt away from the vertical, the force of gravity, which wants to pull it toward the floor, results in a torque that causes the center of its upper surface to describe a circle in space, centered on an imaginary line pointing straight up from the point on which it spins on the table.

In the Earth's case, think of the ecliptic as a sort of penetrable floor. Owing to our equatorial bulge, gravitational forces from the sun, moon, and other planets would like us to stand up straight: they want our axis to stand perpendicular to the ecliptic. This results in a torque that causes the axis to precess like a top's, except that it takes about twenty-six thousand years to complete the circle. Roughly ten thousand years from now, the North Star—one of the many billions of so-called fixed stars—will be just another twinkling in the sky, at a latitude of about forty-five degrees.

Perhaps the hardest thing to visualize here is the rotating of the orbit itself. If you were to look down on the ecliptic from the direction of the North Star and trace out our elliptical path from century to century, the ellipse would appear to be rotating counterclockwise. Meanwhile, the axis is precessing in the opposite direction: clockwise. These two rotations together lead to what is called the precession of the equinoxes: over the course of centuries, the constellation—or sign of the zodiac—in which the sun rises at the spring equinox will shift. As some of us learned in the sixties, it is presently leaving Pisces and entering Aquarius. Although the precession of the orbit and the rotation of the orbital ellipse proceed in opposite directions, the equinoxes precess more rapidly than the axis: the signs of the zodiac repeat every twenty-two thousand years.

This has implications, incidentally, for the standard naïve understanding of astrology. When that art was invented about two thousand years ago, the January sun was rising in the constellation Capricorn, so a person with a January birthday was said to have been "born under" that sign. But that was about one-twelfth of the precession cycle, or one sign of the zodiac, ago: for the past few hundred years, people with January birthdays have actually been born under Sagittarius.

Moving swiftly away from astrology now, consider the effect of all this on seasonal insolation. During epochs of low eccentricity, when our orbit is nearly circular, it makes no difference where on the orbit the solstices and equinoxes take place, because we're equidistant from the sun year-round. During epochs of high eccentricity, however, the Earth will receive 30 percent more sunlight at the summer solstice—primarily in the all-important northern hemisphere—if the solstice happens to coincide with perihelion as opposed to aphelion.

The last important orbital variable is obliquity, the actual amount that the axis tilts. It turns out that the tilt that leads to axial precession does not hold constant at 23.5 degrees. It oscillates roughly one and a half degrees to either side of that average value, with a period of about forty-one thousand years—an effect unrelated to precession.

If there were no obliquity—in other words, if the axis were vertical—the poles would experience perpetual twilight, days and nights everywhere else would be of equal length, and we would have no seasons. Thus, when the obliquity reaches its low value of twenty-one degrees, the seasons are less extreme than average: winters are warmer and summers cooler; and when obliquity is highest, seasons are more extreme.

It does not seem to require too much of a leap of faith to accept the notion that these three variables, affecting the seasons as they do, might have played a role in the growth and recession of the polar ice sheets.

MILANKOVITCH DIED IN 1958, ONLY A FEW YEARS BEFORE GUY STEWART CALLENDAR, and his position in the development of this theory is roughly equivalent to Callendar's in carbon dioxide theory. In Milankovitch's own time, the geological record was not clear enough to prove or disprove his ideas, and in the end his detailed predictions proved to be slightly off the mark anyway. The stock in what has become known as Milankovitch theory rose and fell over the next few decades as new records were recovered, new analytical methods were invented, and the accepted dates for the various advances and retreats of the ice caps were shuffled about. There was a persistent problem with timing, which is often the case in climatology.

The ubiquitous Wally Broecker entered the picture even before Milankovitch's death. In the early 1950s, as a graduate student in geochemistry at Columbia, Wally helped develop radiocarbon methods for dating the layers in deep-sea cores—of which Lamont, one of Columbia's many research centers, had the largest collection in the world. Wally's thesis work, in fact, set a reasonably accurate date for the very end of the Pleistocene (now identified as the end of the Younger Dryas): eleven thousand years ago. Attempts to extend such nascent radiocarbon techniques farther back in time, however, tended mainly to cast doubt on Milankovitch for the next two decades.

In the mid-fifties, when radioactive thorium dating—a technique that works very well on coral reefs and has a range of 150,000 years—came online, Wally applied it to the examination of the world's fossil reefs. Since coral grows only underwater, the heights of undisturbed fossil reefs mark ancient high points in sea level. The basic idea is that glacial melting would have caused sea levels to rise, so the dates of high points should correspond to interglacials and interstadials. In 1965, Wally showed that the three high points he had found thus far—today, 80,000 years ago, and 120,000 years ago—corresponded to three of the four peaks in Milankovitch's summer insolation curve at sixty-five degrees north. Insolation at such high latitudes, Milankovitch's preferred driver, is affected mainly by the tilt of the axis, obliquity, which has a forty-thousand-year cycle; so this seemed promising, except that Wally was missing a high point that should have occurred about forty thousand years ago.

The plot thickened three years later, when he and Brown University's Robley Matthews found that the high points of a series of terraced reefs in Barbados were spaced by between twenty-three and twenty-five thousand years. This would point to the precession cycle, which dominates insolation at low latitudes. So, the emergent difficulty for simple Milankovitch theory was that you could match either a forty-thousand-year beat or a twenty-two-thousand-year beat simply by choosing an insolation curve from a high or low latitude. Nevertheless, in light of the accumulating evidence, the basic premise seemed compelling.

To add to the confusion, in the same year that Broecker and Matthews brought precession into play, the Czech scientist George Kukla showed that the alternating soil layers in his native land—which the ice caps never reached, but whose climate fluctuated wildly with their presence or absence—bore the signature of a one-hundred-thousand-year beat. Two years later, in 1970, Broecker and his Lamont colleague Jan van Donk found the same cadence in a deep-sea core from the Caribbean.

By focusing on specific insolation curves, which may be dominated by either obliquity or precession depending on latitude, Milankovitch had unwittingly ignored eccentricity, which happens to have the one-hundred-thousand-year beat that might explain these last two pieces of evidence. Looking deeper into the theory, Kukla and a colleague from Brown University named Kenneth Mesolella soon realized that the dips and swells in eccentricity line up well with the Czechoslovakian soil changes and came up with what a physicist might call a "hand-waving" argument* to explain how eccentricity's small insolation effect might indirectly have caused dramatic changes at the poles.

All in all, there was a little too much play in both the theory and the data. You could mix and match in order to make the two agree, and there was no way of telling whether an agreement was pure coincidence or not. Part of the problem was that none of the relevant records examined by that point had reached far enough back to cover a significant number of one-hundred-thousand-year cycles.

The solution would come from the bottom of the sea.

BROECKER AND VAN DONK HAD USED YET ANOTHER OXYGEN ISOTOPE RATIO TO TEASE out the one-hundred-thousand-year cycle in their Caribbean seabed core—a measure that had been sorted out by a young British geophysicist named Nicholas Shackleton. In a story similar to Charles David Keeling's, in which extreme attention to detail had resulted in a big payoff, Shackleton spent ten years modifying standard mass-spectrometry techniques to make them capable of producing accurate measurements on the tiny samples required to achieve sufficient resolution in seabed cores. (The reason the samples need to be small is that seabed cores are much shorter than ice cores: one meter of sediment might correspond to as many as three hundred thousand years.) And after developing his novel method for measuring ^{18}O and ^{16}O in the tiny foraminifera shells in the sediments, Shackleton made the scientific discovery that this isotope ratio, amazingly enough, measures the size of the rather larger polar ice caps—a discovery that has implications for ice cores as well.

Consider the bank of water vapor that eventually ends up as snow. The first step in its journey is to evaporate from a body of water, usually the ocean. Since the heavier of the two isotopes, ^{18}O, has a stronger preference for the liquid state, evaporation tends to enrich the isotope mix in ocean water, thus leaving it "heavier." Conversely, the ice caps, which grow over centuries through snowfall, act as repositories for the "light" water that originally evaporated. The bigger the caps, the heavier the ocean water that remains. Forams take on the enriched ratio of the ocean water in their

*A hand-waving argument is one that is based not on solid data or calculations but on an extensive and often poetic description, generally accompanied by much gesticulation. Although this is a venerable tradition in every branch of science—except, perhaps, mathematics itself—physicists in particular tend to frown upon it. Milankovitch was hand-waving when he insisted on the importance of summer insolation at sixty-five degrees north. Wally Broecker's great conveyor is another excellent example; and since I suspect you'd rather not plow through mathematical formulas here, I'm hand-waving my way through this entire book.

shells while they're alive and leave a chronological record of it by dying and dropping to the ocean floor.

Dansgaard's proxy for air temperature is based on the isotope ratio in the evaporated *water vapor* (H_2O). However, the enriched or depleted isotope ratio in seawater enters the *oxygen* of the atmosphere (O_2), the stuff we breathe, by a second route: plant photosynthesis (which happens to remove carbon dioxide at the same time). Todd Sowers, who gave me my first lesson in climatology on Sajama, was among those who showed that it takes about a thousand years for the isotope ratio of seawater to mix into atmospheric oxygen. So, as the ice sheets grow, the isotope ratio in atmospheric oxygen rises, and as they shrink it drops—and its changes over time will be recorded by the *air bubbles* in ice cores.

In the same way that Lonnie used the methane from GISP2 to construct a timescale for Guliya, others have curve-matched with the oxygen isotope ratio in the air bubbles in the Vostok ice core, which has become the standard for this. To build a timescale for a core from another part of the world, they plot the isotope ratio in that core's air bubbles against depth (which is the same as time), then "stretch" and "squeeze" the time axis until the peaks and valleys for the unknown core match up with the peaks and valleys from Vostok, which has a solid chronology.

Another implication would seem to be that the ratios in the bubbles and in the little shells should track one another. But there is a fly in the ointment: they don't, quite.

Remember that in SU81-18 (the Portuguese sediment core that Lonnie used to time Huascarán) the ratio in the forams was taken as a measure of sea surface temperature. In fact, the ratio in forams changes with *both* surface temperature and ice volume. If you're looking at just the last twelve thousand years, during which the ice caps have been small, the ratio will reflect simply temperature, but temperature and ice sheet volume will be tangled together before that time. It is possible to disentangle them, however, if you have just one good air bubble record from an ice core anywhere on earth, as with Vostok, because the ratio in the air bubbles changes only with ice volume.

CLIMAP WAS THE BRAINCHILD OF YET ANOTHER FINE LAMONT SCIENTIST NAMED JAMES D. Hays. Its original goal was to reconstruct climate in the North Atlantic and North Pacific for the past seven hundred thousand years. Two years later its mandate was expanded to include the mapping of earth's surface temperatures at the end of the Pleistocene (an effort that met with qualified success, as we've seen) and the characterization of Pleistocene climate changes: in other words the Wisconsinan glacial stage and its three

Midwestern siblings. The first problem was to sort out a chronology that would work for records as old as seven hundred thousand years, and this they accomplished, but we won't go into that here. Suffice it to say that about seven hundred thousand years ago the Earth's magnetic field was reversed—a compass would have pointed south—and the signature of much of the change in between can be found all over the world.

One of the reasons the Lamont-Doherty Earth Observatory has evolved into one of the preeminent geological research centers in the world is that its philosophy from the outset has been to conduct just the sort of scientific "fishing expedition"—almost literally—that the granting agencies frown upon and of which Roger Revelle, his colleague Harmon Craig, and indeed Lonnie Thompson have always been so fond. For decades Lamont's research vessels have been directed to pull up a core from the ocean floor once a day, no matter where they are.

As Hays and his colleagues, who included Shackleton and Brown University's John Imbrie, began to train their sights on the ice ages, it was Hays who realized that it would take an unusual core to test for the presence or absence of all three of Milankovitch's orbital effects: the core would need high resolution in order to provide a test of the "short" twenty-two-thousand-year precession cycle, but it would need to reach a long way back—four hundred thousand years at least—to test for the long, one-hundred-thousand-year, eccentricity cycle.

To yield high resolution, a seabed core must come from a place where relatively large amounts of sediment drift to the ocean floor in a year, the odd challenge there being small animals who live on the floor and burrow into the surface, blurring out short-term variations. It turns out that the highest sedimentation rates occur in the southern oceans, so Hays began combing through the Lamont collection for southern cores. He found only one that met his specifications, retrieved at about forty-five degrees south, in the Indian Ocean. It turned out to have the required resolution, but it went back only three hundred thousand years. Having exhausted Lamont's treasures, he shifted to a collection at Florida State University in Tallahassee in which the cores came mainly from the Antarctic, and soon found one from a location close to the first. This one reached just far enough back, 450,000 years, but, unfortunately, its top section had been lost during drilling. That was okay, though, because they had the top section from the first core. By matching the contemporaneous features in both, they managed to splice them together and construct a full 450,000-year record.

Shackleton brought his isotope method to the table. Imbrie, among other things, brought a fearsome capability in mathematics, which proved to be the key to linking orbital motion to the geological record without resorting to hand-waving. He simply ignored the mire of seasonal and geographical detail in which Milankovitch had stuck fast and searched only for similarities in what we might call "shape" between the three orbital oscillations and the spliced seabed core's random-looking isotope curve. The technical term is *spectral analysis*. It's like separating a piano chord into its individual notes, and it so happens that the basic mathematics was first elucidated by Fourier, the Frenchman who also played an early role in greenhouse theory. In essence, then, Imbrie and his colleagues treated orbital changes as an input and the climate record as an output, "without," as they wrote, "identifying or evaluating the mechanisms through which climate is modified by changes in the global pattern of incoming radiation." Put another way, they simply dropped Milankovitch's fixation on sixty-five degrees north.

When Imbrie broke the chord of climate into its individual notes he found that all three orbital variables came into play—and, ironically, that the one that had the smallest effect on insolation had the greatest effect on climate. The major glacial stages—Wisconsinan, Illinoian, Kansan, and Nebraskan—were born and died with the one-hundred-thousand-year beat of eccentricity; to confirm the causal connection, peaks in glaciation always coincided with periods of low eccentricity. Half the variation in Pleistocene climate could be attributed to this, a quarter to the forty-thousand-year obliquity cycle, and a tenth to the twenty-two-thousand-year precession cycle. "It is concluded," wrote Imbrie, Shackleton, and Hays, "that changes in the earth's orbital geometry are the fundamental causes of [Pleistocene] ice ages."

As usual, however, the answer to one question gave rise to another. Why was eccentricity so powerful? Imbrie teamed up with his son (science seems to be a family affair in that household) to conjure up an explanation involving good old sixty-five degrees north and a model of ice cap expansion and retreat. Despite its overlay of mathematics, however, it still seemed like hand-waving.

Finally, in the year 2000, twenty-four years after the CLIMAP breakthrough, Shackleton undertook a combined analysis of a seabed core from the eastern equatorial Pacific and the Vostok ice core and showed that deep-sea temperature, orbital eccentricity, and atmospheric carbon dioxide (which he obtained from the Vostok air bubbles) marched in lockstep all through the Pleistocene, while ice volume lagged behind. Thus the Imbries seem to have been mistaken: the ice sheets grew or shrank *in response* to temperature; they could not have been the amplifier; it must have been our old friend carbon dioxide. As eccentricity rose, an as yet undetermined feedback process caused the gas to stream into the atmosphere; as eccentricity fell, it somehow seeped out; and the resulting greenhouse changes led to swift changes in temperature. But the full explanation still eludes us, for we're not sure what that feedback process was.

Thus science marches on, and we are afforded yet another example of the dragon-like power of the greenhouse.

BESIDES REPRESENTING AN EXTRAORDINARY HUMAN EFFORT, SPANNING TWO DECADES, in one of the most inhospitable places on earth, the drilling at Vostok has produced one of the richest scientific treasure troves of all time. Previously,* when I mentioned the three exquisite papers that resulted from the analysis of the second, 160,000-year Vostok core, I pointed out that the analysis revealed tracking between carbon dioxide and temperature and that the magnitude of the carbon dioxide swings could account for the magnitude of the temperature swings. But I failed to mention the main point: that the *timing* of the temperature and carbon dioxide swings lines up very well with the orbital explanation for the ice ages.

The third, 420,000-year Vostok core, recovered in 1998, extends that confirmation back through the last four glacial cycles: the Wisconsinan, Illinoian, and so on. It also shows, ominously, that both methane and carbon dioxide, the main amplifiers of temperature in the geological record, are now at their highest levels in that span and the carbon dioxide curve is particularly alarming: since the start of the fossil fuel era, it has spiked to a level nearly twice as high as any it has reached in almost half a million years.

NOW, THE RELEVANCE OF ALL THIS TO LONNIE'S WORK ON GULIYA IS THAT THE STRONG interstadial warmings there, which coincide with the global methane spikes recorded by GISP2, are spaced by twenty-three thousand years: roughly the cadence of the precession cycle. And it makes sense that this relatively low latitude site would react strongly to changes in precession, for precession controls insolation at low latitudes, as Milankovitch knew. So, seasonal changes in tropical insolation, brought about by precession, seem to have started the interstadials by heating tropical air so that it took on more moisture; increased moisture then led to the growth of wetlands, which caused the methane spikes; and the coordinated greenhouse effect of increased moisture and methane then heated the tropics and the subtropics. Thus the tropics, rather than some obscure current in the North Atlantic, seem to have driven interstadial change—a type of change, moreover, that was hardly felt at the poles, as evidenced by the Greenland and Vostok isotope records.

*See page 160.

TEMPERATURE TRACKS CARBON DIOXIDE AT VOSTOK, ANTARCTICA

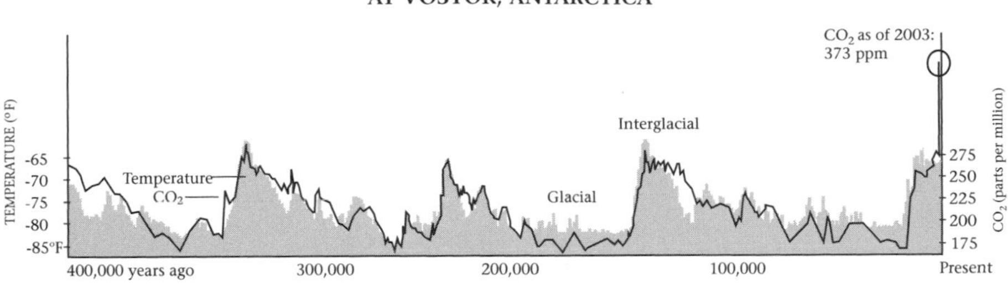

Temperatures inferred from isotope ratios in the Vostok ice core, along with carbon dioxide levels measured in the trapped air bubbles in the same core, show that Antarctic temperatures have tracked global CO_2 levels for the past four 100,000-year glacial cycles. The twentieth century spike in CO_2, measured by Charles David Keeling, shows that the level is now nearly twice as high as it has ever been in 400,000 years.

Although this idea merits only one sentence in the Thompsons' Guliya paper, it represents the first glimmering of Lonnie's next big idea. After he experienced another revelation late on a December night in 1999, we began referring to it as his "asynchrony theory," because it may explain why the advances and retreats of tropical and subtropical glaciers seem to be uncoupled or "asynchronous" with advances and retreats at the poles.

Lonnie also points out that the tiny changes in insolation that initiated the drastic temperature changes attending the interstadials at Guliya are entirely equivalent to—and, in fact, weaker than—the present, man-made increase in greenhouse forcing, and that this bodes ill. Asked if the tropics amplify small climatic signals, Lonnie replies, "I think that's what we're seeing. We saw it in the natural system; I think we'll see it in the human-driven system, too."

In other words, the changes he is seeing in high tropical ice are a harbinger of massive global warming to come.

·17·

SOLVING THE MERCER PROBLEM

Lonnie had not stopped exploring, of course. As he and Ellen analyzed, pondered, and revealed results from the ice he had mined on Guliya and Huascarán, he continued to search for new drilling sites in the Andes, Tibet, and elsewhere—while minor, largely political developments on the global warming front took place in the background.

In the spring of 1994, as it happened, he and Keith Henderson had been home just three days from their first reconnaissance to Franz Josef Land when a call came in from Al Gore's office inviting Lonnie to a "White House Science Breakfast" to discuss the implications of a new report in *Science*.

A highly regarded Dutch glaciologist named Johannes Oerlemans had reviewed historical records for forty-eight glaciers on five continents and concluded that "the retreat of glaciers during the last 100 years appears to be coherent over the globe." This wasn't exactly a surprise, but Oerlemans had also developed a physical model of the glaciers' temperature sensitivity, with which he had estimated that the earth was warming at the rate of about 1.2°F per century. This fine work has been confirmed by subsequent studies: Oerlemans's number missed the IPCC's 2001 estimate of global warming in the twentieth century by only one-tenth of a degree.

This was right up Lonnie's alley, of course, and he happened to have some fresh news from the high Arctic to serve for breakfast: one of Franz Josef Land's four glaciers was also retreating.

The heads of eight federal agencies, Secretary of the Interior Bruce Babbit, the president's science advisor John Gibbon, a few other scientists, and, of course, the vice president were in attendance. Lonnie likes to tell a story from that day to illustrate Mr. Gore's genuine understanding of climate science:

"I remember him stopping me early in my presentation to say, 'Lonnie wait a minute. Let me explain to the others what oxygen isotopes are.' He then proceeded to tell them about ^{16}O and ^{18}O. I was impressed that the vice president of the United States knew this, and I still tell the tale to my environmental geology class to show them why it is important to learn about more than just our discipline."

But in spite of Gore's efforts—not to mention the fact that the greenhouse effect, indeed, the greenhouse effect of carbon dioxide alone, was probably the single most important theme in nearly every tale of climate change, past and present, that scientists were telling at the time—the message was not reaching the public. Jim Hansen had succeeded in placing it on the table during the hot summer of 1988, but other forces had quickly swept it away.

According to an annual report from the early 1990s by the Western Fuels Association, a large coal-mining and electric-power-generating cooperative with facilities in the Midwest and mountain states, "When [the climate change] controversy first erupted at the peak of summer in 1988, Western Fuels Association decided it was important to take a stand. . . . [S]cientists were found who are skeptical about much of what seemed generally accepted about the potential for climate change. Among them were [Pat] Michaels [of the University of Virginia], Robert Balling of Arizona State University, and S. Fred Singer of the University of Virginia. . . . Western Fuels approached Pat Michaels about writing a quarterly publication designed to provide its readers with critical insight concerning the global climatic change and greenhouse effect controversy. . . . Western Fuels agreed to finance publication and distribution of *World Climate Review* magazine [with Michaels as editor]."

A brief review reveals that the publication records of these three gentlemen are weighted much more heavily toward the polemical side of the global warming issue than toward serious research in climatology.

Western Fuels also spent $250,000 to produce a video entitled *The Greening of Planet Earth,* which claimed that global warming would be a *good* thing because it would increase agricultural output—a dubious claim at best. According to author Ross Gelbspan, "Insiders at the [first] Bush White House said it was Chief of Staff John Sununu's favorite movie—he showed it that often."

Gelbspan has done a fine job documenting the lengths to which the fossil fuel industry has gone to undermine the sound science of the greenhouse; that subject will not be covered here. Suffice it to say that it is not easy to restrain an industry with gross annual revenues on the order of two trillion dollars, operating facilities virtually everywhere on the planet, and a unified economic agenda that transcends all national boundaries and interests.

In 1988, in the United States primarily, coal companies for the most part, oil companies as well, initiated an enormous public relations campaign to "reposition global warming as theory rather than fact"—the stated goal, in fact, of a campaign launched by the Information Council on the Environment (ICE), the creation of a second consortium of coal and utility companies. It was a tactic not unlike that of creationists vis-à-vis evolution or tobacco companies vis-à-vis cancer.

Gelbspan observes that an ICE campaign launched in 1991 was,

> . . . according to strategy documents that were later exposed in the press, . . . aimed specifically at "older, less educated men" and "young lower-income women." The geographic targets of the campaign included areas where electricity came from coal and districts whose congressmen served on the House Energy Committee.
>
> The effectiveness of the campaign can be seen in the results of two *Newsweek* polls, conducted in 1991 and 1996. In 1991, 35 percent of the people polled said they believed that global warming was a serious problem. By 1996 the number had dropped to 22 percent.

For almost twenty years now, a small, handpicked cadre of so-called experts has been trotted out for interviews on television, radio, and in the print media to present "the opposing view." There are about ten in all, and their names crop up everywhere. They hold advisory positions or sit on the boards of what are essentially advocacy groups posing as scientific think tanks, many of which are incestuously interrelated and have exceedingly fuzzy origins and sponsorship. They publish newsletters and position papers, respond predictably to any relevant piece of science or opinion as it appears in *Science, Nature,* or the *New York Times,* weigh in equally predictably on policy decisions, hold "scientific" conferences, and so on.

Dr. Michaels continues to enjoy a close relationship with Western Fuels. His *World Climate Review* has now morphed into the *World Climate Report,* published by the Greening Earth Society (GES), which was founded by Western Fuels on Earth Day 1998. In fact, according to the Union of Concerned Scientists, "GES and Western Fuels are essentially the same organization. Both used to be located at the same office suite in Arlington, Virginia"—although one could not possibly tell this by looking at the GES Web site or its publications. And with little in the way of true research to occupy his time, Dr. Michaels has made a virtual career of appearing at congressional hearings.

This campaign has succeeded in providing equal footing for a view that was completely at odds with that of the majority of working climatologists even when it began, and with the overwhelming majority since about the mid-1990s.

The poisonous seed of environmental partisanship that had been planted by Ronald Reagan in the 1980s only grew and proliferated during the Clinton-Gore years. Virtually no progress was made on a domestic greenhouse policy, although Clinton and Gore did manage to set a dialogue in place with the international community and build bridges that might be used to set policy if a future administration and Congress are ever so inclined.

Some progress was made by the international community, however. In 1996, the IPCC issued its second assessment. In the first, they had merely stated that climate had warmed in the twentieth century. Now, pointing especially to the close match of the observed *patterns* of temperature change—both geographical and over time—with the predictions of global circulation models based on the measured carbon dioxide buildup, the IPCC concluded that the warming was "unlikely to be entirely natural in origin."

The panel had found, essentially, that the human influence could be detected but was still too small to be quantified. This led to the report's most memorable line, a clear product of the "science by democratic consensus," by which the contrarians had hoped to sabotage the effort from the outset: "Nevertheless, the balance of evidence suggests that there is a discernible human influence on global climate."

Thus the most peer-reviewed scientific report in history backed its way into a recognition that there might be a problem. I believe it is fair to say that serious scientific debate about the existence and potential danger of human-induced global warming died with that statement. Realizing the gravity of the moment, the contrarians rose to new heights of invective in their contention that the process was politically skewed. (And most would concede that point; it was skewed in their favor.)

Given the complexity of the problem, science had come up with a remarkably unequivocal answer in a remarkably short time, only seven years—the basic reason being that the greenhouse effect is quite powerful. As the Vostok ice cores and even the orbital explanation for the ice ages had shown by then, the greenhouse effect of carbon dioxide has been the basic amplifier for nearly all the changes in temperature that have ever been studied.

The political divisiveness of this time actually had a direct effect on Lonnie: the infamous budget standoff between Bill Clinton and Newt Gingrich in early 1996 (ultimately won by Clinton) added so much uncertainty to the budgeting process at NSF, which has the character of a roller coaster even in normal years, that Lonnie's grants came through too late for an expedition that year. Thus the annus mirabilis of the year I met him, 1997, was not something he had planned. The grants for Franz Josef Land, Sajama, and Dasuopu came through all together near the beginning of that year, and he figured he ought to spend the money right away because inflation was rampant in Russia, Bolivia, and China, and according to the rules of grantsmanship, he might lose the money if he didn't spend it by year's end.

This was an astoundingly productive time. The results from so many places were beginning to add up and extend Lonnie's intuition; his asynchrony insight was just beginning to percolate; and, as I look through the notes I scrawled in the snow cave on top of Sajama, I find that he had a clear enough head in that thin air not only to rattle through the highlights of his career, tossing out accurate references as to where I might dig deeper as he went along, but also to make a bold prediction about what he was going to find in the ice he was fishing up as we spoke.

He was quite sure that they were digging back into the Wisconsinan glacial stage, of course. More impressively, however, he predicted that he would find exactly the opposite phenomenon in Sajama's deepest ice than either he or anyone else had found in the glacial stage ice in any previous ice core. There in the snow cave, Lonnie told me he believed Sajama would show that Bolivia's climate was wetter during glacial times than it is now; Sajama's Wisconsinan ice would be clear, not dark and dusty. And he turned out to be right.

According to what was now becoming the inevitable paper in *Science*—which the Thompsons pushed out rather swiftly, at the end of 1998—Sajama's ice was not only eight times less dusty during the glacial than it is now, there was more of it as well: the layers are thicker—and the climate was colder, of course. The isotope ratio drops about as much as it does in Greenland and Antarctica and at Huascarán. Furthermore, the fact that it was wetter at Sajama and drier at Huascarán—while the ratio dipped equally in both places—argues against the so-called amount effect. Hard to explain those contradictory bits of evidence with anything except a temperature change.

ONE WAY TO GET A HANDLE ON THE INCREASED GLACIAL PRECIPITATION IN BOLIVIA IS to invoke the north–south El Niño seesaw: when Huascarán is dry, Sajama should be wet, and vice versa, because Sajama sits on the southern end of the seesaw, not far south of Quelccaya. This would also lend support to the notion of the four-hundred-year metronome that once paced pre-Incan civilization. But it isn't really an explanation.

For a few years after the expedition Lonnie toyed around with a certain hand-waving argument involving the so-called Hadley Cell; there is little doubt that this prominent feature of the planet's air circulation has a lot to do with it, but he has now admitted that he and his colleagues don't really understand how.

The Hadley Circulation resembles El Niño's Walker Circulation, except that it transports moisture north and south, from the tropics to the midlatitudes, rather than east and west along the equator. If you were to look at the Earth from, say, the moon on any given day, you would generally see a band of clouds circling the planet, parallel to the equator, at a tropical latitude determined by the season. The band is narrow over the ocean and spreads out over land, and it lies at the latitude directly under the sun at high noon on that particular day. At the summer solstice, the northern edge of the band reaches its northernmost latitude, the Tropic of Cancer; at the equinoxes it crosses the equator; and at the winter solstice its southern edge reaches the Tropic of Capricorn. The clouds are formed mainly by water evaporating from the oceans due to heating by the sun.

This cloud band marks the rising arm of the Hadley Cell and is similar to the rising arm of the Walker Circulation in the west, over Indonesia. It is also known as the Intertropical Convergence Zone (ITCZ). Where there are clouds, obviously, there is a higher chance for rain. Thus the passages north and south of the ITCZ over tropical regions is what causes their rainy seasons and monsoons.

Now, the spinning of the Earth exerts a Coriolis (pseudo-)force on this rising arm of air: if the air is north of the equator it will circulate north; if it is south it will circulate south. In parallel with the Walker Circulation, these currents drop most of their moisture in the form of rain as they rise and dry out significantly by the time they reach the troposphere. They descend back to Earth somewhere near thirty degrees north and south, thus carrying aridity to the midlatitudes in the same way that the Walker Circulation carries dry air east to coastal Peru.

El Niño ties into all this by delivering a tremendous amount of moisture to tropical South America and strengthening the Hadley Cell there—which results, ironically, in drying out the descending arm at Sajama and Quelccaya.

The second grand effect of the Hadley Cell in the present world, in addition to producing the monsoon, has been to create wide bands of desert in the midlatitudes, both north and south of the equator. Sajama stands at the northern edge of a large desert on the Altiplano. A few shallow salt lakes and the largest salt flat in the world, the Salar de Uyuni, lie to the south. The mirror image of this in the northern hemisphere would be the Great Basin, which covers portions of western Utah, eastern Nevada, and southern Idaho. And just as Sajama was wet at the Last Glacial Maximum, the Great Basin was filled with twenty-thousand-square-mile Lake Bonneville, as deep as a thousand feet in places, between roughly fourteen thousand and thirty thousand years ago. That lake has now receded to leave the Bonneville Salt Flats and many shallow salt lakes, most prominently the Great Salt Lake in Utah. A recent study has shown, similarly, that the Salar de Uyuni was also full of water at various times during the glacial and that the nearby lakes were larger as well.

BY THE TIME LONNIE WAS DRILLING ON SAJAMA, WALLY BROECKER'S SPECULATIONS HAD shifted from the past influence of his great ocean conveyor belt to its implications for the future. In the monthly newsletter of the Geological Society of America, he had recently asked, "Will Our Ride into the Greenhouse Future Be a Smooth One?"; and within a few months—days before the meeting in Kyoto that would result in the first international agreement to limit greenhouse emissions—his beautifully named "Thermohaline Circulation, the Achilles' Heel of Our Climate System: Will Man-made CO_2 Upset the Current Balance?" would appear in *Science*. He would simultaneously issue a vivid press release for consumption at Kyoto.

Both papers were essentially editorials. They presented no new information or data, only Wally's latest ruminations. He tied the tumultuous events at the end of the Wisconsinan to the toggling on and off of the conveyor beyond a shadow of a doubt and in the "Achilles' Heel" paper openly implied that global warming stood the paradoxical chance of inducing a major *cold* spell. Greenhouse-induced melting of far-northern glaciers and the increased transport of tropical water vapor to high latitudes had the potential, he argued, to flood his North Atlantic "switches" with dangerous amounts of freshwater, shut the conveyor down, and trigger as drastic a cooling as the Younger Dryas. If so, in his own words,

> Iceland would become one large ice cap. Ireland's climate would be transformed to that of Spitzbergen [at nearly eighty degrees north latitude]. Winters in Scandinavia would become so cold that tundra would replace its forests. The Baltic Sea would be permanently ice covered, as would much of the ocean between Greenland and Scandinavia. Further, the impacts of such a mode change would not be limited to the northern Atlantic basin; rather, they

> would extend to all parts of the globe. Rainfall patterns would dramatically shift. Temperatures would fall. The atmosphere would become dustier. . . .

He also suggested that the warming might lead to the same sort of highly disruptive "flickering" that had occurred at the Dansgaard-Oeschger transitions of the Wisconsinan, when climate whipsawed back and forth between warm and cold states lasting as few as two years. "Is it possible that about 100 years from now, when our descendents struggle to feed the 15 or so billion Earth inhabitants, climate will jump to a less hospitable state?" he asked. "It is difficult to comprehend the misery that would follow on the heels of such an event!"

These outlandish speculations, which only someone of Wally's stature could possibly have gotten into *Science,* proved to be enticing fodder for the popular press (the news staff at *Science* itself even jumped on the bandwagon). Such contrarian logic was perfect for the *Atlantic Monthly,* which published a starry-eyed cover story in January 1998 entitled "The Great Climate Flip-Flop," as part of a series looking ahead at the twenty-first century. The author, William Calvin, taking it for granted that we're facing an ice age, actually proposed that an international effort be mounted to study the problem, mainly through computer modeling, and to come up with artificial ways of preventing a flip. "Perhaps computer simulations will tell us that the only robust solutions are those that re-create the ocean currents of three million years ago before the Isthmus of Panama closed off the express route for excess-salt disposal," he wrote. "Thus we might dig a wide sea-level Panama Canal in stages, carefully managing the changeover."

I suspect one or two climate modelers might become excited at the prospect of playing with those big new computers, but I cannot imagine the collection of freethinking climatologists I have met over the years ever reaching enough of a consensus to propose such a project, much less politicians taking it seriously enough to go along.

But Wally has nothing if not an active mind. By January 1999 he was distancing himself from the false prophets who had construed doomsday scenarios from his modest conjectures. Even if a flip were to occur, he now believed, it was unlikely we would freeze as much as our ancestors did during the Younger Dryas, because temperatures would have to rise by seven or nine degrees in order to flip the switch, and since it would be much warmer at the beginning of this flip than the last one, we would end up only slightly colder than we are today. Nevertheless, he wrote, "[t]he fact that we are unable to provide satisfactory estimates of the probability that a conveyor shutdown will occur or of its consequences is certainly reason to be extremely prudent with regard to CO_2 emissions. The record of events that transpired during the last glacial period sends us a clear warning that by adding greenhouse gases to the atmosphere, we are poking an angry beast." (This time he included a cartoon of a bespectacled kid poking a dragon with a stick.)

Then, in the summer of 2000, he came full circle. Putting together three observations—(1) that there have been no great flip-flops in the ten thousand years of the Holocene; (2) that the Holocene has seen much less northern sea ice than the chaotic Wisconsinan did before it; and (3) Gerard Bond's evidence that the conveyor has been switching on and off all through the Holocene with the same, sun-driven 1,500-year beat it had during the Wisconsinan—Wally guessed that sea ice amplifies the effects of the conveyor, so that without it a flip would not cause significant change. "Were this scenario to be the correct one," he wrote, in what I believe to have been his final words on the subject (although he has continued to stir the pot in social and media circles), "then a greenhouse-induced reorganization of deep water flow would be less of a threat, for coming on the heels of a significant polar warming, there would be even less sea ice than now. As I am the one who first raised a warning signal, it is a relief to be able to declare that it may have been a false alarm."

BUT WALLY HAD "SHAKEN THE WORLD" YET AGAIN. HIS NOT-ALTOGETHER-INNOCENT speculations had taken on a life of their own. While he and his North Atlantic colleagues continued to proselytize for the great conveyor and the central importance of Greenland to their professional colleagues (many, in fact, continuing to speak of an imminent cooling), the Great Climate Flip-Flop evolved into an urban myth. At cocktail parties and the like, folks with a casual interest in popular science, who knew very little else about climatology, came to display a vague knowledge of the strange current up by Greenland that might flip us into an ice age. Even the military bought into it.

In late 2003, the Pentagon released a study predicting with more certitude than Wally ever did that global warming would cool the northern hemisphere and cause concurrent disaster nearly everywhere on earth (with the curious exception of North America). "Abrupt climate change"—a buzz phrase that had somehow become identified with the great conveyor, even though the only abrupt changes we have actually seen in the past two decades have originated with the manifestly tropical phenomenon El Niño—was portrayed as an imminent national security threat on a par with terrorism. In the proposed scenario (characterized as unlikely but plausible) the conveyor flips in about 2010. In addition to northern cooling, the report foresees "mega-droughts . . . in key regions in Southern

China and Northern Europe" and advises "that the duration of this event could be decades, centuries, or millennia and it could begin this year or many years in the future."

The report was actually written by the staff of a private, evidently liberal or libertarian consulting group, which has also constructed futurist movie scenarios for Hollywood; but it was commissioned by an eighty-three-year-old Pentagon legend named Andrew Marshall, who, according to *Fortune* magazine (the first media outlet, inscrutably, to have been provided with a copy of the report), "is known as the Defense Department's 'Yoda'—a balding, bespectacled sage whose pronouncements on looming risks have long had an outsized influence on defense policy. Since 1973 he has headed a secretive think tank whose role is to envision future threats to national security." In the early seventies, Marshall became a close associate of Donald Rumsfeld, defense secretary under George W. Bush, and joined Rumsfeld in the zealous promotion of ballistic missile defense. These are hardly liberal credentials.

While the climatic details of the Pentagon report are greatly out of touch with current scientific thinking (and give us yet another reason to question the perspicacity of our so-called intelligence services), its vision of the chaos that is likely to ensue as climate change limits the earth's capacity to feed and provide water for the burgeoning human population is very much worth considering. "[I]t seems undeniable that severe environmental problems are likely to escalate the degree of global conflict," the report reads. "Nations with the resources to do so may build virtual fortresses around their countries, preserving resources for themselves. Less fortunate nations, especially those with ancient enmities with their neighbors, may initiate in struggles for access to food, clean water, or energy. Unlikely alliances could be formed as defense priorities shift and the goal is resources for survival rather than religion, ideology, or national honor. . . . Once again warfare would define human life."*

Although President Bush ignored this report (as he had previously ignored—and some say altered—others from the National Academy of Sciences and the Environmental Protection Agency), he and his vice

*Worth considering, but, again, not particularly new: ten years ago, in his much-discussed *Atlantic Monthly* essay, "The Coming Anarchy," Robert Kaplan observed that it was already "time to understand The Environment for what it is: the national-security issue of the early twenty-first century." He predicted that environmental scarcity—particularly that of water—exacerbated by both global warming and population growth, would give rise to a bifurcated world, rampant with criminal anarchy: "Think of a stretch limo in the potholed streets of New York City, where homeless beggars live. Inside the limo are the air-conditioned post-industrial regions of North America, Europe, the emerging Pacific Rim, and a few other isolated places, with their trade summitry and computer-information highways. Outside is the rest of mankind, going in a completely different direction."

president and fellow oilman Dick Cheney might have enjoyed it had they bothered to read it: the "geo-engineering" solution proposed here goes William Calvin one better by suggesting that we add *more* greenhouse gases to the atmosphere in order to offset the cooling!

When these absurd notions made the news, their ultimate originator, Wally himself, saw fit to write a letter to *Science* dismissing the report. "Exaggerated scenarios serve only to intensify the existing polarization over global warming," he wrote. "What is needed is not more words but rather a means to shut down CO_2 emissions to the atmosphere." Well said.

And in a rare instance of Hollywood and the Pentagon thinking along the same lines at the same time, a movie that owes equally as much to Wally's speculations and is only slightly more fanciful appeared about six months after the report. *The Day After Tomorrow,* directed by Roland Emmerich, who also brought us *Independence Day,* presents global warming (actually an ice age, arising in a matter of days) as a menace to equal the alien warships of Emmerich's previous go-round. Tsunamis and "super storms" threaten to destroy civilization, and New York City is covered in ice. At the time of the film's release, Emmerich told the *New York Times* that it was "a popcorn movie that's actually a little subversive": he hoped to wake the public to the dangers of the greenhouse buildup. And perhaps he was right. A few months after its release, a group of British and South African scientists described a (perhaps tongue-in-cheek) survey they had conducted at four cinemas in southeastern England in an attempt to assess both the level of concern moviegoers felt about global warming ("measured as how much out of a hypothetical £1,000 people wanted to give to climate mitigation versus four other good causes") and their knowledge of it. A series of questions was put to audiences on their way into the movie and (to a different group) on their way out. The scientists found that the outgoing group had significantly "less realistic expectations" about future climate; however, in spite of this "dumbing down," they allocated about 50 percent more of their hypothetical money to mitigation.

Fair enough, but somehow I continue to believe that only the even-handed presentation of the scientific truth will finally succeed in changing policy.

For the record: in 1998, just as the flip-flop fad was reaching its height in scientific circles, Lonnie stated calmly in an e-mail that he believed it was "fundamentally flawed" to reason that warming could cause an ice age. Furthermore, even *with* a conveyor shutdown the 2001 IPCC report sees northern Europe becoming warmer this century—though, admittedly, less so than in the global mean.

. . .

THE FACT IS THAT LOOSE ENDS HAD TURNED UP IN THE TIDY STORIES OF ICEBERG ARmadas and draining lakes in the far north long before Wally provided this distraction. Harvard's Jim McCarthy believes it would have been "an overstatement" even in the early 1990s "to say that most people thought [the great conveyor] was the engine" driving climate change. He "wouldn't have taught a course on the subject without discussing it," but he would have called it an "important factor," rather than "the engine driving everything." "Every now and then it looks like a really critical piece of this grand puzzle has been understood and that it will explain more than it does after closer scrutiny," McCarthy observes.

By the time Lonnie had turned his attention to Sajama, the tropical revolution begun in part by Huascarán had cast many doubts on North Atlantic supremacy. Sajama would cast another.

AWARE THAT HUASCARÁN'S RELEVANCE TO THE "TROPICS VERSUS POLES" DEBATE WAS greatly diminished by its reliance on a chronology from the northern hemisphere, the Thompsons went to extra lengths to develop independent timing for Sajama, and its tropical location helped them there. The winds that nearly lifted my tent off the ground as I was sleeping on the summit (or not sleeping, as the case may be) sometimes waft bits of wood, grass, and other organic fragments onto the snow. This doesn't happen at the poles, which are thousands of miles from the nearest blade of grass.

Mary Davis discovered quite a few organic fragments as she melted Sajama's core segments with distilled water, dressed in her bunny outfit in the dust lab clean room. She remembers starting in horror as she watched one particular black dot through her microscope: the surrounding ice melted, the focus improved, and she found herself staring into the eyes of a perfectly intact fly—whose age, according to subsequent carbon dating, turned out to be 5,600 years. Of the fifteen organic fragments in that particular core, the oldest was a piece of wood, five feet above bedrock, which came in at 24,950 years. The dating of these, along with layer counting, Huaynaputina's A.D. 1600 ash horizon, and flow modeling, allowed the team to tie down the chronology quite well.

That done and the isotopes and other markers analyzed, the Thompsons found—as they had at Huascarán, Guliya, and Dunde, and as others had found in Greenland—that a sudden and temporary cold spell had set in at Sajama just as the Wisconsinan seemed to be ending. The Thompsons refuse to tag this event with the name "Younger Dryas," however. They call it a "deglaciation climatic reversal (DCR)," which began fourteen thousand years ago "as the climate shifted abruptly to colder conditions that were similar to those attributed to the North Atlantic Younger Dryas."

The Younger Dryas is perhaps the purest story in North Atlantic myth: climate was warm and getting warmer, the conveyor was running nicely, thank you; then Lake Agassiz broke its dams, the conveyor shut down, and the earth slid back into an ice age. But the Thompsons find that temperatures in Bolivia begin to drop about 1,100 years *before* the first signs of the Younger Dryas crop up in the layers of GISP2. Temperatures rise again either simultaneously with those in Greenland or even a few hundred years *later*. If they began to drop at the southern end of tropical South America more than a thousand years before they did in Greenland, it is difficult to attribute this particular global change to the flip of a switch in the North Atlantic.

Owing largely to Wally Broecker's great influence and promotional skill, the phrase "Younger Dryas" has become intimately associated with his great ocean conveyor belt. To use the phrase in relation to the cold spell that gripped the Andes (and Tibet) at about the same time, the Thompsons feel, attributes far too much power to Wally's favorite mechanism (which is rather a vaguely defined concept anyway). They would like to avoid the entire semantic tangle. "The abrupt onset and termination of a Younger Dryas–type event [at Sajama]," they wrote, "suggest atmospheric processes as the probable drivers."

AS THEY WERE PENNING THESE WORDS, OTHER FINE THINKERS WERE "AUDITION[ING] the tropical Pacific for the lead role in producing both orbital (forced) and millennial (internal) climate cycles," as Mark Cane put it in an elegant perspective just two months before the Sajama results were published. Ray Pierrehumbert even produced a tropical counterpart to Wally's angry beast, naming his own perspective on the subject "Climate Change and the Tropical Pacific: The Sleeping Dragon Wakes."

Cane is one of the more thoughtful, even-keeled, and rigorous members of this new "Tropical School." His background is applied mathematics and meteorology, and his Ph.D. advisor at MIT was none other than Jule Charney, who had participated in von Neumann's legendary Meteorology Project and subsequently given his name to an influential greenhouse assessment.

After early success as a member of the MIT faculty, developing the first computer model capable of predicting El Niños,* Cane moved to Lamont (thereby joining his temperamental opposite, Wally Broecker) and proceeded to use his model to predict the 1986 El Niño—providing advance

*See pages 192–95.

notice that saved thousands of lives. About ten years later, at the height of interest in the North Atlantic, he shifted his focus from meteorology to paleoclimatology.

In the summer of 2000, assuming he must have heard quite a bit about the North Atlantic in his years at Lamont, I asked him if he had "converted" from there to the tropics. He replied that it was more like being lost and then found; he had never bought the arguments for the great conveyor: "The crucial piece of evidence was the idea that a lot of these changes were pretty much global," he said. "It's hard to do that from anywhere but the tropics."

The center of the planet is in the best position to broadcast changes to both hemispheres, because most air and ocean currents flow from the equator to the poles, which makes it difficult for a change to cross from one hemisphere to the other.

"The other thing about the North Atlantic is—well, I think of it as the Mercator fallacy. You look at too many Mercator projections, showing the North Atlantic as this huge expanse, and you forget that it's actually not a very large part of the surface of the earth."

A Mercator projection is the most common way of mapping the spherical surface of the earth onto a flat page. It distorts it in such a way as to magnify the polar regions relative to lower latitudes. If you inspect a globe, however, you will understand just how minuscule the oceans south of Greenland are compared with El Niño's nine-thousand-mile playground in the tropical Pacific. It is difficult to imagine how so little water could possibly affect the oceans all over the earth, much less the entire atmosphere—to which the oceans are coupled only at their surface, anyway. Cane points out that climate is mainly (and perhaps obviously) an atmospheric phenomenon. And anyone who has lived through the 1980s and '90s knows that El Niño has direct and essentially immediate consequences nearly everywhere in the world.

Lonnie points out that the polar ice caps are so large and so cold that they tend to make their own weather, while the tropics and subtropics not only comprise half the earth's surface, which makes them more representative, but house about 70 percent of the human population, which makes them more relevant.

Cane agrees: "Greenland and Antarctica are atypical places in the world by a huge margin. They're not necessarily that well connected to global changes. When the temperature in a Greenland record drops by ten or fifteen degrees fairly suddenly, is it because the whole world changed or because there was a wind shift at very high latitudes? The tropics are more likely to be typical of a greater swath of the globe."

• • •

THERE IS A SUFI TALE OF THE HOLY FOOL NASRUDDIN, WHO LOST A KEY IN THE MIDDLE of a street one night and promptly went to look for it under the nearest lamppost, for at least there he could see. The story illustrates the hopelessness of understanding God through the intellect alone. Over the last five or seven years, more and more climatologists have come to see the North Atlantic as a similarly misplaced lamppost.

Mark Cane points out that "most of the records are coming out of the oceans, and most of those from the North Atlantic, so most of the theories about climate are centered in the North Atlantic." He then remembers a comment he heard in a lecture by a retired climatologist named Andy Macintyre: "He's a character, and I suppose he's old enough to say such things without getting into trouble—but his simile was that it's like when you're first learning about sex: you go for the most responsive system. Only later do you develop any taste. Needless to say, not everybody was pleased by this particular analogy, but I thought that might be part of it, too."

By the turn of the millennium, a migration from the poles to the tropics had clearly begun. Many groups began taking their field studies to the coral reefs and deep ocean floor of the Pacific, and some also found records in the Andes to complement Lonnie's. This required the development of new methods, for nothing like Heinrich Layers are to be found on the immense floor of the Pacific. The changes are more subtle there, and the chemistry of the forams, which have proven so useful in the North Atlantic, does not tell as obvious a story.

Cane is also a great baseball fan (he once claimed that the Dodgers' move from his hometown of Brooklyn to Los Angeles when he was a kid had "precipitated the decline of American civilization"). In a perspective in *Science* in 2000, he characterized a group that had just pulled up a new coral record from the South Pacific as having "wisely followed the dictum of Wee Willie Keeler* to 'hit 'em where they ain't.' "

Many others were soon following that dictum.

IN EARLY 2003, A TEAM LED BY KATHERINE VISSER OF THE U.S. GEOLOGICAL SURVEY REported on a seabed core from Indonesia's Makassar Strait. The oxygen isotope ratio in the forams in this core varies with both polar ice volume and local water temperature, as it does elsewhere, but Visser's team also exploited a new temperature measure, based on the ratio of magnesium to calcium in the shells. That meant they could disentangle temperature from ice volume in a single record (without resorting to Vostok's air bubbles)

*Keeler was a professional baseball player who won the batting crown in 1897 and 1898.

Approaching the Quelccaya ice cap, Peru, in 1983, after a three-day walk. (*Bruce Koci*)

The team that spent three months in the Peruvian backcountry on the "Dream Expedition" to Quelccaya in 1983. Left to right: Dave Chadwell, Phil Kruss, Mike Strobel (kneeling), Lonnie Thompson, Eugenio Angeles C., Benjamín Vicencio M. (kneeling), Bruce Koci, Keith Mountain. (*Keith Mountain*)

On the Gobi Desert, approaching the Qilian Shan from the north in 1986. Lonnie's team would drill both that year and the next on the summit of the wide, flat Dunde ice cap, just visible at the far right of the photograph. (*Keith Mountain*)

Chinese team members "man-hauling" equipment to the 17,500-foot summit of the Dunde ice cap in 1987. (*Bruce Koci*)

Huascarán, the highest mountain in Peru, from the southwest. The Garganta Col lies between the two summits. The route to the col ascends the complex and dangerous glacier on the left. (*Lonnie Thompson*)

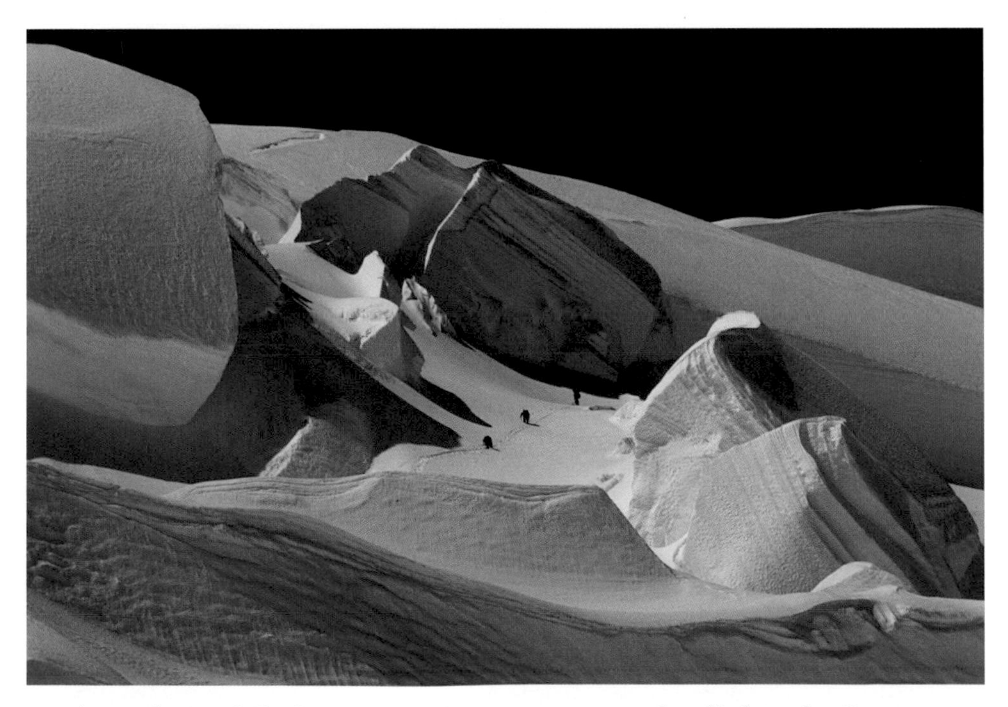

A roped trio of climbers weaving between seracs on the climb to the Garganta Col. (*Bruce Koci*)

The drilling camp on Huascarán's 19,500-foot Garganta Col, where Bruce, Lonnie, and Vladimir spent more than fifty consecutive days in 1993. Note solar panels on right. (*Bruce Koci*)

The 6,000-foot west face of Nevado Sajama. Our route followed the left skyline to Lonnie's drilling camp on the 21,500-foot summit. High Camp was located just above the snow-topped molar of rock about halfway up. (*Mark Bowen*)

"An island in the sky." The solar-powered drill on the circular summit of Sajama. (*George Steinmetz*)

The team in the snow cave that served as dining room, living room, and lounge on (or just under) the summit of Sajama in 1997. They worked for twenty-eight consecutive days up there. Left to right: Bruce Koci, Patrick Ginot, Lonnie Thompson, Vladimir Mikhalenko, Victor Zagorodnov, Benjamín Vicencio. (*George Steinmetz*)

Porters with full ice core boxes descending the northwest ridge of Sajama. (*George Steinmetz*)

Keith Mountain hand drilling on the northern ice field on our reconnaissance of Kilimanjaro in 1999. Lonnie Thompson and Julias Minja, right, measure and weigh the core segments as Keith produces them. (*Mark Bowen*)

Drilling on Kilimanjaro's southern ice field, about a hundred yards from the summit of the mountain, near the end of the deep drilling operation in 2000. Keith watches as Vladimir steadies the drill and Victor extracts the screw from the drill housing. The "anti-torque skates," which prevent the housing from spinning as the screw cuts its way into the ice, are mounted on the left (top) end of the housing. (*Mark Bowen*)

Mary labels a cylindrical tube containing the newest segment of ice core, as Victor prepares the screw for reinsertion into the housing, and Lonnie communicates by radio with Keys Hotel in Moshi, which he faces to the south. The "inversion layer" of dust, smoke, and haze from below marks the horizon. (*Mark Bowen*)

Looking south at the Himalayan crest during the search for the Dasuopu ice cap in 1996. According to Bruce, "One of the trucks got stuck on the way in, so we were 'forced' to spend the night out on the plains close to the mountains. It was a magic evening, as you can see." Dasuopu lies on the hulking massif of Xixapangma, center, the world's fourteenth-highest mountain (8,012 meters or 26,286 feet). (*Bruce Koci*)

The eastern end of Kilimanjaro's northern ice field, as of February 2000. "This is the way a glacier dies." (*Mark Bowen*)

and definitively reveal the relative timing of changes in both. They found that the water in the Makassar Strait was six or seven degrees colder at the Last Glacial Maximum than it is today, confirming Huascarán and its coincident studies; more important, they found that the temperature of the Indonesian ocean began to rise two or three thousand years before the northern ice sheets began to shrink.

Working nine thousand miles to the east at about the same time, another group demonstrated that the glacial began to relax in Peru about five thousand years before it did in Greenland.

Meteorologists refer to the tropical ocean as a "firebox," because the greatest portion of the sun's energy streams in there, and the ocean stores that energy as heat. This is the fuel that powers the enormous cyclones and hurricanes that wander like rogue elephants into the higher latitudes in summer, cutting wide swaths of destruction if they happen to hit land. A rise of only one or two degrees can make the difference between a strong hurricane season and a weak one.

And the Makassar Strait happens to lie in the hottest spot in the tropical firebox: the Indonesian warm pool, the largest reservoir of warm water on the planet and the source of that crucial player in the Walker Circulation and El Niño known as the Indonesian Low. Recent El Niño events are vivid evidence that this critical location is at least as "tippy" as Wally Broecker's switches up by Greenland. What's more, in his "search [for] a mechanism capable of propagating the effects of [the great conveyor] across the globe" (to quote from his Huascarán comments), Wally himself calls upon El Niño to do the actual *work* of changing climate.

"Water vapor is the atmosphere's most powerful greenhouse gas," he writes elsewhere. "If you wanted to cool the planet by 5°C [9°F] and could magically alter the water-vapor content of the atmosphere, a 30% decrease would do the job. . . . Water vapor is supplied to the atmosphere primarily in the tropics . . . [s]o if we invoke a change in the atmosphere's water-vapor inventory, we must look to the tropics—in particular, to the western tropical Pacific. . . ."

He then argues that the flipping of his putative switches in the North Atlantic basin might somehow affect the upwelling of the Humboldt Current off equatorial Peru and, thereby, tickle El Niño—although he does admit that "[t]his aspect of my argument is particularly speculative." Equatorial Peru is, after all, not only sixty latitudinal degrees south of his switches, but on the other side of two connected continents that reach nearly from pole to pole.

Mark Cane, who knows something about the causes of El Niño, believes the ocean is "a secondary player." His understanding of its physics tells him that the phenomenon is triggered mainly by events in the atmosphere: "[The ocean] feeds cold water to the equator in the first place, so I wouldn't say it has nothing to do with it, but the changes that happen along the coast of Peru are a consequence [of El Niño], not a proximate cause."

In fact, Cane's preferred explanation for past climate changes turns Wally's on its head. Cane believes the Indonesian warm pool is the main player in climate change and that changes in its position and temperature may have caused the interglacial warmings, the violent rumblings of the late Wisconsinan, and, with significantly less consequence than either of those, the switching on and off of the conveyor!

"It is well established that El Niño years are warmer [on average]," he writes. "One influence of El Niño [is to warm] northern North America. An interesting if somewhat anecdotal confirmation is based on Hudson Bay Trading Company records of the date that the ice goes out on Hudson Bay. In El Niño years, the ice goes out early; we extrapolate, perhaps outrageously, to the idea that moving the [Indonesian Low] to the central Pacific [as in El Niño] will help to melt the Laurentide Ice Sheet, while keeping it in the far western Pacific will favor ice sheet growth."

Cane has run computer simulations that show that the severity and frequency of El Niño are controlled by the twenty-two-thousand-year precession cycle. This would make sense because El Niño is highly seasonal, and precession, which is felt most strongly in the tropics, changes the seasonal intensity of incoming sunlight. The Earth presently finds itself closest to the sun in early January and farthest away in early July; eleven thousand years ago, at the start of the Holocene, the timing was reversed.

The tropics versus poles debate also reaches farther into the past.

IT TURNS OUT THAT ABOUT THREE MILLION YEARS AGO, AROUND THE TRANSITION FROM the Pliocene to the Pleistocene, a change in climate dried and may have cooled the Horn of Africa. A green woodland landscape, mostly rain forest, was transformed into grassy, brown savannah. Nicholas Shackleton showed in 1984 that this African change coincided reasonably well with the onset of a major glaciation in the northern hemisphere. The Isthmus of Panama also closed around then. This probably initiated the Gulf Stream, which would have added to the existing currents of the great conveyor. So, not surprisingly, a strengthening of the conveyor was credited with changing the climate of Africa.*

*There seem to be two flies in the ointment here: recent thinking favors the view that the isthmus closed four or five million years ago, much earlier than the change in Africa; and a strong conveyor should *warm* northern climates, not cool them.

But in 2001, Cane and MIT's Peter Molnar suggested that it may have been a tropical change that simultaneously dried Africa and cooled the Arctic. Their thinking went as follows: Australia and New Guinea sit on a tectonic plate that is presently moving north toward Eurasia and the larger islands of the Indonesian archipelago at the blistering pace of about forty-five miles every million years. Calculating back—paying due heed to current knowledge in plate tectonics and continental uplift—Cane and Molnar showed that three and more million years ago there was probably a wide passage between New Guinea and the eastern islands of Indonesia. In those early times, then, comparatively warm water from the South Pacific would have streamed through this passage into the Indian Ocean. Then, about three million years ago, this "valve" closed, resulting in the present pattern in which somewhat cooler Pacific water from north of the equator flows south through the smaller Makassar Strait (between the islands of Sulawesi and Borneo) then west into the Indian Ocean.

Before the valve closed, Cane and Molnar's thinking goes, the earth would have experienced a permanent state of El Niño, which presently brings rain to the Horn of Africa, partly by warming the Indian Ocean, just as the ancient, more southerly flow of Pacific water would have done. The subsequent closing of the valve would have mimicked La Niña, which brings drought to the Horn today. Noting again that El Niño also tends to heat the far north, Cane and Molnar guessed that the closing of the valve may also have cooled northern North America and enhanced glaciation.

This particular climate change, incidentally, seems to have played a role in human evolution. The fossils of the earliest hominids, excavated in Tanzania and Kenya, indicate that our family tree diversified at about the time the valve closed. *Australopithecus* was forced out of the trees and onto the savannah, where he had to walk on two legs in order to survive. The first fossil remains of our own genus, *Homo,* characterized by much larger skulls (and, by inference, brains) than their *australopithecine* ancestors, date from about 2.5 million years ago, as do the earliest-known stone tools.*

Quite a few climatologists believe global warming might again induce permanent El Niño. When I asked Cane if this was nonsense, he replied, "I don't know that it's right, but it's not nonsense. Right now the best

*Lonnie's biggest take-home message from a meeting of the International Union for Quaternary Research in Durban, South Africa, in 1999, came from a talk by Yale paleontologist Elizabeth Vrba, whose research told her that our brains have tended to grow during cold periods. He ended his own talk, which included much evidence that the earth is now getting warmer, with the line, "No one's brain is going to be growing in the near future."

models we have are showing that, and if I had to make a guess, I think I'd have to agree."

Incidentally, in spite of disagreeing with Wally about the importance of the great conveyor, Cane commends him as "the leading person in this field, no question about it. Everybody knows it. He has lots of ideas—a lot of them are wrong; usually he moves along more fluidly from one to the next—but I have enormous respect for him as a scientist. One of the nice things about science—I think it's a nice thing—is that it's done by human beings. Einstein couldn't tolerate quantum mechanics no matter how successful it was, right? And he was smarter than Wally, so . . . [chuckle]."

IT DOES SEEM THAT THE ARGUMENTS FOR THE GREAT CONVEYOR HAVE BEEN BASED more upon fixed belief and received wisdom than upon critical thinking. Carl Wunsch, a physical oceanographer at MIT (rather a rigorous profession at rather a rigorous place), laid many of them bare when he asked, "What Is the Thermohaline Circulation?" in *Science* recently. "A reading of the literature on climate and the ocean suggests at least seven different, and inconsistent, definitions," he observed.

Even the most basic belief in the North Atlantic myth cycle—that the great conveyor is responsible for the mild winters of western Europe—took a body blow recently. This bit of folklore was enshrined a century and a half ago by the American oceanographer Matthew Maury, who wrote, "One of the benign offices of the Gulf Stream is to convey heat from the Gulf of Mexico, where otherwise it would become excessive, and to disperse it in regions beyond the Atlantic for the amelioration of the climates of the British Isles and of all Western Europe. . . . It is the influence of this stream upon climate that makes Erin the 'Emerald Isle of the Sea,' and clothes the shores of Albion in evergreen robes; while in the same latitude, on this side, the coasts of Labrador are fast bound in fetters of ice."

In 2002, using real measurements of the heat stored and transported by the atmosphere and the Atlantic Ocean, complemented by simulations using two different global circulation models, a fine study by a group led by Richard Seager of Lamont and David Battisti of the University of Washington (and including Mark Cane) found that the simple reason winters in western Europe may be as many as thirty-five degrees Fahrenheit warmer than those in eastern North America is that the prevailing winds at high northern latitudes blow west to east. Thus Europe has a typically maritime climate, its warmth maintained year to year by the annual summer heating of Atlantic surface water, while eastern North America's is typically continental. This accounts for half the temperature difference between the eastern and western shores of the Atlantic. The rest is explained by a "standing wave" in air circulation, set up by the Rocky Mountains. The long mountain range running north–south on the western half of North America steers westerly Arctic winds southward to "bind Labrador" and its neighbors on the eastern seaboard in their "fetters of ice." Over the Atlantic, the winds curve back north, thus arriving in Europe mainly from the warm southwest rather than the west itself or the much colder Arctic. The sole contribution of the actual *flow* of the Gulf Stream and its fellow currents in the great conveyor to the *difference* in temperature is to warm northern Scandinavia by restricting sea ice cover in the Norwegian and Barents Seas. Seager and Battisti find that the conveyor adds "a few degrees" of heat to the northern hemisphere, but that it adds that heat equally to both sides of the ocean. "In retrospect," they write, "these conclusions may seem obvious." Indeed.

Wally Broecker accepts the conclusions (and apologizes for his "previous sins") but remains steadfast in his belief that the North Atlantic is the main trigger point of climate change.

THIS IS A SCIENTIFIC LANDSCAPE THAT HAS ALREADY SEEN ONE DECADE OF EXPLORATION and will probably see many more before it is sufficiently revealed. After all, the explanation for the ice ages remains incomplete after more than a century. Strong belief dies hard. As one of Lonnie's colleagues observes, "Science proceeds one funeral at a time."

Much new terrain has been revealed since Lonnie first emphasized the importance of the tropics to me as we gazed out together on what seemed an endless tropical expanse from the summit of Sajama.

CLIMAP is currently being updated, for instance, and the first results from GLAMAP (Glacial Atlantic Ocean Mapping) seem to be saying that the Atlantic as a whole cooled less during the glacial than the CLIMAP estimate, while the lower latitudes cooled much more. This turns CLIMAP's main conclusion on its head: the tropics may have cooled more than the poles did during the glacial. Furthermore, as one paleoceanographer not involved in GLAMAP observes, "The extent and timing of equatorial changes may be consistent with an idea that dynamics of the tropics trigger global climate change."

Recent work has also indicated that past changes in the tropics have coincided with those in the *Antarctic*. Thus a Southern Ocean School has appeared—populated mainly by those who have drilled Antarctic ice or Southern Ocean mud, oddly enough.

And in 1998 Wally Broecker proposed the idea of a "bipolar seesaw," whereby conditions have purportedly warmed in the Arctic as they have cooled in the Antarctic, and vice versa. (Naturally, his explanation involves the great conveyor.) So the crosshairs focused on both poles for a while, and it was soon revealed that changes north and south have indeed been out of sync, but that it is hard to tell which takes the lead. Since there is so little evidence to go by, speculation is rife, and the polar enthusiasts—particularly those for whom the conveyor is climate's be-all and end-all—are now asking which *pole* is the primary trigger. Seems like the old "drunk under a lamppost" routine, with two awkwardly placed lampposts instead of one.

There is plenty of room for disagreement here. The most recent of these events took place before mankind learned to use a hoe; and it is probably easier to predict the future than the past, as the saying goes, because we'll never really know what happened.

Dan Schrag, an originator of the Snowball Earth hypothesis, has the refreshingly global perspective that one might expect of a person associated with so sweeping an idea: "When Mark Cane says the tropics are driving things, guess what Mark Cane studies? When Lonnie says the tropics are important, Lonnie studies the tropics. When Wally Broecker says the North Atlantic is important, Wally studies the North Atlantic. When people who work on Antarctica say the Southern Ocean is important . . . the answer is it's all important."

In the year 2000—the same in which he won a so-called genius award from the MacArthur Foundation—Schrag wrote an "opinion" in *Nature* entitled "Of Ice and Elephants":

> One reason for concentrating on the North Atlantic is that many of the highest-quality, best-dated records originate from this region. But, like the story of the wise men and the elephant, could we be seeing only part of the climate system? The story tells of an emperor who sends his wise men to investigate a gift from a foreign land, an exotic animal called an elephant. But the wise men have grown blind from reading their books, and after each feels a different part of the strange beast, they return to the emperor in total confusion.

So which end is the head? Furthermore, in the vast geography and incomprehensibly long life of this planet, so many climate-affecting features have changed—from the location and size of continents, which steer ocean and air currents, to the life-forms that inhabit it (today, most prominently, ourselves)—that the beast itself has evolved. The head may have moved. Different regions may have triggered different climatic events at different times; change has probably spread gradually around the planet in a unique way each time; and in some cases two or more regions may have collaborated to produce the change in the first place. Time will tell; and as well-dated records from all latitudes and longitudes continue to roll in, a more sophisticated and unified understanding is bound to emerge.

While Lonnie remains a tropical partisan for now, his main concern is to retrieve as global a set of high-quality records as he can—while they last. "I think in five years we are going to be surprised at how little we knew about our climate system," he says.

He was saying the same thing five years ago. Scientists may put up a rational front, but the best of them know that progress is not linear and that there is no such thing as a "scientific method." The asking of a simple question may open a door to a new world, where that question is no longer relevant and new, more beguiling questions arise. This is what drives the best scientists. It certainly drives Lonnie. Furthermore, climatology is a good place for an explorer to be right now, for it is yielding the thrilling paradigm shifts that scientists live for on almost a yearly basis.

JOHN MERCER TOOK THE FIRST STEPS INTO THIS PARTICULAR SCIENTIFIC LANDSCAPE when he began beachcombing peat in the glacial moraines of Patagonia in the 1950s. He was still at it twenty years later when he took a greenhorn from West Virginia along on his first trip to the Andes. Recall that the age of Patagonia's moraines told Mercer that the Younger Dryas was probably limited to the northern hemisphere, or that if a fitful cooling *did* set in as the Wisconsinan loosed its grip on South America, it started about five hundred years before the Younger Dryas began in the North Atlantic region. This so-called Mercer Problem could still be found in textbooks in the mid-1990s.

To weigh the similarities between scientific exploration and wandering lost in uncharted terrain, consider that the present tension between the tropics and the North Atlantic may be nothing more than a new way of stating the Mercer Problem, with a vastly expanded view of the surroundings. In 2000, when Lonnie himself had been on the trail for an additional twenty-six years, he, Ellen, and Keith Henderson—pointing to Sajama's evidence for a "deglaciation climatic reversal" that began in Bolivia a thousand years before the northern Younger Dryas and ended at either the same time or slightly later—finally claimed that they had solved the problem named for Lonnie's mentor.

18

ENDURANCE AND TRAGEDY ON DASUOPU

So they went from the Andes to the Himalaya with two weeks between.

The journey began in Kathmandu, Nepal. Lonnie, Mary, Keith Henderson, and Shawn Wight, a twenty-six-year-old graduate student on his first trip abroad, flew in from Ohio; Vladimir flew in from Moscow; Benjamín, Maximo, Alberto, and Jorge, the mountaineers who had just provided support on Sajama, flew in from Peru (missing a flight connection along the way and sending Ellen, back at "mission control Columbus," into crisis mode for a while). In Kathmandu, Lonnie hired four Nepalese Sherpas, led by the well-known Lopsang Sherpa, whose formidable résumé included climbing and guiding on Everest. The bulk of their equipment had been shipped via Beijing to Yao Tandong at the Lanzhou Institute, from which he and twenty-six companions transported it west by truck. On the mountain itself, a few Tibetan yak herders and their yaks would join the team.

Demonstrating his usual impeccable timing (and lack of interest in sightseeing), Bruce Koci arrived the day before they set out by bus, early one morning, on the "highway" connecting Kathmandu and Lhasa. In view of subsequent events, the hazards encountered on the northward journey are worth keeping in mind.

Within a few hours of leaving Kathmandu, they reached the first landslide to completely block the road.

"Had to abandon the bus," says Mary, "and, of course, there were huge traffic jams on both sides. People are used to dealing with this so they just gather up their bags, traipse over the debris, and hire a bus on the other side. We're carrying our stuff back and forth over this landslide, and rocks are still coming down off the mountain. Every time we hear a rumble everybody freezes in their tracks until it's over and then starts walking again."

They abandoned their second bus in Kodari, the last town in Nepal, where they hired a truck to shuttle them across the fifteen miles of disputed territory between there and the border with China. Few of the available vehicles had decent tires or brakes, and the one they finally judged safe enough was filled with lime, so their clothes and belongings were soon coated in caustic gray dust. Above Kodari, the road became muddy and deeply rutted and began winding through mountainous green rain forest. The air turned gloomy. "Not rainy, but dank," as Mary remembers it. "Humid, dank, and foggy."

They passed customs at the "Friendship Bridge," within walking distance of the town of Zhangmu—and they did walk, for their truck was not permitted across the border. The bridge collapsed about a year and a half later.

Zhangmu was a tough place. Mary remembers "huge dead rats in the muddy streets and dead animals everywhere," and Shawn noted in his diary that "*Zhangmu* must be Chinese for shithole." They were rescued after two nights by a pair of students from Lanzhou who had driven south in Land Cruisers to get them.

North of Zhangmu they began to enjoy the scenery. The road traversed steep forest-covered mountainsides, at the occasional edge of a vertical rockface with a thundering river in the point of a V-shaped valley below, as in a Chinese scroll painting. Mary describes "absolutely beautiful green hills, towering mountains, fantastic scenery; but we had to drive through a couple of waterfalls, so the people on the back of the truck got soaked, and it's not warm up there."

They crossed some high passes, including one that may have reached seventeen thousand feet, before descending into the twelve-thousand-foot valley that holds the town of Nyalam, which is an interesting place climatically. It lies by the edge of the Tibetan Plateau, at the boundary of monsoon rainfall, so the southern half of the town is green, and the northern half is brown. The team experienced their first headaches and stomach upsets there; the process of acclimatization had begun.

The next day they crossed a final 17,000-foot pass to gain the Tibetan Plateau and soon left the road to travel west. North of the Himalaya now, they turned south and reached their 17,500-foot base camp before dusk. On the reconnaissance in 1996, it had taken a few days to find a suitable site, but this year the Chinese had arrived and set up camp in advance.

Their objective, the Dasuopu ice cap, lay on the northwestern flank of

Xixapangma, which, at 8,012 meters, or 26,286 feet, is the least high of the world's select fourteen eight-thousand-meter peaks. They could see Everest forty miles to the east.

Having gained more than five thousand feet that day, they put on the brakes for a while, remaining in base camp for a full week, acclimating, waiting for Yao to arrive with the second of their two shipments of equipment, and sorting the gear they did have. This was the first time Shawn had experienced acute mountain sickness, and he kept a detailed record of his discomfort in his diary—the usual symptoms: headache, nausea, sleeplessness, listlessness . . .

On the twentieth of August, they left for advance base camp, which would become the nerve center of the expedition. At this point, according to his diary, Shawn seemed to be acclimating more easily than Mary or Keith, although much would later be made of the fact that Lonnie carried Shawn's pack for a while that day. Lonnie perceived Shawn as being overly enthusiastic—inclined to exert himself more than one should on the hikes—and wanted to prevent him from paying for it later.

"You reduce risk," he says. "The less effort you expend, the less stress you put on your body, the less chance you're going to have altitude problems later. That's in Houston's book."

Lonnie is speaking of Dr. Charles Houston, the towering figure in the field of high-altitude physiology, who conducted the first medical study of acclimatization as a naval flight surgeon in the late 1940s and documented the first case of high altitude pulmonary edema twenty years later. Lonnie had also studied the work of Dr. Peter Hackett, whose handbook *Mountain Sickness* is published by the American Alpine Club. In past years, he had brought in experts to teach his group how to stay healthy at altitude, although he had not done so this year.

The team waited at this intermediate, 19,100-foot camp for thirteen days.

Mary says, "We were encouraged—especially Shawn and Keith, because this was their first time—to hike around as much as possible, acclimatize, put on as much weight as we could, eat. Eating wasn't easy for Shawn and Keith. They hated the food. It was Chinese food, and of course our idea of Chinese food is nothing like the real thing. But everybody was doing pretty well there, and I got bored real fast, because I'd already spent a lot of time at [advance base] in 1996."

She had developed a cold on the reconnaissance expedition, which had been exacerbated early on by the climbing of endless flights of stairs on a tourist visit to the Potala Palace in Lhasa, inhaling the fumes of the innumerable yak butter lamps on the Buddhist shrines. She had hidden the cold from Lonnie, for he had threatened to send her home if it persisted; it had evolved into bronchitis on the mountain, so she had spent a few tedious weeks at advance base.

On September 1, this year's party climbed to what I will call Camp III, their second-highest camp, at 21,500 feet. This is almost exactly the height of Sajama's summit—and was quite high enough to bring on a new wave of acute mountain sickness.

Bruce points out that "there's a lot of humor involved, when you look around and see how many people are feeling as bad as they are. Everybody's laughing. You see somebody leave, walk fifty meters, charf cookies all over the place, and say, 'My God, I got another twenty times that to go!'"

But Shawn wasn't providing much amusement: on the day he reached Camp III he wrote in his diary that it was the best he'd felt since entering China.

Bruce referred to Camp III as the "parabolic mirror," because it was situated in the cup of a snow bowl and got extremely hot during the day. The sun baked the shirts on their backs, making them smell as though they'd just come out of a clothes dryer. Ironically, the heat also made it necessary to lug electric freezers to the high camps in order to cool the blue ice they would need to keep the ice cores cold. The indomitable Sherpas carried the freezers up on backpack frames, with tumplines strapped across their foreheads and back around the box.

Up to this point, the Sherpas and Peruvians had climbed ahead of the scientific team to set up the camps in advance, so that the scientists could simply flop into a sleeping bag and wait for dinner when they later arrived, but at this point Lonnie and Vladimir, the two with the most high-altitude experience, were seized by a desire to work. They moved up the very next day with Benjamín and a few Sherpas to set up the drilling camp at twenty-three thousand feet: Camp IV.

It was no accident that the camps were situated almost precisely 6,500 vertical feet, or 2,000 meters, apart. That also came from Houston's book.

They had been living above seventeen thousand feet for three weeks when Mary, Keith, Bruce, and Shawn moved up to the drilling camp, two days after Lonnie and Vladimir.

According to Mary, "Shawn was fine. He hadn't been displaying any symptoms; and as a matter of fact, he was as active as anyone up there. He was helping us dig out the underground kitchen. He was helping to lay out the drill line. He was helping stock the kitchen with food. He was helping to fix meals."

Having suffered mightily on Sajama less than two months earlier, the group used chain saws to carve out a luxurious snow cave this time around. Stairs descended from the entrance. "Then," according to Mary, "there was a short hallway and, facing back, off to the right, was a cooking alcove, where Jorge set up the stoves and pots and did the cooking. Shawn and I dug out shelves in the walls for storing food. We sorted out the food and stored it by desserts, entrées, lunches—row after row of shelves. Off to the left was the dining room—a pretty big room—ten or twelve feet on a side." They even carved out benches and tables.

During lunch in this refuge, at about three-thirty in the afternoon on his third day at Camp IV, Shawn's arms went numb and his fingers began tingling. Lonnie's first thought was diminished oxygen, so he asked Shawn to step outside and walk around. Within about an hour the numbness and tingling disappeared, but after another hour he developed a headache and lost the ability to recall names. Lonnie noted in his field book that it might be the first symptoms of high altitude cerebral edema and took him straight down to Camp III, where Shawn still could not summon names.

"He kept calling me Chief," Lonnie remembers. " 'I'm sorry Chief. You shouldn't have to come down.' "

So they continued down to advance base camp, where, according to Lonnie, "Shawn could remember everything, was in great shape, had a Coca-Cola, went to bed.

"Got up the next morning—Yao was with me—and we talked about whether he should go on home; but Shawn was fine. He ate breakfast, no problem; said, no, he didn't want to go home. He actually wanted to go right back up with us. But I said, 'No. You stay here a few days. See how you're doing.' "

So Lonnie and Yao left instructions with one of Yao's Chinese colleagues to keep an eye on Shawn and climbed back to the high drilling camp the next day. There were about twenty people with Shawn at advance base, five of whom spoke English, including one Chinese scientist who had shared an office with Shawn in Ohio.

On his sixth day there, Shawn awoke to coughing fits and severe pain in his chest and stomach cavity. The following day his campmates summoned a Chinese physician from a climbing group camped nearby, who gave him painkillers and antibiotics, believing he was passing a kidney stone. On the evening of September 14, Shawn sent a letter up to Lonnie telling him he disagreed with the doctor—he thought he had a hernia—and that he had decided to leave for Kathmandu.

The Sherpas carried that note up to Lonnie the next morning, whereupon he sent Yao back down with three hundred dollars and detailed instructions on what Shawn should take with him. Lopsang Sherpa would accompany him all the way to Kathmandu, which meant Lopsang would be lost to the expedition: like everyone else, he had only a single-entry visa to China.

But when Yao reached advance base, he found Shawn immobilized with a swollen and discolored right leg. The two men concluded correctly that Shawn had developed a blood clot, and the next morning Yao sent a note up to Lonnie telling him to come down immediately.

The ice core work had been proceeding well, thanks in large part to a new twin-cylinder engine that Bruce had incorporated into the drill, which was about ten times more fuel-efficient than the one they had used on Guliya and much better matched to the job. (Since it was cloudy, they used generators rather than solar power.) They had just completed the first core to bedrock when Lonnie received Yao's note. He instructed the team to move the drilling dome to the crest of a col at 23,622 feet (from which they could have jumped into Nepal), then descended with all the Peruvians except Benjamín—which left Benjamín, Mary, Bruce, Vladimir, Keith, and a few Chinese students to continue drilling.

When Lonnie reached advance base, after dark, he was concerned that Shawn's blood clot might flow to his heart or brain and lead to a more serious collapse. On the other hand, the books he had read on high-altitude medicine suggested that it is sometimes wise *not* to move a person with a blood clot. "You give them lots of liquid and keep them at altitude; let the clot dissolve on its own," he says. A discussion between Lonnie, Yao, Lopsang, the Peruvians, and, of course, Shawn, however, resulted in a decision to build a stretcher out of backpack frames and take Shawn to a hospital. Kathmandu was out of the question, as it lay in a different country, on the other side of the highest mountain range in the world. They would head for Lhasa.

The next morning, seven people—Lonnie, Yao, Maximo, Alberto, Jorge, and two Chinese porters—carried Shawn to base camp, where the trucks were parked.* When they arrived, again after dark, they were faced with another decision: should they leave immediately or wait for morning? They would be driving off-road in a featureless, dark brown landscape, and the driver was afraid they might drop off a cliff. (Luckily, they had a good driver; most would work only a certain number of hours a day.) They left at five A.M. the next morning and reached Xigaze, the second-largest city in Tibet, at nine that night.

*Having assisted in a few evacuations myself, I once pointed out to Lonnie that seven is not a lot of people to carry a person in a stretcher eight miles at seventeen thousand to nineteen thousand feet.

"Yes. Very strong people, though," Lonnie responded. "Part of strength comes in mental commitment to things."

The first thing they did was to carry Shawn into an English restaurant that was about to close and fill him up with mashed potatoes and other non-Chinese food. They knew they might have to drive all night to Lhasa, and the driver also needed rest and food. Shawn was still experiencing chest pain, so they went to the local hospital, where they managed to get his chest and abdomen X-rayed. There were no doctors on hand, but the equipment was state of the art, and the technicians were "sharp," according to Lonnie. The X-rays showed no abnormalities to this admittedly unqualified group, and no hard copy was made.

Shawn's party drove through the night to Lhasa and delivered him to the emergency room at the best hospital in Tibet at five the next morning, exactly twenty-four hours after they had left base camp.

Recognizing the importance of supporting Tibet's profitable tourist industry with good medical care, the Chinese government has built a fine hospital in Lhasa. Lonnie's expedition also had a "logistics man" stationed in the city, a member of the Communist Party, who made sure that Shawn received special attention. There were ten buildings in the hospital complex, and Shawn was admitted to the one reserved for party officials, the mayor of Lhasa, and other VIPs. Three hours after his arrival, he was lodged in the "Ambassador Wing," in the care of four doctors, including the director and assistant director of the entire hospital complex.

On the day he was admitted, a Friday, Shawn asked Lonnie to get in touch with his father, Brad Wight, a heating-and-cooling contractor in Ashtabula, not far from Cleveland, to make sure his medical insurance for the fall quarter was paid up. This was not a call for help. Shawn had been estranged from his father for years. He had put himself through college and his first few years of graduate school without his father's help.

Lonnie left a message on Wight's cell phone, apprising him of the situation, telling him his son was okay, and relaying Shawn's request about his insurance. He added that the situation was well in hand and that there was no need to call back. Indeed, he purposefully did not leave the hospital's number, for Shawn's doctors had specified that he should not take calls. There was no phone in his room, and he was not to be moved. The doctors wanted him to rest. The lack of a phone had nothing to do with the sophistication or modernity of the hospital, by the way: none of the rooms in the Ambassador Wing had phones, because the VIPs who usually stayed there often did so to escape their hectic lives. There was a secretary in the center of the wing who screened calls and passed on messages.

By Sunday morning, Shawn's chest pain was gone, the swelling had diminished, and his leg had returned to a healthy color. He was reading and doing crossword puzzles, and Lonnie was bringing him yak burgers and pizza from the nearby Holiday Inn. X-rays taken at this hospital showed no evidence of a kidney stone and no abnormalities in his chest cavity.

At breakfast that morning, Lonnie and Yao discussed their concern for the thirty-seven people they had left on the mountain. Since this was Yao's home country and he spoke the language, it seemed logical for him to remain in Lhasa to deal with Shawn's medical and travel arrangements and for Lonnie to return to the expedition. The two men bought Shawn a first-class plane ticket home, and Lonnie gave Yao five thousand dollars for other expenses that might arise. Lonnie says that when he discussed this plan with Shawn, he asked him if he wanted Keith or Mary to come to Lhasa and return with him to the States, but that Shawn replied, "No. If they came they wouldn't be able to finish their work."

So Lonnie left Lhasa. Things might have turned out very differently had he stayed one more day.

BACK IN OHIO, SHAWN'S FATHER WAS UNDERSTANDABLY CONCERNED AND, DESPITE Lonnie's message, had been trying to reach his son since Friday. He had called various administrative offices at Ohio State over the weekend, but found no one who knew his son's whereabouts. (Had he known more about Shawn's life at the university, he might have thought to call Ellen Thompson, either at home or at work. She was most likely working in her office and would certainly have been checking her messages.) On Monday morning, China time—which would have been sometime Sunday in Ohio—Wight reached the U.S. consulate in Chengdu, the capital of Sichuan Province, which borders Tibet to the east, and the staff there located Shawn. He reached Shawn by phone on Monday morning, Ohio time, which is either late Monday night or very early Tuesday morning in China.

Having missed Lonnie by hours, Wight then proceeded to override the plans that had been made on the spot by the people in Tibet, from half a world away. He contacted an infectious disease specialist at Ohio State, and the two held a conference call with the doctors in Lhasa, mediated by an interpreter. On the basis of that call, the specialist concluded that the Chinese doctors were not treating Shawn properly and recommended that he be moved to a hospital in Hong Kong. Wight ultimately decided to evacuate Shawn on an American Express charter flight; however, the charters were not allowed to land in Tibet; the nearest possibility was Chengdu, about a thousand miles east of Lhasa. So Yao bought five seats in the front of a commercial jet, to accommodate Shawn's stretcher, and flew with him and two of his doctors to Chengdu. (Ohio State paid for everything.)

Four days after Lonnie had left Lhasa, Shawn flew on another charter flight to Hong Kong, accompanied by a representative of the American embassy in Chengdu. That would be at least one day before the doctors in Lhasa had said he should be moved at all. His father was waiting for him in Hong Kong.

IN THOSE FOUR DAYS, LONNIE HAD RETURNED TO DASUOPU WITH FRESH SUPPLIES AND climbed all the way to the twenty-three-thousand-foot drilling camp—a risky move, because acclimatization is quickly lost when one descends. In 1981, he had induced pulmonary edema in just this way: after working at 18,600 feet on Quelccaya, he had descended to sea level in Lima and then proceeded directly to the top of 20,700-foot Chimborazo, the highest mountain in Ecuador, to drill a short core by hand. (Incidentally, the American doctor who treated his edema told Lonnie never to go to altitude again.)

The drilling team was very happy to see Lonnie, for they had been without a leader for nine anxious days, and Benjamín had taken ill the day after Lonnie had left with Shawn. "Thank God Benjamín could move down by himself, because there weren't enough of us to carry him," says Mary. Benjamín never returned to Camp IV.

"We had no leaders in the field," she continues, "and if Benjamín had had to be evacuated, I didn't know if I carried the authority to make the drivers take him out. Bruce was the boss of the drilling, but he didn't want to be the boss of the expedition, and neither did Vladimir nor I. So we decided that if a decision outside the drilling had to be made, it would be made by the three of us, and Bruce would deliver the orders. Being the oldest, his authority would have carried more weight with the Chinese."

Luckily, there had been no further health problems and the drill had continued to hum. They had retrieved a second core while Lonnie was gone, and with him back they drilled a third at the crest of the col, near the site of the second.

CONDITIONS WERE SEVERE. THEIR METEOROLOGY STATION WAS READING −20°F AT night, and the wind, which is the real killer, was blowing constantly. They were not only 1,500 feet higher than Sajama—which I had found absolutely frigid—but it was also about twenty degrees colder, and it snowed constantly.

The snow cave was a godsend, although drifting snow soon transformed its entrance stairs into a sort of slide. Upon arriving for breakfast in the morning or for dinner after a long day's drilling, they would simply sit down at the entrance and slide in. To get out to their tents at night, on the other hand, they had to crawl on their hands and knees up the ramp and out into the subzero temperatures and blowing snow.

At the very end, their only food was some microwave popcorn, which Mary had been teased for bringing along in the first place, because no one believed it would pop at twenty-three thousand feet. But the resourceful Jorge succeeded with a simple frying pan.

In the end, Mary lived and worked above twenty-three thousand feet for thirty days, Keith and Bruce for thirty-two days, and Vladimir for thirty-three days. These seem to be records for high-altitude endurance: as far as I can tell, no one else has ever stayed so high for so long in one continuous stretch.

After leaving the ice on October 14, Mary and Lonnie accompanied Yao to Lanzhou, where they split the first core with the Chinese, gave them the second, and shipped the third back to Ohio. The rest of the party returned to Kathmandu (taking the same lime-filled truck from Kodari back to the landslide that they had used on the way in). By October 20 both the people and the ice had reached Columbus.

They found Shawn Wight listed in "satisfactory condition" at the University Hospitals in Cleveland.

ACCORDING TO THE HONG KONG HOSPITAL RECORDS, SHAWN HAD BEEN RECOVERING from the blood clot when he arrived. His lungs had also been X-rayed, and nothing out of the ordinary had been found. At some point during the two weeks he was in Hong Kong, however, an opportunistic and rarely pathogenic bacterium named *Alcaligenes xylosoxidans* had infected a cavity near one of his lungs. The Hong Kong doctors had recommended that the resulting abscess be drained immediately, but Shawn's father had refused to let them do it and flown his son home to Cleveland.

Lonnie says that when he returned to Ohio, Wight made it very clear that he did not want him contacting Shawn. However, Mary, who was under no such constraint, called him right away and discovered that he had just undergone surgery to remove part of the lung. She says that Shawn was upbeat anyway and looking forward to returning to Columbus—that he "could hardly wait" to finish his master's thesis, a study of volcanism in the southern hemisphere based on Ellen's Plateau Remote core. Once that was done he planned to pursue a doctorate, with Lonnie as his thesis advisor.

Shawn had never shared much about his family life with his coworkers, but they eventually learned that his childhood and high school years had been extraordinarily painful. His parents had divorced when he was three, and their relationship remained intensely antagonistic more than twenty years later.

When he was released from the Cleveland hospital sometime near the first of November, Shawn and his mother, Jacqueline Clark, who had traveled from her Texas home to be with him, placed a call to Lonnie to ask if Shawn might take up his old position in the lab when his health returned. Lonnie says, "There was never any question about that." He remembers Shawn as being apologetic about his father's intervention overseas, which Shawn had believed he had not needed.

Shawn had been out of the hospital for only three days when his health took yet another turn for the worse. This time everyone from the Byrd Center went to visit him immediately. But the bacterium that had infected his lung was incredibly tenacious. His organs began to fail, and he lapsed quickly into a near-coma, in which he could communicate only with faint facial expressions. Mary visited him four times during this period. She was present with his mother and father when Shawn Wight died on November 26, 1997.

In its autopsy report, the Cleveland hospital listed the cause of death as high altitude pulmonary edema.

A MEMORIAL SERVICE WAS HELD AT OHIO STATE, AND SHAWN'S ASHES WERE LATER strewn in Arches National Park in Utah, the place he most loved. In Texas, two days after the memorial service, Jacqueline Clark filed for custody of Shawn's estate only to find that her ex-husband had beaten her to it. This is a prerequisite for filing a lawsuit on behalf of the deceased. Clark warned Lonnie that a suit was in the offing and stated in an e-mail that she did not want anyone getting rich over the death of her son.

Roughly a year later, in January 1999, about a week before Lonnie, Keith Mountain, and I left for the reconnaissance of Kilimanjaro, Brad Wight filed a twenty-one-million-dollar lawsuit against Ohio State University and Lonnie Thompson, claiming that their negligence had led to the death of his son. Lonnie was later dropped as a defendant. Despite her earlier protestations, Jacqueline Clark later joined as a plaintiff.

There was always the possibility of settling out of court; however, Lonnie wanted the case heard. "It came down to reputation," he says. "The problem with any settlement is that it's seen as a settlement. I wanted a decision based on right and wrong—clear cut and no doubt about it. It was important that I, after thirty-four expeditions, be able to defend my actions so that I could justify training future graduate students to go to remote parts of the world. Our greatest challenge as teachers is not so much to teach students how to use today's technology, but how to think so they can solve tomorrow's problems. We need to inspire them to set themselves apart as risk-takers, dreamers, and doers. I've had a great life. I want to make sure the next Lonnie Thompson can have a great life, too."

The two sides did get together once, in the library of the Byrd Center, to discuss settlement. The plaintiffs requested $3.7 million; the university offered to pay their legal expenses and put seventy-five thousand dollars in a Shawn Wight Memorial Fund, to support future graduate study and fieldwork. No agreement was reached. As it happened, in the midst of the weeklong trial, which took place in June 2000 (after we returned from deep drilling on Kilimanjaro), Lonnie and Ellen received word that the prestigious *Journal of Geophysical Research* had accepted a paper based on Shawn's thesis work. The Thompsons dedicated the paper to Shawn's memory.

Since the trial had implications for field programs nationwide, other universities watched it closely; and, at first, the local news media, including Columbus television, paid attention as well. Going in, their coverage had a certain "David and Goliath" slant: the wronged family (and no one would deny that they had suffered a tragedy) against the famous professor and the big university. It was not a jury trial; it was heard by a judge, and it began on a Monday morning. The plaintiffs presented their case first, calling Lonnie as the first witness, and it would be fair to say that the trial was over by the time he left the witness stand. His integrity, honesty, intelligence, impeccable professionalism, and concern for Shawn shone through instantly. (He believes the plaintiffs' lawyers were unprepared, having expected to settle.) And as the true story came out, the television crews withdrew. Only one remained for Friday's closing arguments.

Although the legal proceedings do not seem particularly pertinent here, the decisions Lonnie made at altitude do, somehow. The leading experts on high-altitude physiology provided expert testimony at the trial: Charles Houston on behalf of the plaintiffs; Peter Hackett on behalf of the university. Other medical experts appeared as well.

None believed in the accuracy of the autopsy prepared at the Cleveland hospital. One cannot die of high altitude pulmonary edema two months after leaving high altitude. The infection and its complications were the most likely cause of Shawn's death. The judge refused to accept the autopsy as evidence.

Dr. Hackett testified that he believed Lonnie had acted more conservatively than necessary even in taking Shawn down from the drilling camp when he could not recall names. The fact that Shawn was capable of descending by himself indicates that he did not have cerebral edema. Hackett believed he had incurred a type of migraine in which the rupture of small, isolated blood vessels in the brain leads to temporary memory loss. Such ruptures usually repair themselves within hours (as they seem to have in Shawn's case). In any event, Lonnie had responded as he should have if it had been edema. You descend until the symptoms disappear. There is no magic altitude; you keep going until they're gone.

The plaintiffs also questioned Lonnie's acclimatization strategy. When Dr. Houston took the stand, however, one of the university's lawyers convinced him that Lonnie had ascended more slowly than Houston himself recommends in his own book. After the trial, Houston wrote to an internist who had been prepared to testify for the university that "once he was apprised of all the facts," he realized "he had been duped to get on a losing horse."

It was not a horse race, of course, although the internist may have been right in characterizing the trial as an exercise in "fatuous gymnastics." It was the agonizing end to a story that went from bad to worse from the moment it began. Everyone I've spoken to has expressed compassion for Shawn and his family first and foremost.

In March 2001, the judge ruled that Ohio State University had not been responsible for Shawn Wight's death.

PART VI

KILIMANJARO

IF . . . THE READER GATHERS THAT WE HAVE DONE SOMETHING MORE THAN MERELY "TRAVEL," I SHALL FEEL AMPLY REWARDED FOR THE TIME, THE TOIL, AND THE MEANS I HAVE SPENT IN THE EXPLORATION OF GERMAN EAST AFRICA.

—HANS MEYER

· 19 ·

SCRATCHING THE SURFACE

Before breakfast on our first morning in Africa, I wandered out to the dirt road in front of Key's Hotel to admire our objective. The highest mountain in Africa was not as "unbelievably white" as it had been for Hemingway, but it did seem nearly "as wide as all the world."

The Kilimanjaro massif consists of three dormant volcanoes standing in an east–west line, which together comprise more than one thousand cubic miles of rock and dirt. The massif rises alone from a flat plain at the eastern edge of a rash of volcanic bubbles known as the Great Cauldron, which emerges by the Rift Valley to the west and fans east along the northern edge of the Maasai Steppe. From my southern vantage point that morning, the massif was fifty miles wide and thirty miles deep, and I couldn't really appreciate its size just by looking at it. Perhaps I do now, after having climbed it a few times and lived on its summit for about a month. Outside of the moon and stars, it is probably the largest single thing I have ever seen.

The peak that most people think of as "Kilimanjaro" is actually 19,344-foot Kibo, the central, highest, and newest of the three volcanoes. Next in order of age is the rugged tower of Mawenzi, the plug of lava exposed by the erosion of the volcanic ejecta that once enveloped it, which stands seven miles to Kibo's east, across a fourteen-thousand-foot plateau known as the Saddle. The oldest of the three is Shira, which lies nine miles to Kibo's west and seems more of a foothill, since it is only thirteen thousand feet high (although geologists believe volcanic pressure once pushed it as high as eighteen thousand feet). Shira's five-mile-wide crater was transformed into a circular plateau when it was filled with lava during the enormous eruption that created Kibo, six thousand feet above it. That must have been quite a sight.

Kilimanjaro is one of a handful of mountains claimed by competing parties to have the greatest vertical relief on the planet. Alaska's Denali is another; so are Nanga Parbat and Rakaposhi in Pakistan; there must be more in the Himalaya. Whether it wins the contest or not, it's pretty tall. What thin veneer remained of Hemingway's famous snows gleamed in the air almost seventeen thousand vertical feet above my head.

This southern side of the mountain is sometimes known as the breadbasket of Tanzania, and it has an energetic, festive air—unlike the harsh wasteland below Sajama. In contrast to the Andes and the western United States, where the mountains act as water towers, storing the precious substance as snow and metering it out through the dry seasons, Kilimanjaro makes rain almost daily. Kibo is rarely visible at sunset: by midmorning the massif is usually shrouded in clouds that may climb to two or three times its height; and as we would learn by living in the summit crater, these cloud towers are generally hollow, like smokestacks. In the dry season at least, which is when we were there, the sun almost always shone in the summit crater, and even though massive clouds swirled around us every afternoon, the sky was almost always blue directly overhead.

Since Kilimanjaro sits at three degrees south latitude, the Hadley Cell, or Intertropical Convergence Zone, drags the monsoon through the region twice a year. On its southward passage the ITCZ brings the "short rains," from mid-October to mid-December, and on its way back north it brings the "long rains," from mid-March through the end of June.

The rich land in the mountain's rain shadow, to its west and southwest, has attracted farming peoples for centuries. During the vast Bantu migration that swept sub-Saharan Africa more than five centuries ago, a tribe now known as the Wachagga, or Chagga, settled the region. Their fields and irrigation ditches quilt the plains and climb the mountainsides nearly to the nine-thousand-foot contour that defines the boundary of Kilimanjaro National Park.

Most of Moshi's people are Chagga, and it is a country town. Nearly everyone walks or rides bicycles to work or school. I watched smiling women in multicolored *kangas* and *kitengas* stride silkily by with bunches of fruit or laundry in bright red, green, or blue plastic bags balanced on their heads and sinuous necks. A boy in rags sat in the shade of a tree, holding a stick with a string dangling from one end, tending three scrawny goats who nosed the red dust. Groups of nuns passed by, dressed completely in white, along with troops of uniformed schoolkids in clean white shirts, dark shorts, and knee socks with horizontal stripes—the local version of the old school tie, sported also by the occasional businessman strolling along in jacket and tie.

LATER THAT MORNING WE MET WITH PHIL NDESAMBURO IN HIS SPARE CORNER OFFICE on the first floor of Key's. Phil is a charming and charismatic man—with a ruthless business ethic. His eyes lit up as he came to understand the scope of Lonnie's plans, and he quickly offered to be our agent in Tanzania, both for this short probe and for the full-scale drilling that would follow. We liked him immediately. He had a warm, old-fashioned way of holding my hand as we walked the grounds in conversation. On that innocent, low-budget reconnaissance, his charm managed to outpace his rapacity.

That afternoon we planned our climb with Phil's top guide, Julias Minja. Guides are required in the park, and they undoubtedly prevent many untoward adventures and deaths; however, they also provide significant income to the tour operators and the park. More than half the fee we paid for Julias and three-quarters of that for the porters went directly to Phil. The porters survive mainly on tips.

Kilimanjaro is billed as the highest mountain in the world that can be climbed by a regular tourist, and I've been told that it generates more foreign currency than any other enterprise in the country. A nonresident pays an outrageous eighty-five dollars a day in park fees for himself, plus five dollars a day for each of the three porters/guides he typically employs—so about a hundred dollars a day. According to the booking warden at park headquarters, 18,327 people attempted the mountain in 1997, generating nearly thirteen million dollars in park fees alone, and only about half made it to the top.

More popular a spot than Lonnie's usual drilling site . . .

WHEN JULIAS LEARNED THAT KEITH, LONNIE, AND I HAD ALL CLIMBED HIGHER MOUNTAINS before, he decided to take us up this one from the west, by a combination of routes prized for their beauty and solitude. This would be more difficult, but it would allow us to avoid the hordes on the standard Marangu route, which ascends from the Saddle and is known to the guides as the "Coca-Cola route," because 90 percent of the Kilimanjaro hopeful choose that way—or have it chosen for them.

This was a standard reconnaissance: Lonnie wanted to inspect the three snowfields on the summit, decide which to drill, sound the deepest, and hand-drill one or two short cores to determine if there was any climate signal at all in these fabled snows. If so, he would generate preliminary data for a grant proposal for deep drilling the following year. Aside from the fact that we would camp on top, this was not significantly more complex than the usual Kilimanjaro trek. We took care of the details in short order, then chatted with Julias over popcorn and beer.

He is Chagga by heritage and was thirty-eight at the time. His father had been a Kilimanjaro guide, and Julias himself had climbed the mountain more than two hundred times at that point. Though he never finished high school, he is a born leader and a thinking man. As the conversation ended, I asked if he had noticed any changes in the glaciers in the fifteen years he'd been guiding.

"The glaciers are shrinking, because the earth is getting warmer," he said without hesitation.

"No need to call in the experts here!" laughed Lonnie.

"Maybe he should testify before Congress," I chimed in.

THE NEXT MORNING, WE RODE IN A FLEET OF 4×4S THROUGH THE OPEN COUNTRY WEST of Moshi, the air black from the smoke of burning sugarcane fields on both sides of the road. Kilimanjaro dominated the northern sky, its upper slopes hidden by cumulus clouds that beetled in inverted tiers to form a mirror image of the massif in the sky. We soon turned north, off the main road, and entered the farmland in the mountain's rain shadow. To our left, dry Maasai country stretched west to the Serengeti Plain. Tractors tilled the soil, sending up high tails of red dust. Smoke rose from more burning cane fields. Solitary Maasai shepherds tending small herds of goat or cattle stood stock-still in the shimmering heat with their arms draped in Ts on the sticks across their shoulders. A rough dirt road led us straight across a flat plain, Kilimanjaro to the right, the blue pyramid of Mount Meru, Africa's fifth highest, to the left, small parasitic cones of the Great Cauldron bubbling into the distance before us.

This land was still in the grip of the La Niña that had followed the El Niño that had hit Sajama a year and a half before—shortly after Lonnie drilled there. And the people who live here do not need Mark Cane's computer models to know that El Niño brings rain and his sister brings drought to this region. The short rains had never come this year—in fact, there had been no significant rainfall for nine months—and famine raged in central and western Tanzania. The prime minister had recently warned that three hundred thousand people in the Dodoma region faced starvation within five months, while the Kilimanjaro region, sustained by the moisture bestowed daily by the mountain, was managing just to struggle by. The tilling and burning all around was in expectation of the long rains, which were due in about a month.

Both extremes of the El Niño/Southern Oscillation create misery in the

Horn of Africa. The recent El Niño, which had come on the heels of yet another drought, had induced one of the most devastating rainy seasons in memory—a sobering prospect if global warming indeed brings the permanent El Niño that many climatologists anticipate. Some parts of Tanzania received forty times their average annual rainfall. Floods obliterated roads and entire villages, and the country's tenuous systems for delivering food and health care were disrupted and in large part destroyed. The health care system would have been taxed anyway, for increased rain nearly always leads to outbreaks of vector-borne disease.

The 1997–1998 El Niño had brought tens of thousands of cases of cholera and malaria and eighty-nine thousand cases of Rift Valley fever to the countries of the Horn, the fever alone resulting in nearly a thousand deaths. At the height of the chaos, a relief worker for the World Health Organization described one village where, "I told the mothers not to feed their children the meat of dead and diseased animals. But they said they had nothing else, and I couldn't disagree." And after watching malaria claim forty-three lives in forty-eight hours in a small-town hospital in Kenya, a district commissioner said, "In the wards, up to three patients were sharing a bed. In one of the male wards, two elderly patients were in the same bed with the corpse of another patient who had been dead for several hours."

Like the Aymara in Bolivia, tens of millions of poor people in Africa, who lead a marginal existence even in the best of times, stand to suffer more in a warmer world.

IN THE MOIST FOOTHILLS DIRECTLY WEST OF THE MOUNTAIN, THE PLANT LIFE GREW lush and the towns took on the aspect of rural Appalachia. Footpaths the color of red Georgia clay wound between clusters of rectangular log cabins. Women in bright prints walked toward a stream in single file with plastic buckets perched on their heads. A bit farther on we entered the second of Kilimanjaro's five ecological zones, the rain forest band, and waited for a while at the park entrance in the town of Londorossi, which is a timber and match concession. Julias told us that no permanent structures were allowed there; the town is made entirely of wood.

A four-wheel-drive road led from Londorossi through a forest of unique pine, dripping combs of long, waving needles. This opened on the moorland zone, covered with low, brown and olive heaths and heathers. With views of wide hills to the front and sides and Meru and the Serengeti behind us, we crawled up the side of the Shira Plateau and stopped in a chill fog by a barrier gate at twelve thousand feet. I felt the first vagueness of altitude and searched in my pack for a coat. Climbing a steep hillside, we gained the plateau and surged and wallowed on a muddy road to Simba Cave, our camp for the night (not entirely reassured to learn that *Simba* means "lion" in Swahili, the lingua franca of East Africa). A golden jackal sprang from the heather and bounded across the road just as we arrived, and Julias told us that eland, leopard, lion, and elephant visit the plateau in the rainy season. Kibo hid to the east behind the inevitable afternoon clouds.

The next day we walked a few miles across a moonlike landscape of black spherical boulders in all sizes, spilling across a slanting surface of copper sand. Golden brown heathers and tan tussocks of grass grew between the boulders, and slate-green or rust-orange Spanish moss hung from the boulders and the heathers—a product of the frequent mist on this side of the massif. We passed a stand of *Senecio kilimanjari* tracing the bed of a creek: yucca-like trees, some as high as thirty feet, with thick shaggy branches ending in bushy green fists.

The porters took up residence in a fourteen-thousand-foot Moir hut—constructed of plywood and shaped like a Gemini space capsule—while Keith, Lonnie, and I retreated to the peace and privacy of our tents. Time to slow down. We took a short acclimatization hike the next day and stayed at Moir Camp another night.

On the day of the hike, we returned early enough for Lonnie and me to enjoy afternoon tea in the warm sunlight on a brightly printed *kitenga* spread on the sand. Munching on popcorn and roasted nuts prepared by one of Julias's brothers, we discussed the important things in life and found that we agreed on most.

"Nice place for a chat," I said, waving toward Meru and the Serengeti.

"Well, it is," said Lonnie. "That's another thing about human beings. I don't know if there's any truth to it, but when I've looked at religions around the world—where they got their messages and their insights—it seems that it was often on a mountain. And the religions that come out of isolated and remote mountain areas also tend to be contemplative. Mountains are interesting places. . . . There are a lot of words in the world of science and everyday affairs, but words don't work on mountains; you have to experience them. You can't argue with them or compromise; it's hard reality. Just being in these places pulls something up from deep inside a person. Makes for clearer thoughts. Not sure why. Maybe just the physical struggle of getting up there sharpens your perceptions."

"Like the American Indians doing vision quests on mountaintops?"

"Right. I've often wondered if there's not something to that."

Keith wandered through, muttering in his standard unprintable argot about the bones of an elephant he had found nearby. He grabbed his camera and disappeared.

"As the highest mountain in Africa, Kilimanjaro is the only place on the continent that has any chance of holding an ice core record," Lonnie continued. "But who knows? We may be too late already. This whole thing hinges on something being up there, and just from looking around I'd guess that all the ice on this mountain will be gone in ten to twenty years. I'm worried too much melting may already have taken place. Guess we'll find out fairly soon . . . But if there is a record I bet we'll see El Niño. The prevailing wind comes from the east, the Indian Ocean, and Kilimanjaro is the first thing it hits."

Even a century ago, Gilbert Walker and his contemporaries were aware of the relationship between the African and Asian monsoons. As *The Imperial Gazetteer of India* put it in 1908, "[I]t is now fully established that years of drought in western or northwestern India are almost invariably years of low Nile Flood. The relation is further confirmed by the fact that years of heavier rain than usual in western India are also years of high Nile Flood." The strength of the flood is determined by the summer monsoon (the long rains) in the headwaters of the Blue Nile, which lie in the mountains of Ethiopia, not far to the north of Kilimanjaro; and the annual spilling of water onto the Nile floodplain has sustained Egyptian agriculture for nearly seven thousand years, as the Nile Valley itself has received virtually no rain in all that time.

And for more than five thousand of those years Egyptians have kept records of the Nile flood. William Quinn, a now-deceased friend of Lonnie's, once compared those records with recent historical accounts of El Niño and showed that they were connected.

"Bill thought the frequency of the El Niños here may have changed in the last thousand or so years, just as it has in the Andes," said Lonnie, lounging on an elbow. "He said the cycle of seven good years followed by seven bad years, which Joseph prophesied in the Bible, was not an unusual thing in biblical times. It may actually have been an El Niño cycle. In fact, Joseph may have been the first climatologist.

"We were having a beer together down in Lima, Peru, when Bill first told me that. He was just starting his investigation. Then he came over to Egypt to see how the Nile flood was actually measured."

The first Nilometers, or river gauges, were put in place in 3500 to 3000 B.C. They evolved from simple marks on riverside cliffs to more elaborate and accurate wells that siphoned off the river water. The level was marked on the walls of the well or on a central pillar within.

"The keepers of the early Nilometers were Copts," wrote Quinn. "Joseph himself is said to have built the first Nilometer at Bedreshen on the west bank of the Nile near the remains of a wall which was said to be part of the 'Granary of Joseph' at the foot of which the Nile used to flow."

According to the Book of Genesis,

> . . . Pharaoh dreamed: and behold, he stood by the river.
>
> And, behold, here came up out of the river seven well favoured kine and fatfleshed; and they fed in a meadow.
>
> And, behold, seven other kine came up after them out of the river, ill favoured and leanfleshed; and stood by the other kine upon the brink of the river.
>
> And the ill favoured and leanfleshed kine did eat up the seven well favoured and fat kine. So Pharaoh awoke.
>
> And he slept and dreamed the second time: and, behold, seven ears of corn came up upon one stalk, rank and good.
>
> And, behold, seven thin ears and blasted with the east wind sprung up after them.
>
> And the seven thin ears devoured the seven rank and full ears. And Pharaoh awoke, and behold it was a dream.
>
> And it came to pass in the morning that his spirit was troubled; and he sent and called for all the magicians of Egypt, and all the wise men thereof: and Pharaoh told them his dream; but there was none that could interpret them unto Pharaoh.

It was the Hebrew "dreamer" Joseph who finally interpreted them:

> "Behold, there come seven years of great plenty throughout the land of Egypt: And there shall arise after them seven years of famine; and all the plenty shall be forgotten in the land of Egypt; and the famine shall consume the land."

Quinn dubbed this prophecy "probably the first indication of persistence in the hydrologic time series" and pointed out that "[i]t appears that the seven-year famine foreseen by Joseph set in about 1708 B.C." Having also found that long periods of low Nile flood occur more frequently in cold times, he guessed that Joseph lived in a cold time.

"Bill also looked at the historical records of El Niño in coastal Peru and Ecuador, going all the way back to the days of the conquistadors," Lonnie continued. "He believed part of the reason the Spanish [Francisco Pizarro and his motley band] conquered the Incas so easily was that an El Niño set in the year their ship put into port north of Lima. That meant there was water and grass in the desert, so their horses could eat as they walked into the mountains. Had they come in a regular year, they would never have gotten across that band of desert by the coast."

MIDWAY THROUGH THE NEXT DAY OUR PATH MERGED WITH A STREAM OF TREKKERS ON the popular Machame route, and our splendid solitude came to an end. We walked east in a long line on the path that circumnavigates Kibo at the fourteen-thousand-foot contour, through a "high desert" zone wearing every shade of brown, black, and gray. Shin-high chocolate-colored sage grew sparsely on the volcanic dirt.

Holding back patiently, Julias shepherded us into sixteen-thousand-foot Arrow Camp, where our three tents joined fifteen more, while fifty or more porters huddled in, around, and under a cluster of boulders and metal sheets, scrounged from a destroyed hut. Finally in the high country, breathing crisp alpine air in a world of lichen, snow, and rock, we crawled into our tents to sleep off the altitude until dinner.

Arrow Camp lies on a flattened spine in the lower, nearly level section of the Western Breach, a wide bowl rising two thousand feet to Kibo's crater rim. Looking up, the Breach is bound on the left, or north, by a long ridge spiked with anthropomorphic gendarmes, and on the right by a soaring, dead vertical, diamond-shaped wall. Below the camp, the spine narrows to a ridge and sends up a rocky spur before dropping off sharply.

After dinner, I climbed out to the tip of the spur and looked up at the Breach, towering clay red in the setting sun. It held very little snow, and what remained seemed seasonal, sustained by the shade of the diamond wall. Arrow Camp is named for a glacier of that shape, which Julias can remember descending to the level of the tent sites. It was once a source of water for the camp. Now it is virtually gone.

I turned out to gaze at a scarlet and blue-gray cloud-ocean under a clear, darkening sky. The arrow tip of Mount Meru poked through to the west. A few cloudy wisps floated up to my level, and huge, dark gulfs revealed patches of the volcanic plain below—in a silence as vast as the view.

THE FIRST HISTORICAL MENTION OF AFRICA'S GREAT SNOW MOUNTAIN DATES FROM THE second century A.D. According to Ptolemy's *Geography,* the mountain stood inland from a market village, twenty-three days south by sail from Ras Hafun, a town that still exists on the Somali coast. The Arabs who colonized East Africa from the sixth to the sixteenth centuries built towns on the coast and developed a rich trade in ivory, gold, rhinoceros horn, and, most profitably, slaves through caravans to the interior. Once the Chagga settled the southern slopes of Kilimanjaro, the region became a center for the Arab slave trade and a stopping point for caravans. There were more than twenty tiny kingdoms in the five hundred square miles of Chagga, and from the accounts of the first European visitors it would seem that the main pastime of the local chieftains was to form temporary alliances with one set of neighbors in order to launch slave raids on the others. They kept some prisoners as wives or slaves for themselves and sold the rest to the passing caravans.

The next reference to Kilimanjaro after Ptolemy is found in *Suma de Geographia,* published in 1519 by the Spaniard Fernandez de Encisco, who learned of its existence on a visit to the port city of Mombassa, in conversations with caravan leaders. "[W]est of this port stands the Ethiopian Mount Olympus, which is exceedingly high," he wrote. It would have been a landmark for the caravans, of course. In Swahili, *kalima* means "mountain," evidently, and *njaro* means "caravan." Some called it simply *Njaro.* To the Maasai it is *Ngaje Ngai:* "House of God."

By the second half of the nineteenth century, the search for what Sir Richard Burton called "the coy sources of the White Nile" was in full swing, and the stories of those who explored the interior of the "Dark Continent"—with names such as Stanley, Livingstone (and Burton)—made headlines throughout the Western world. (They still make great reading.) According to Ptolemy—who got pretty much everything right—the Nile's longer, western arm flowed from large lakes in the center of the continent, near a range he called the Mountains of the Moon.

The first Europeans to see Kilimanjaro were the Swiss missionary Johann Rebmann and his German colleague Johann Krapf, a persistent pair who originally intended to establish a necklace of missions east to west across the continent but, with thirty years of trying, made it only about one-tenth of the way. They didn't have much luck spreading Christianity, either. When the ravages of malaria finally forced Krapf to quit Africa after eighteen years—his wife and infant daughter having already succumbed to the disease—he and Rebmann had failed to convert a single soul.*

They founded a first mission near Mombassa, a second ninety miles inland, and resolved to found a third at "Jagga," which lay, according to Swahili traders, at the foot of a high mountain called Kilimansharo, which was shrouded in silver and "full of evil spirits (Jins)." Rumor held that

*Three years later, however, when Krapf received a letter from Rebmann announcing that a woman near one of their missions had begun to pray, he observed prophetically, "God allows great things to arise out of small and insignificant beginnings, as He has promised:—'A little one shall become a thousand.' " Today, more than 90 percent of the Chagga people practice Christianity.

"people who have ascended the mountain have been slain by spirits, their feet and hands have been stiffened, their powder has hung fire, and all kinds of disasters have befallen them."

In 1848, Krapf having been laid low with a recurrence of malaria, Rebmann journeyed to Chagga alone. On the morning of May 11, he wrote afterward,

> The mountains of Jagga gradually rose more distinctly to my sight. At about 10 o'clock (I had no watch with me) I observed something remarkably white on the top of a high mountain . . . and while I was asking my guide a second time whether that white thing was indeed a cloud and scarcely listening to his answer that *yonder* was a cloud but what that white was he did not know but supposed it was *coldness*—the most delightful recognition took place in my mind of an old European guest called *Snow*. All the strange stories we had so often heard about the gold and silver mountain Kilimansharo in Jagga, supposed to be inaccessible on account of evil spirits . . . were now rendered intelligible to me, as of course the extreme cold, to which the poor natives are perfect strangers, would soon chill and kill the half-naked visitors.

He strolled on toward Chagga, equipped with only an umbrella, and stayed for a few days in a village near the base of the mountain. In three subsequent visits to Chagga over the years, he viewed the snows many times, learning that *kibo* means "snow" in the Chagga language and that when the locals had bottled some of the "silver-like stuff" and brought it down, it had turned to water. His attempts to establish a mission, however, failed completely.

Krapf saw the mountain about a year after Rebmann's sighting, on an expedition of his own on which he also happened to catch the first European glimpse of Africa's other great snow mountain, Mount Kenya.

BUT THE FIRST REPORTS OF SNOW IN EQUATORIAL AFRICA INVOKED NEARLY AS MUCH derision in the late 1800s as the reports of its imminent demise do, for different reasons, today.

The most virulent attacks on the veracity—and indeed sanity—of Krapf and Rebmann were launched by one William Desborough Cooley, a "critical geographer" and Fellow of the Royal Geographical Society, who had gained a reputation as an authority on equatorial Africa without ever having bothered to leave England. After analyzing many old texts (discounting Ptolemy, evidently) and interviewing a single Arab caravan leader on a visit to London, Cooley was disinclined to accept Rebmann's eyewitness account: "I deny altogether the existence of snow on Mount Kilimanjaro. It rests entirely on the testimony of Mr. Rebmann . . . and he ascertained it, not with his eyes, but by inference and in the visions of his imagination." (The missionaries, by the way, couldn't care less about the controversy. "Why should I endanger my mission for the sake of science?" wrote Krapf. "Let Geography perish.")

In 1861, Carl Claus von der Decken, a German baron with serious amounts of time and money on his hands, decided to settle the matter once and for all. At the end of June, he and a young British geologist named Richard Thornton left Mombassa at the head of a company that included more than fifty porters, slaves, and servants. They sighted the snowcapped mountain on July 14, thirteen years after Rebmann, and proceeded to make the first European attempt on the summit. Weather turned them back at about eight thousand feet, still within the "primaeval forest."

Thornton put Kibo's height at between 19,812 and 20,655 feet (ever so slightly high) and was first to realize that the massif was volcanic in origin and put the relative ages of the three caldera in the correct order. He died of malaria two years later, on an expedition up the Zambezi River with David Livingstone, a few days short of his twenty-fifth birthday.

Von der Decken undertook a second summit attempt in November 1862 with a German scientist named Otto Kersten. They got to about fourteen thousand feet and claimed to have experienced a heavy fall of snow.

From his desk in Britain, Cooley congratulated the "Sporting Baron" on "the opportuneness of the storm," but found it "easier to believe in the misrepresentations of man than in such unheard-of eccentricity on the part of Nature." According to Hans Meyer, whom we'll meet in a moment, Cooley's "unwarrantably fierce attacks upon the worthy [missionaries] . . . combined with a subsequent passage at arms which he had with Von der Decken, won for this otherwise estimable savant a certain degree of unenviable notoriety." While Cooley went to his grave denying the possibility of snow in Africa, von der Decken's African discoveries earned him the Gold Medal of the Royal Geographical Society in 1863. He was murdered two years later, at the age of thirty-three, on an expedition to Mount Kenya.

IN THOSE DECADES, THE STARTING POINT FOR MOST EUROPEAN ADVENTURES ON KILIMANJARO was the same as ours more than a century later: the tiny kingdom of Moshi, which was ruled by a fearsome and capricious one-eyed chief named Mandara. A wide-eyed Methodist missionary named Charles New paid a visit in 1871, showered Mandara with enough gifts to secure two shots at the summit, and became the first European to make it as high as the snow line—probably about thirteen thousand feet at that time on the side he climbed it. New fell in love with the beauty of the place and returned two years later to found a mission in Moshi. However, in the words of Sir Harry Johnston, a clever British explorer who spent six months in the region in 1884, "Mandara received the missionary New, and on his second visit robbed him courteously of many valuables." New departed ill and exhausted, with no supplies or gear, and died before reaching Mombassa. This explains the ten-year hiatus between his visit and Sir Harry's.

Johnston was a tiny man—exactly five feet tall—with a high-pitched voice and a lively sense of humor, who led an astoundingly prolific life. He was at various times an explorer, a colonial administrator, a painter, a botanist, a comparative zoologist, and an author. (He wrote more than sixty books about Africa.) He was also a natural linguist, fluent in Arabic, Italian, Spanish, French, Portuguese, and, to varying degrees, thirty different African languages. Of course, he was also a great raconteur. In the dinner talk he gave to the Royal Geographical Society about his own visit to Moshi, he explained that he used guns and other weapons only for their theatrical value and provided a few moments of high comedy with his description of the night he repelled an attack by Mandara's warriors simply by setting off fireworks. He may sometimes have been given to exaggeration, however: his claim to have climbed above the Saddle on Kilimanjaro was subsequently discounted owing to inaccuracies in his description of the terrain, and there is no corroborating evidence for the claim in his autobiography that he was acting as a British secret agent on his ostensibly botanical mission to Kilimanjaro.

On the other hand, some colonial intrigue was definitely afoot at the time of his visit. Germany's "Iron Chancellor," Bismarck, having hitherto maintained a strict policy against colonial expansion, developed an interest in East Africa in the 1880s, when he saw a use for it in the complex diplomacy of Europe. ("My map of Africa lies in Europe," Bismarck famously declared.) Mandara took advantage of colonial competition (and made off with a few extra baubles) by signing treaties awarding suzerainty to both the British government and the German East Africa Company within days of each other, but control of the Kilimanjaro region reverted to Germany in the end. Thus, in 1887, a second immeasurably wealthy German aristocrat, Hans Meyer, decided it was his "national duty" to climb the highest mountain in the German Empire.

On his first try, he and a companion made an attempt on Kibo from a camp in the Saddle. His companion quickly succumbed to exhaustion, and Meyer was required to "press forward alone." He was finally stymied by a "solid wall of ice."

The next year, he organized a grandiose expedition "with a view to exploring the German sphere of influence throughout its entire breadth." So little was known of the African interior just 120 years ago that when Germany and Britain agreed to divide East Africa in 1886, they simply drew a line running northwest from the coast, with no defined endpoint. The land to the north became British East Africa, now Kenya; the land to the south became German East Africa, then Tanganyika—and later Tanzania, when Tanganyika merged with the island of Zanzibar.

Meyer planned not only to climb Kilimanjaro but to explore Lake Victoria (the true source of the White Nile, though this was not known at the time), the Ruwenzori Mountains (probably Ptolemy's Mountains of the Moon), and Lakes Edward and Albert in present-day Uganda. A few weeks in, however, he was attacked by a marauding Arab sheik, and Meyer and his companion, an Austrian geographer, were "overwhelmed and made prisoners, loaded with chains, and thrown into a dark hut, where we were left to lie for some days, ignorant of what fate might be in store for us." They finally paid a ransom to the sheik and were escorted to the coast. "The expedition was totally ruined, and all of our European equipments, with the goods intended to last a caravan of 230 men for two years, were lost."

On his third try, in 1889, he didn't take any chances. He limited himself to a climb of the mountain and a geologic survey. His company included an eight-man Somali bodyguard, sixty-two porters in the charge of two Swahili headmen, and a gymnastics teacher and experienced alpinist named Ludwig Purtscheller. Even the porters were armed, although Meyer confiscated their weapons at night to forestall desertion or mutiny.

Climbing Kilimanjaro was different in those days; the trek from Mombassa took a few weeks. A Pangani guide walked at the head of Meyer's procession, carrying the German flag, and the boss walked second, mapping and charting the terrain as they went. They followed a strict daily schedule, stopping at midday to set up camp, whereupon Meyer and Purtscheller would "fortify their inner men" with a meal, and Meyer would light his pipe and sketch a rude map of the day's journey. Then he would venture into the field to take photographs, with a Somali to carry his camera, while Purtscheller gathered botanical and geological specimens. "On my return to camp," wrote Meyer, "it behoves me to mount the judge's chair, and to mete out condign punishment to evildoers at the hands of the Somál, ten to twenty lashes being the quantum for ordinary offences." Only three porters dared desert.

In Moshi, Meyer found Mandara more irritating than threatening and put up with his antics for only a few days. He then proceeded east to Marangu and struck a deal with Mandara's bitter enemy, Mareale, whom he saw as a more dignified and courteous man. While Mandara had a weakness for European suits, Mareale loved fine cloth, so Meyer bought him off with a sewing machine. Thus, the first ascent of Kilimanjaro began in Marangu; Marangu became home to the national park headquarters, and certain elements in Moshi seethe with resentment to this day.

LONNIE, KEITH, AND I WERE CLIMBING THE MOUNTAIN ON THE OPPOSITE SIDE FROM Meyer. However, Meyer was not one to do things halfway: after his successful ascent, he led his caravan around to our western side in order to complete his geological survey. Viewing Kibo from the west, he saw a "great glacier, which issues from a stupendous fissure with precipitous walls, by which the cone is here cloven from head to foot [undoubtedly the Western Breach]. The glacier is formed partly by the ice which issues from the Kibo crater through the great notch on its western side, and partly from névé [compacted seasonal snow] in the fissure itself. In a vast sheet over 1,500 feet thick, the glacier descends like a cascade from a height of 18,700 feet to below 13,100 feet."

In other words, a little more than a century before I sat on a spur looking up at a bare, red Western Breach, it had been filled with a glacier that had reached nearly three thousand feet below my toes.

NOWADAYS THE GUIDES METE OUT CONDIGN PUNISHMENT ON THE CLIENTS. AT MIDnight on the one night we were to spend at Arrow Camp, all of our fellow campers—even a troop of French ladies in their fifties—were herded into the darkness for their one summit attempt. They had spent less than twelve hours above sixteen thousand feet. I'm amazed that even half make it to the top.

AS THE MORNING SUN CLEARS THE DIAMOND, LONNIE, KEITH, AND I PLOD UP A STEEP, unbroken slope in loose sand. Julias pads along like a cat, unaffected by the altitude, waiting calmly whenever we flatlanders need a break.

It hits me almost immediately: nausea, listlessness, apathy, a tendency to doze off when I sit, dark voids between superficial thoughts; I mutter a few words and watch them blow away on the wind. Through the mental fog, I am dimly alarmed at Lonnie's vicious cough. He tells me it's exercise-induced asthma from the marathons he used to run and vows to see a doctor for some new drug when he gets home.

The ground becomes steeper—broken rocks on loose sand, an even chance of sliding backward on every step. I try to climb at a constant pace, but Lonnie climbs in spurts, hunching athletically into the hill for a minute or so, then bending down with his hands on his knees to gasp and hack, while Keith scrambles along in uncharacteristic silence, as amiable as ever. Near the top, we mince across loose piles of rock between rickety towers, which arise and disappear in the windblown mist. Some steep sections require hands, and on a few of these the consequences of a fall appear serious. Around the fire at Moir Camp two nights before, Julias told us that *Kilimanjaro* comes from the Chagga word *Kilemankiaro:* "journey without end."

But when it does end, it ends abruptly. Climbing up through a notch with the aid of our hands, we emerge at the edge of the summit crater and file across one of the more unusual landscapes in my experience: a desert with ice caps. Ginger-colored sand dunes rise before us. The sky is blue except for a few puffy, windblown clouds unfurling like flags from the sides of the chestnut rock towers that ring the crater's edge and give it the appearance of a broken teacup. A fragment of aquamarine ice with steep cookie-cut sidewalls, the tiny Furtwängler glacier, melts on the sand before us, sending an anemic trickle of water through the notch we have just climbed.

(From the summit, on his first ascent, Meyer observed an "enormous cleft, through which the ice that at this point covers the bottom of the crater issues in the form of a glacier." In other words, the crater was the breeding ground for the great glacier Meyer later observed, from below, in the Western Breach. Not only has the lower glacier disappeared, but the one large glacier Meyer saw in the crater has now fragmented and nearly disappeared. The crater floor is now all but entirely sand.)

There are actually three concentric craters, and we are walking on the uneven plain of ash, rock, and seasonal snow between the outer two. The innermost Ash Pit, which is the youngest of the three, emits the occasional puff of sulfurous smoke from a few small fumaroles. The great dunes in front of us form the Inner Cone, a doughnut of fine volcanic ash, four hundred feet high and half a mile wide; we walk along an arc between that doughnut and the chestnut shards along the edge, which are all that remain of the outer and oldest crater. We finally stop at a group of boulders under the shard that holds the highest point in Africa, Uhuru Peak. In thick, windblown mist we set up tents and dive into them, even the porters intent only on their own well-being.

I lie on my back in my tent with too little energy to unpack my sleeping bag, but eventually summon enough strength to cover myself with whatever clothes I can reach and sink into a near faint. Sometime later I awake with a headache and down three Tylenols with silty, smoky water—melted from ice over a fire by the cook. After one of my usual sleepless nights at altitude, I rise for breakfast, inscrutably restored.

LONNIE WAS IN THE MOOD FOR EXPLORING. AFTER BREAKFAST WE FISHED OUT THE drilling and sounding equipment and set off with a few porters for the largest of the remaining glaciers, the northern ice field, directly across the crater from our camp, about half a mile away.

The moment the two geologists caught site of the ice field they began marveling at its odd, contradictory appearance: cathedral-like peaks and spires on the eastern end attested to its slow demise, while the vertical walls to the north and south indicated that it was "polar," or healthy. They were pleased, however, by the vivid horizontal strata on the vertical wall facing us, which promised a good climate record farther in.

Lonnie's attention was immediately drawn to a three-foot band of dark ice that ran for a short stretch along the base of the wall. Thinking of the band of dark glacial-stage ice that he had discovered near bedrock on Huascarán, he said, "We might make a discovery right here."

I should point out that neither he nor Keith would have looked far out of place on Shackleton's *Endurance* expedition in 1914. They wore khaki pants and identical gray wool balaclavas topped with cute little pom-poms. Keith, who hadn't changed his clothes since the plane ride, wore a blue-and-red-striped rugby shirt. Lonnie wore a plaid flannel shirt and, when he put on his baseball cap, could easily have been mistaken for a sheep farmer from his native West Virginia.

The scientific method here called for Lonnie to whack away at the ice with an old wooden ice axe the length of a cane and scatter chips into a plastic bag held by Keith, who knelt in a position of penitence with bowed head so as to keep his pom-pom below the arc of Lonnie's backswings. They took one sample from the black ice and another from the clear ice above it, hoping that a difference in the oxygen isotope ratio might signal a transition to the glacial. When he would test the samples in the lab a few weeks later, however, Lonnie would find no difference—so no great discovery that day.

THAT MISSION ACCOMPLISHED, WE WALKED EAST ALONG THE MARGIN OF THE GLACIER, passing at one point the mummified carcass of a jackal curled on the sand in a bony orange ball. Farther along, we found a break in the wall, which seemed to offer access to the top, so I climbed up with my two modern ice axes, dragging a rope, and prepared to belay Lonnie and Keith up behind me. As testimony to the fact that they don't do this for the fun of it, neither even knew how to tie in; the Peruvians take care of those sorts of details for them. So I yelled down some instructions, averted my eyes, and brought them up. There was much grunting and gasping for air, and a few patches of knuckle skin were left on the ice. Meanwhile, Julias and three porters climbed up unroped and without crampons. Then, together, we hauled the duffels of equipment up from the sand.

The top surface of the glacier was scored with odd, stunted *nieve penitente,* which not only made it frustrating to walk on but also brought a look of concern to Lonnie's face: if the surface was roughed up like this every year, the annual signal had probably been obliterated. He also shook his head at a small lake, refrozen from meltwater, on the surface nearby. We made a feeble attempt to carry the gear to a suitable drilling site for a while, then gave up and returned to camp.

BACK ON THE GLACIER THE NEXT MORNING, WE DID THE JOB WE'D COME HALFWAY around the world to do. First we measured the thickness of the ice with a bootlegged sounding device, consisting of an oscilloscope, a signal generator, and a few long wires—an instrument Lonnie claims to have built for less than forty bucks; then, in one of the more impressive feats of dogged exertion I have ever witnessed, Keith drilled ice core by hand.

Although Bruce Koci's composite hand auger is elegant in its own way, it resembles nothing so much as a giant corkscrew. The business end is a hollow tube, about three feet long and six inches in diameter, with sharp teeth on the bottom end and threads on the outside. A handle at the top imparts the shape of a T. As the intrepid driller twists it, it bites into the glacier, and, ideally, an ice core climbs the inside.

But this was far from ideal ice; it was clear and hard—in fact, we feared we had hit another frozen lake. In order to get the drill to bite, you have to grab the ends of the crossbar, lean on it by bending at the waist, and wrench it clockwise. Once I managed to get it engaged, it took all my strength to twist it farther in, so I could drill for only two or three minutes before collapsing on the snow in exhaustion; but Keith could drill for ten or fifteen minutes at a stretch and kept it up for most of the day. By now his pants were torn, filthy, and bloodstained—the last from the knuckles he'd bashed ice climbing the previous day—and the intense high-altitude sunlight, amplified by reflection from snow and sand, had burnt away a few layers of his fair skin, so his hands and face were mottled and puffy.

Whenever he made two new feet of downward progress, he would grab the crossbar like a weight lifter doing a clean and jerk and grunt in the effort of breaking the set of the drill in the ice—usually needing four or five tries. When the drill finally twitched free, the porters and I would run in to help him pull it out of the hole. Then he would set it on his shoulder in a pose suggestive of the savior carrying the cross to Golgotha and lurch a few steps toward Lonnie, who knelt on the snow nearby. Keith would spill the ice shards from the tube, and Lonnie would arrange them in line, by depth, concentrating fiercely and writing detailed notes. With Julias's help, Lonnie would weigh the shards and measure their size. (Since they would melt before reaching the lab, this was his only chance to get their density, which he would need in order to estimate annual snowfall.) Then he would bag and label them.

And from this time forth Lonnie would guard the ice like a lioness guarding her cubs, sleeping with it in his tent at night and keeping the porter who carried it (in a venerable old Samsonite suitcase) in his sights all the way down the mountain.

As the hole got deeper, Keith added extensions to the drill, and when it grew past a certain length, he and the porters had to gather in a tableau reminiscent of the marines raising the flag at Iwo Jima in order to tip it vertical so it would drop back into the hole.

Although we had hoped to drill ten meters, we gave up after six, with the sun sinking fast, then packed up and raced, finally rappelling to the crater floor just as the western sun touched down on the rim. Lonnie and the porters charged toward camp, while I filled my pack with heavy ice and hung back with Keith, who was utterly and justifiably exhausted. We slogged together across yellow sand streaked with sharp, lengthening shadows cast by the many boulders strewn about. The clouds and ice turned pink and faded into blackness before we reached home.

UNFORTUNATELY, KEITH FORGOT TO FILL HIS WATER BOTTLE BEFORE CLIMBING INTO BED that night, so he showed up late and ghastly looking, severely dehydrated, for breakfast the next morning. All the same, he lobbied enthusiastically to climb back up on the ice and find a better place to drill. But in a sign that Lonnie has mellowed in the decades since Quelccaya, he replied, "Look. We've got the first oxygen isotopes from Africa. That's news in itself. Let's take a look at the summit and go down."

A mild thirty-five-minute walk led us to the top of Uhuru, the summit of Africa, and since trekkers are inevitably dragged up there much earlier, for sunrise, we had the pleasure of soaking in the view for nearly an hour in the sole company of a rant of white-necked ravens, soaring in a blue, windless sky. Ever the scientist, Lonnie kept exclaiming about the vivid horizontal strata on the wall of the southern ice field, a hundred yards away.

One of our secondary objectives was to assess the prospects for flying the balloon that had failed on Sajam, and they looked good. Two of our three mornings in the crater had been breathless, and what little wind there was had blown steadily in the direction of the airport, which we could now see in the distance on the plain below. Perhaps it would be possible to float next year's ice directly to the airstrip.

As the freeze-dried jackal suggests, one doesn't exactly "live" at nineteen thousand feet; one slowly dies. To descend into oxygen-rich air with the scent of rain on the wind is to return to one's senses, and a gradual descent of Kilimanjaro's long, gentle Mweka route might be the best way on earth to savor that basic wealth.

At first we reversed the route along the spine of Uhuru that Hans Meyer and Ludwig Purtscheller had taken on the first ascent. Our path diverged from theirs at Hans Meyer Point, where we took one last look east at the great sand-filled crater, spotted here and there with blue fragments of ice, then turned and gazed west under gray clouds at the immense slope of sand that stretched down to the brown and maroon Saddle, five thousand feet below, then up to the rocky peak of Mawenzi, seven miles away. To our right, a few tiny patches of ice clung to the southern slopes of Kibo.

ON HIS FIRST ASCENT, IN 1889, LOOKING UP FROM THE SADDLE AT THE SPOT WHERE WE now stood, Hans Meyer described Kibo as an "ice-dome." Three main flows poured from the summit in his time: northwest (the northern ice field), west through the Breach, and south toward Moshi. They were fed by the vast fields of ice and *névé* in the crater, which were replenished by seasonal snows and preserved by steady cold.

"We had reached a height of 14,200 feet," Meyer wrote, "and were still about a mile and a half from the actual cone, which towers upward for another 5,510 feet, its base measuring about four miles in breadth. . . . On the left [south] side . . . the ice descends almost to the base of the cone in long tongues full of great rents and chasms. In the middle, that is, on the side facing us, a broad sheet of ice comes downward, filling up the valley between two long high ridges of rock."

He and Purtscheller climbed the ice dome on the left of the side that faced them, using the standard technique of the day, which was to chop a series of steps with their shoulder-length alpenstocks and mince from one to the next in the "climbing irons" they had strapped to their boots. (Meyer's had been stolen with a great amount of other gear before they reached Africa, so he managed in hobnailed boots.)

They climbed roped, and Purtscheller, with his superior knowledge of glacier travel, led the way. The slope was too steep to address directly, so they were forced to zigzag. "The task was no easy one: each step cost twenty strokes of the axe." When the slope eased, they stopped to rest. Over lunch they agreed to name the body of ice on which they were sitting the Ratzel Glacier, in memory of a friend of Meyer's who had taught geography in Leipzig.

In the end, they labored on the Ratzel for three and a half hours before attaining the edge of the summit plateau, at which point they realized they could not make the summit that day and decided to go down. Even with the help of gravity, it was another two hours before they finally stepped down off the ice.

The Ratzel Glacier has now shrunk to a small block at the crater rim, and the three large flows that existed in Meyer's time have dissolved into dozens of small fragments.

After a day's rest, he and Purtscheller climbed the steps they had previously cut and made for the highest point on the rim (having reached it, by luck, on its highest shard). As leader and financier of the expedition, Meyer took the liberty of perching first on the small pile of rocks that formed the very summit. He removed the topmost rock and placed it in his pack (to present later to the kaiser). Then, "taking out a small German flag, which I had brought with me for the purpose in my knapsack, I planted it on the weatherbeaten lava summit with three ringing cheers, and in virtue of my right as its first discoverer christened this hitherto unknown and unnamed mountain peak—the loftiest spot in Africa and in the German Empire—Kaiser Wilhelm's Peak. Then we gave three cheers more for the Emperor, and shook hands in mutual congratulation."

The peak was rechristened "Uhuru," meaning "Freedom," when Tanganyika gained independence in 1961 (from the British by then). A venerable metal plaque has replaced Meyer's flag. It reads, WE THE PEOPLE OF TANGANYIKA WOULD LIKE TO LIGHT A CANDLE AND PUT IT ON TOP OF MOUNT KILIMANJARO WHICH WOULD SHINE BEYOND OUR BORDERS GIVING HOPE WHERE THERE WAS DESPAIR, LOVE WHERE THERE WAS HATE, AND DIGNITY WHERE BEFORE THERE WAS ONLY HUMILIATION.

It is no longer possible to repeat the ascent of these two explorers. The steps they cut in rock-hard ice would lie a hundred or two feet in the air, above a dusty gully between the present Mweka and Marangu routes. On summit day for either of those, the aspirant of today will need only warm clothes, food, water, and a pair of comfortable approach shoes. Running shoes will do. You won't need to step on snow.

AND AT THE MOMENT, I WOULD SAY, KEITH MOUNTAIN WAS MISSING THE SNOWS OF KILimanjaro more than anyone else on earth. For in the scurry to leave camp, he had again missed filling his water bottle and had now fallen even deeper into the listless hell of dehydration. Since I wasn't running on quite the deficit he was, I hung back to share my small supply with him. Even so, lower down on the dusty trail, his desperation drove him into a cave on his hands and knees, where he held a water bottle under a slow drip from a small icicle, which he then broke off and brought back sucking like a Popsicle. Wasn't doing him much good.

But the slope finally eased; we passed the first thin tussocks of grass and green mosaics of rosette, and entered the moorland, all gray, brown, and green. The bushes grew higher as we descended; tiny flowers flashed out the first sparks of yellow from the shade of round boulders; the heather grew tree-size, mingled with stands of protea bearing large bloodred bulbs, and again cast-down Spanish moss in slate and orange shawls.

Joining roughly a hundred tourists and twice that many porters among the junipers at Mweka Camp, Lonnie, Julias, and I celebrated with the inevitable Kilimanjaro lager—on ice, no less—purchased from a wily entrepreneur in a dilapidated metal hut, and kept one waiting for Keith, who finally stumbled in with his ever-present smile, but a tired one.

When we arrived at Key's the next afternoon, the first thing Lonnie did was to spread the plastic bags of ice on the floor of his room so it would melt faster. Two mornings later, he sneaked into the kitchen without telling us and commandeered the staff into helping him pour the samples into plastic bottles and seal them with wax.

As our bodies strove to recover, we lapsed into a slothful pattern of deep sleep and ravenous eating, and wasted a lot of time watching CNN in Key's dark bar. It was there that Lonnie and Keith first opened up with their dark view of the greenhouse future.

THE EVIDENCE HAD CONTINUED TO BUILD ALL THROUGH THE NINETIES. NINETEEN ninety-five had set a record as the hottest in the 120-year history of instrumental measurement. Then the prediction Al Gore had made at the Senate hearing in 1992 proceeded to come true: 1997 eclipsed 1995, and just weeks before Keith, Lonnie, and I left for Africa we learned that the year just past, 1998, had set another record—the third in four years—assisted by a slight nudge from El Niño.

But things were not moving quite so rapidly on the policy front. In December 1997, shortly after Ellen had accompanied the vice president on a trip to Montana to highlight the global warming threat with the receding ice of Glacier National Park as a backdrop (Lonnie being off at Dasuopu), Gore flew to Kyoto, Japan, to join negotiators from more than one hundred other nations in hammering out the language for the first international agreement to limit carbon dioxide emissions. The Kyoto Protocol called for developed nations to scale back their total emissions to a level 5 percent below that of 1990. For the United States, which had increased its already disproportionate share in the intervening decade, this meant a reduction of about 13 percent. Months before Gore even left for Japan, however, the Senate, objecting mainly to the fact that Kyoto required no commitment from developing nations, had voted 95–0 against ratifying any treaty of the kind.

"Kyoto is just a first step," said Lonnie in the bar. "It's too little too late anyway, and even a five percent reduction in CO_2 will be difficult if not impossible to get through the U.S. Congress. But just look at all the scientific facts we have linking smoking to cancer, yet Congress won't pass a bill to help keep cigarettes out of the hands of our young people. Add the fact that China and India aren't even part of the protocol and that they'll probably be the largest emitters in ten years, and I think we'd better start getting used to the idea of living in a hotter world.

"There is a lot about the climate system that we don't understand—our knowledge is not complete, and of course it won't ever be—but my personal opinion is that scientists are by training and nature conservative and that we have probably underestimated our impact. Fifty years from now—I hope I'm wrong—I think you may be living in a world where you don't go outside between one and four in the afternoon. Of course, we in North America and Europe will cope with this. But these people here"—he waved out the window at the multitudes walking by on the road in front of Key's—"they don't have a choice; they're not going into any air-conditioned office. As usual, the third world will pay the price.

"There are people who would argue that looking at the past doesn't make a lot of sense, that there are no analogues; we've never had a creature take off and change the composition of the atmosphere the way humans have in such a short period of time. But I think there are analogues in our past, and they're saying we ought to be very careful.

"There was a time about 3.5 billion years ago when there was no oxygen in the atmosphere, and a kind of anaerobic bacteria occupied all the oceans of the world. They produced oxygen just by living, the same way we produce CO_2, and they multiplied until they occupied every part of the earth. But the oxygen they gave off was poisonous to them, so they eventually changed the atmosphere to the point that they killed themselves off. Now you only find them in the restricted pockets where anaerobic conditions still exist.

"Well, we like to think we're smarter than bacteria, that humans do more and think more about the big picture, that we realize we're in a big ocean and we're part of a big planet. We would like to think we're smart enough to see that and to adjust our lifestyles accordingly, but I don't think it's true. I think humans are like every other organism: they try to maximize the system to their advantage, take every resource they can use to make whatever it is they're trying to produce, and they will keep doing it until that resource is no longer available to them. Our economic system is based on that: maximum production. And every country in the world wants to be like the Western countries—same lifestyle, same air-conditioning, same TVs. We have fine universities, we train people to think; but actions speak louder than words, and as long as we stay on this path I don't think we're any smarter than the bacteria. We're behaving the same way they did. You can do that until you exceed the boundaries of the system, and then it's going to collapse."

"You mean the whole system will fall apart?" I asked.

"Oh no, the system will keep working. I'm very optimistic about the system. The system will take care of itself. This is like a cancer growing on the surface. The planet will react in such a way as to stop that cancer."

"The earth will stay healthy?"

"Yes. It might be big storms; it might be wiping out Bangladesh or Africa; the world will go on; and there will be creatures that will multiply in that new world. Plants like CO_2; maybe the world will be dominated by plants. Whenever a creature exceeds its resource base, its population collapses—think of lemmings—and I think that's ultimately what will happen to humans. I'm not a religious man, but I can't help thinking it's a little funny that this is all happening at the beginning of a new millennium.

"You should be prudent about these things. If you have a system that's changing and you have a good indication you're part of that, a prudent person says, 'Hey, let's slow this down so we can buy time until we better understand the system.' The Kyoto Protocol? Absolutely! No-brainer! We can meet our quotas through conservation, which we should be doing anyway to conserve this valuable natural resource. Okay, so in 2010 you find out that these scientists are off the mark. You haven't lost a thing. The oil is still there; the gas is still there; the human race still needs energy. A prudent person would do that. Prudent bacteria would do that."

The bacteria image has stuck with me. I sometimes flash on it when I'm stuck in a traffic jam in a big city.

BACK IN COLUMBUS, LONNIE WASTED NO TIME ANALYZING THE SAMPLES. SIX WEEKS AFter our return from Africa, he and Ellen gathered their group to decide whether or not to go back (though I believe he had actually decided before we left the rain forest). It turned out, however, that there wasn't much to be learned from the short core Keith had expended so much energy wresting from on high. Exasperated when the meeting closed without a decision, I followed Lonnie into his office and asked, "Well? Are you going back or what?"

He tossed his head and offered a confidential nod. "Yeah. It's a gut feel. You go by gut feel on these things—especially this one, because it's a complicated record. The thing is, I have a feeling there *is* a record up there. It's probably not the record I think it is, but I'm willing to bet there is one."

·28·

DRILLING DEEPER

The strange desire to be in an interesting place at the moment that defined their particular millennium drove five hundred people to the summit of Africa for sunrise on January 1, 2000, one year after our reconnaissance. Presumably, about the same number tried and failed. Of those thousand-odd people, I am told, five died of edema that day. But Lonnie's team spent it with their families for the most part: Phil Ndesamburo had warned him of the impending bulge in tourism, so he scheduled our arrival for the fifteenth.

Feeling privileged to be joining the inner circle on a true deep-drilling expedition, I met up with Mary Davis, Vladimir Mikhalenko, Keith Mountain, and Victor Zagorodnov in Amsterdam—Lonnie having flown ahead to set the Tanzanian bureaucracy in motion and Bruce Koci being detained at the South Pole (where I had just been working with him).

This was not the improvisational experience of the year before. Lonnie met us in the atmospheric darkness of the airport with a fleet of 4×4s from Key's and then sat in the front seat of my vehicle as we rode to the hotel. Once we were moving, he stretched his arm along the seat back and turned around to share some news he couldn't contain: he'd just had another of the epiphanies that scientists live for.

As is often the case with these things, it had occurred under pressure, as he had been preparing his first talk about Dasuopu for the December meeting of the American Geophysical Union. Since Dasuopu was the highest and coldest glacier he had ever drilled—he had dropped a probe down the holes and measured 7.2°F at the ice-bedrock contact—it could not have been "losing time" at its base, so he had expected to find glacial-stage ice down there.

They had kept the ice cold all the way from Tibet to Ohio, so they were confident that the trapped air bubbles were well preserved; and, to take an initial stab at the chronology, they had sent samples to three different labs to obtain methane profiles.* The first two profiles indicated that the cores went back only about ten thousand years—not even as far as the Younger Dryas.

"I was ready to argue that there was something wrong with the labs," Lonnie said. "How can you go to the highest, coldest place on earth and not find glacial ice? It doesn't make sense."

The third profile came in just two weeks before he was scheduled to give his talk—and confirmed the first two. That sealed it; it was a true result.

"We'd already noticed a problem with the isotopes and the dust," he explained. "They also said there was no glacial ice—so I began thinking, 'Maybe there's something strange going on here . . .' "

It hit him, as usual, late one night.

IN ALL THE EXCITEMENT SURROUNDING THE DISCOVERY OF GLACIAL-STAGE ICE ON Huascarán, one puzzling aspect had been overlooked: why did the record go back only nineteen thousand years? How could the high, cold Garganta Col have been completely bare at the Last Glacial Maximum, when temperatures there were almost certainly lower than they are today? The same argument holds for Sajama, where the record ends twenty-five thousand years ago. Furthermore, once it was born, Huascarán's ice cap persisted through what is known as the Holocene optimum, between about nine thousand and six thousand years ago, when temperatures were higher there (and in many other places) than they are today. Now, on Dasuopu, where it had almost certainly been far below freezing for at least the last one hundred thousand years, the record didn't even span the Holocene.

But we're thinking backward; let's turn the timescale around: Sajama's ice cap, at a latitude of eighteen degrees south, was born twenty-five thousand years ago. Huascarán's, at nine degrees south, was born nineteen thousand years ago. Dasuopu's, at twenty-eight degrees north, just outside the tropics, was born (as more precise estimates would later show) eight thousand years ago.

*Since methane, like carbon dioxide, mixes quickly through the atmosphere, one can match the peaks and valleys of the methane curve from an unknown core with the peaks and valleys from GISP2 and infer dates via GISP2's layer counting. Todd Sowers, whom I met on Sajama, played a role in constructing GISP2's methane "standard curve" (pages 12–13).

The evidence for strong interstadials that Lonnie had only recently uncovered at Guliya, thirty-five degrees north, was also percolating in his mind as he prepared his Dasuopu talk. Those periodic warmings, remember, interrupted the Wisconsinan glacial stage with the twenty-two-thousand-year beat of the orbital precession cycle. When I had visited him in Columbus less than two months before his talk, he had speculated that the closer one moved to the tropics, the stronger the precession effect would become. In his actual tropical cores, however, he had never seen a precession signal, because none reached far enough back into the glacial. This was a large part of the reason, in fact, that he had gone to the intermediate latitude of Dasuopu in the first place.

What he had realized late one night in December 1999 was that the cyclical precession of the earth's axis had led to the births of all the tropical and subtropical glaciers he had studied.

OWING TO THE FACT THAT THE EARTH PRECESSES LIKE A TOP AS IT FOLLOWS ITS ELLIPTIcal path around the sun, the latitude that will experience the height of summer just at perihelion—the moment we are closest to the sun—will cycle back and forth between the Tropics of Capricorn and Cancer with a period of about twenty-two thousand years. Examining calculated insolation curves for various latitudes, Lonnie realized that Sajama's ice cap was born just at a peak in summertime insolation at its particular latitude; Huascarán, nine degrees farther north, was experiencing a peak as its glacier was being born six thousand years later; and the Tropic of Cancer, not far south of Dasuopu, was experiencing a peak as Dasuopu was giving birth to a glacier eleven thousand years after that.

One might think that *more* summertime sunlight would cause ice caps to melt; however, an ice cap requires two ingredients for growth: both cold and moisture. What with Huascarán and its coincident discoveries, climatologists still believed that all Lonnie's tropical and subtropical mountains had been *colder* during the Wisconsinan—and, furthermore, that tropical temperatures have tended to track temperatures at high latitudes as the eons have rolled by: they're "synchronous." So Lonnie's new hunch was that precession hasn't affected tropical temperatures so much as the amount of atmospheric moisture available for the building and sustenance of high tropical snow.

To add to his suspicions, tropical mountains receive their snow in summer, ironically, as the Hadley Cell passes over them, dragging the Intertropical Convergence Zone and the Asian and African monsoons along with it. And when summer insolation peaks over a particular region, the cell will be stronger and deliver more snow than usual. (Remember that the lakes in the now dry Salar de Uyuni grew to maximum size just as the ice cap was born on Sajama, nearby.) Conversely, when insolation dips, the ice caps (and lakes) will be starved of water and tend to waste away. So these mountains seem to have been bare since their previous insolation minima, twenty or forty thousand years ago, and required the significant moisture increase associated with a new peak in insolation to give birth to glaciers once again.

This also means that the small ice caps at low latitudes will not necessarily follow the lead of the great northern ice caps when they decide to grow or shrink: the advance and retreat of high- and low-latitude ice will be asynchronous. So, there in the car in Tanzania, Lonnie and I began calling this his "asynchrony theory."

"The problem has been that we haven't had much between Antarctica and Greenland," he said, "where the same things seem to be happening in the same time frame, by and large. So, our group has been trying not only to get the records before they disappear but also to get enough records from different latitudes—like building blocks—to put the big picture together. This is the first big picture development that's come from this effort.

"It's amazing how science works: you labor and you labor and you learn things that don't fit, that don't make sense, and suddenly you get a new piece of information from some far corner of the world and it makes you say, 'Hey, your paradigm was wrong, you didn't understand how the system worked.' "

As we bounced through the African night, Lonnie also pointed out another implication to all this: he confidently predicted, based on his new asynchrony idea, that the ice cores we were about to retrieve would show that the snows of Kilimanjaro had been born about eleven thousand years ago. As usual with him, science and adventure were turning out to be essentially the same thing.

ACCORDING TO THE USUAL VIEW OF ADVENTURE, THIS TRIP WOULD NOT HOLD A CANdle to Guliya, Huascarán, or Dasuopu, of course. By Lonnie's standards it was quite literally a "walk in the park." On the other hand, it became immediately obvious that it would be much more complicated than any expedition I had ever been on—and, as events would show, even walks in the park can be dangerous.

We had brought the balloon that had failed to fly on Sajama, and this time Lonnie believed that destiny was lending a hand. On the first of what would turn out to be many long evenings in the dining room at Key's, as he, Keith, and I quailed under the weight of a certain Kenny G instrumental, playing over and over again as it had the year before, and as the Kilimanjaro lager began to flow, Lonnie proclaimed that "things happen when they're meant to happen. The balloon wasn't meant to fly on Sajama; it's meant to fly here. Kilimanjaro is an easier place to fly it, and it will attract more attention." He was hoping that an image of a balloon rescuing climate's history from these famous, dwindling snows might become an icon of global warming.

He also found significance in the fortuitous emergence of Jon "JR" Russell, an itinerant British balloonist who had learned of the Sajama debacle through the ballooning grapevine and sent an unsolicited résumé to Ohio State from his temporary perch in Australia, knowing nothing of Lonnie's future plans. Indeed, some synchronicity may have been involved, for it turned out that JR had lived in Tanzania for a while, piloting balloon safaris over the Serengeti, and that he not only held a commercial balloon pilot's license in this country but had helped write the test for getting one. Although he would not arrive for two weeks, he had been monitoring Kilimanjaro's weather and had recently told Lonnie that the prevailing winds seemed to be blowing at a moderate, balloon-able speed straight toward the airport—as we had guessed on the summit the year before.

But the balloon only complicated what was already an unfathomable permitting process. With the commendable goal of preserving the dignity and sanctity of the mountain, park regulations specifically forbade ballooning and other "high-profile" activities, such as mountain biking and skiing, and in the face of the three or four competing bureaucracies that oversaw the park, neither Phil Ndesamburo nor Lonnie had any idea exactly how many permits we would need for this unusual endeavor. There was also an expedition to organize: everything from porters to food to odd items like shovels and water containers, which we purchased one scorching afternoon at the local market. And then there was the ice.

Fortuitously, a warehouse-size cargo freezer had been constructed at the airport in the year since our reconnaissance, and with some luck JR might be able to float straight to it from the summit. (This place really was perfect for the balloon.) Grotesquely, however, when Lonnie and Keith dropped by to inspect the freezer, they discovered the body of one of the New Year's Day edema victims inside. Her relatives were still struggling with the bureaucracy to get her remains out of the country.

After four days, Lonnie had twelve permits, and our five tons of equipment had finally reached the airport. But the same customs officials who refused to let the cadaver out were refusing to let the cargo in. They wanted ninety-one million Tanzanian shillings in bond (more than one hundred thousand U.S. dollars!) against the excise Lonnie would have to pay if he imported the stuff; and although he would supposedly get the bond back when he proved he was taking his equipment back out, that seemed an iffy proposition in this seat-of-the-pants country. Thus a cargo waiver was added to the expanding list of permits we would need from innumerable bureaucratic appendages.

The rest of us helped when we could, but for the most part we lolled away the hours in the hotel, to the unceasing accompaniment of Kenny G.

The tune was wearing extremely thin after six more days. We still had no cargo waiver, and things were looking bad for the balloon. Lonnie now had twenty-five permits, as well as letters from the national ministries of parks and aviation approving everything, including the balloon. But the wardens at park headquarters in Marangu and at the headquarters for all national parks, in Arusha, were still steadfastly refusing to allow a balloon to take off in Kilimanjaro National Park. It seemed that we had permission to fly over the mountain but not to touch down.

The cargo waiver came through that night. We took over the large back veranda of Key's to unpack, inventory, and repack five tons of gear.

It was a good sign when Bruce Koci showed up the night after that, for he has a knack for showing up just as the team heads into the bush. He appeared unceremoniously at the hotel door, hair askew, lopsided backpack slung over one shoulder, in jeans and running shoes, with a faded cartoon of the South Pole marker, resembling a barber pole, on the front of his T-shirt. Sure enough, the next morning Lonnie decided to leave the balloon up to Phil and JR and begin moving up the mountain.

Our fate now passed into the hands of one Ron Kinet (not his real name), who owned the only trucks in the neighborhood capable of transporting our gear and porters to the Shira Plateau. Kinet was a humorous, voluble, well-read, and unpredictable fellow; rumor had it he smoked opium.

Somehow or other, nevertheless, on our thirteenth morning in Tanzania, we seemed ready to go. Kinet's two trucks were loaded and waiting, and a joyful Julias Minja, resplendent in black—gray camouflage on his pants and bright African motifs on his shirt—was presiding over forty enthusiastic porters in Key's front parking lot.

Kinet showed up four hours late, then led the caravan toward town at the wheel of the more impressive of his two vehicles, a battle-scarred, six-wheel-drive dump truck, carrying half the gear, with a porter riding shotgun on top. The second truck followed with the rest of the porters and their scant personal gear. We trailed in two 4×4s.

Kinet shook us off within half a mile to pursue further unknown business in Moshi, so we whiled away a few more hours at the park entrance in Londorossi, watching a clan of colobus monkeys frolic in the trees: lithe longhaired animals with shimmering black bodies, white faces, long white tails, and black capes fringed in white, which hung like wings from their sides and arms.

When Kinet finally appeared, Lonnie greeted him amiably and—perhaps to keep him in sight—let his big truck go first up the next rock-filled section and through the forest of weeping pine, so that we ate its dust and black exhaust as we also climbed into thinner air. On the open moorland below the plateau we came across the porters, standing around with their hands in their pockets by Kinet's other truck: it was low on gas and had to turn around.

Thus we made a magnificent entrance onto the Shira Plateau, the cargo truck leading the way with porters lying on top, clinging to its sides, and standing on its bumpers—on what proved to be the only day I have ever seen on Kilimanjaro that remained crystal clear through sunset and the rising of the moon. In a strong tailwind, the big truck lumbered up the final slope to the plateau in slanting yellow light, kicking a plume of red dust up before it, profiled—porters and all—against the blue sky and the rocky hump of Kibo.

In splendid copper light, the dust blowing crossways, our vehicles crawled like ants across the slate green plateau. A truck tire went providentially flat, permitting us to pause in the center of the caldera and watch Kibo turn red in the setting sun. Then the truck foundered in a muddy wash, and we had a great time pushing it out.

A mild panic set in as we reached the trailhead in the dark, with the cold deepening by the minute. Forgoing dinner, we searched for our tents and sleeping bags and began setting up camp, as Kinet rumbled down the hill in his elephantine vehicle to spend *his* night in a warm bed.

So there we were: so high and so close to the equator that those of us who took the time to look up were treated to the North Star, the Big Dipper, and the Southern Cross all at the same time.

The next day, the first of what would become daily morning radio calls to Key's consisted primarily of caving in to Kinet's demand for more than the previously arranged sum in order to bring the rest of the equipment to the plateau. On the other hand, Kinet did surprise us by actually showing up late that night, his truck wheezing and grunting into camp long after we'd gone to bed. He roused the porters to unload it and then quickly departed, while Lonnie hid in his tent until he'd left and then went out to make sure everything was there.

WE WERE HAPPY TO HAVE FINALLY LEFT KEY'S AND KENNY G BEHIND FOR A SILENT AND spacious natural landscape, and our first job was to acclimate, so we walked just for fun for the first few days—smiles on every face whenever we bumped into each other on the trail or at the same viewpoint to watch the sun set behind Meru.

There was Bruce in his fur-lined "mad bomber" cap with dangling earflaps, robed to his knees in two hooded anoraks that lent him the appearance of a storybook druid or magician. He had bought the inner garment, made of thick New Zealand wool, on his way through that country on one of his annual visits to the South Pole. The outer was a sky-blue Goretex creation by Apocalypse Designs, Fairbanks, Alaska. Beneath these, he wore the outfit he's worn every time I've seen him in the places most of us usually inhabit: blue jeans, a venerable T-shirt (often commemorating some bizarre event in one or the other polar region), and a bulky, untucked, usually unbuttoned chamois shirt. And I have never seen clear evidence that Bruce has combed his hair.

Unlike most of us, he works at his own pleasure, not that of his employers. He has basically walked out of his job at the South Pole in order to get here (and, owing to his critical importance, they will take him back).

"I don't need the money, and neither does Lonnie," Bruce tells me at one point. "We have an agreement that if we run over budget on this expedition our salaries will be the first things to go. If necessary, I'll buy the airplane ticket, too. For someplace like this? I can't duplicate this!"

Having played a major role over the past few decades in the development of scientific ice drilling in virtually all its forms, Bruce is here mainly for the thrill of the chase. He's a consultant, really. Victor, who is employed full-time at the Byrd Center, will be in charge of the drilling.

Bruce's musical tastes range from Tom Waits to Wagner. (He may have dressed up a little on the trip he and his wife took to New York recently, to attend the complete *Ring* cycle at the Metropolitan Opera.) And he is an accomplished photographer with an artist's eye. He has brought four or five single-lens reflex cameras as well as a 4x5 view camera with a massive tripod. He doesn't take "people pictures," only landscapes featuring light, clouds, earth, and sky. He is recognized in the distance by his unique wizened posture, his lopsided backpack, a camera hanging from one hand, his dangling earflaps, and his robelike attire—always alone.

Victor also wears a bomber's cap, but his is an authentic leather Russian bombardier's unit. While Bruce's is basically a joke, Victor's is more serious—as is its wearer. Victor has a pathos about him—moody, warm, and somehow sad—and voices strong opinions during dinner conversations.

Vladimir seemed shy down below, but the mountain has brought him out of his shell. He and I have a common interest in mountaineering, and I have learned that he has climbed some of the more estimable routes in his home range, the Caucasus, where the mountains are basically piles of rubble, very prone to rock fall—and where I once had the most terrifying climbing experience of my life. He'll come out with a phrase of Spanish every once in a while, which he must have picked up on his visits to Huascarán and Sajama. He's quiet and polite, reliable and self-contained. During idle moments he'll sit straight-backed on a boulder and write in his field book on a crossed leg. One sees why Lonnie takes him everywhere.

Mary has many of the same qualities, and she's thrilled to be in the mountains. She can always be counted upon for a friendly conversation.

Keith provides much entertainment, which the boss seems to appreciate more than anyone. Whenever they're together, Lonnie's bell-like laugh rings off the mountain walls.

Funny thing: all six have blue eyes.

AS WE MOVE UP THE HILL, IT BECOMES CLEAR NOT ONLY THAT THIS IS NOT A WALK IN the park but also that Phil Ndesamburo has never before supported an effort of this magnitude.

In a meeting down in Moshi, Julias had suggested fifty porters, but Lonnie had balked, reminding everyone that he *did* have a budget: there was a limit to what he could spend. Phil was charging thirty dollars a day per porter (most of which was going into his own pocket), and porter fees promised to be the largest single expense of this trip. Lonnie had agreed to forty porters, but we now saw that most were inexperienced and nearly all were utterly underequipped. It seemed that Julias had simply put out a call to his friends and neighbors, and in a country where a dollar a day is a very good salary, nearly every male he knew had decided to give it a go. Few had hats, gloves, or boots; most appeared in holey sneakers or street shoes with dangerously smooth leather soles; hardly any had sleeping bags.

These problems came to a head at Lava Tower, which at fifteen thousand feet begins to qualify as high altitude. We experienced our first headaches and bouts of sleeplessness there, while the porters, with their lack of equipment, suffered that much more. Most pled for headache pills whenever they saw us and showed a sensible aversion to sleeping in the cold with no sleeping bag by running down to the Shira Plateau every night. This compounded the organizational difficulties in the morning, for when time came to assign loads, Julias never had as many bodies on hand as he expected.

The arrangement had been for each porter to carry two loads a day from the plateau to Lava Tower, but the night we reached the Tower they declared a slowdown: one load a day. Thirteen reinforcements came up with the next truckload of food, which the expedition was consuming at a prodigious rate. Julias was not the happy-go-lucky mountaineer of the previous year. He was making two round-trips himself most days and listening to complaints from both sides all the while.

Since all this pain and confusion demonstrated vividly just how much five tons of equipment really weighs, I gained a new appreciation for Lonnie's feats on Guliya, Dasuopu, and especially Huascarán.

BUT NOTHING COULD KEEP KEITH DOWN, OF COURSE. ONE OF OUR GOALS WAS TO make a definitive three-dimensional map of Kibo in order to document the size of the glaciers. Lonnie had contracted with an aerial photography outfit in Nairobi to fly over the summit a few weeks hence and take a series of stereoscopic images, and Keith was charged with laying out markers that would be visible in the photographs and with measuring their locations with a scientific GPS instrument, so an accurate three-dimensional map could later be constructed from the photos. He had spent many hours in his room at Key's with wires dangling from his mouth and a desperate look on his face, trying to figure out how to plug the GPS unit into his laptop computer—at somewhat of a disadvantage, because he didn't have the manual—and it looked as though he might soon be wishing he was working with simple paper and pencil, as he had on Quelccaya in 1983, when he had to stick his leg out the tent all night to keep himself awake.

On the day we climbed to Lava Tower, he and Vladimir took a ten-hour detour to some subsidiary peaks to place the first marker. When they shuffled in after dark, just as the rest of us were finishing dinner, Keith collapsed to the ground, raised his eyes to the heavens, and shouted, "Thank you, sweet Jesus, for letting me live another day! . . . See? I don't need much—just keep me alive."

MEANWHILE, JR LANDED IN MOSHI AND GOT RIGHT TO WORK. ON OUR SIXTH MORNING on the mountain, as we prepared to leave Arrow Camp and embark on the "journey without end" up the Western Breach to the summit crater, he took the balloon on a test flight from the magnificent Simba Farm in the flats in the western rain shadow of the mountain, which we had passed on the way in. Lonnie had arranged to store the ice at the farm as it was brought off the mountain—either on foot or by balloon—and part of the deal was to give the Dutchman who managed the farm a balloon ride.

We looked for the balloon through a pair of binoculars before starting our climb, but it was impossible to pick it out in the vastness between ourselves and Mount Meru. Even Arrow Camp, which is more than three thousand feet below the summit, stands almost two and a half vertical miles above the plain.

It took about the same effort to climb the Breach this year as it had the year before, but the addition of so many people—and especially the infectious gaiety of the porters—added significant wind to our sails. Despite inadequate clothing and almost no food, a few of them shuttled loads all the way to the crater and returned to sleep at Shira Plateau that night. I remember stopping on one of the more spectacular outcrops to rest at one point and finding several porters giggling and lazing in the sun, passing a joint around—which I politely refused. (A few days later, our true superhero climbed from Shira to the summit of the northern ice field with an eighty-pound refrigerator on his head, smiling and smoking dope the whole way—no sign of a headache, nor the munchies—and made it back to Shira by nightfall. That's about three vertical miles. He received a tip of one hundred dollars and retired until the end of the expedition, then returned to do the reverse trip for the same fee.)

We pitched camp by the northern ice field, across the crater from Uhuru, near the spot where Lonnie and Keith had hacked out samples of dark and light ice the year before. I remember staggering across the crater, nearly last, under oppressive skies blowing wind and graupel, which darkened the surface as it instantly melted, the stragglers before me leaving evanescent footprints of white, upturned sand.

There was a homey feeling in the cook tent that night as Keith and Mary mixed up a soup with a history stretching back two decades to the first expeditions to Quelccaya: an improvisational concoction of freeze-dried mashed potatoes, cheddar cheese, and mixed vegetables that hit the spot. It was bland enough to soothe our queasy high-altitude stomachs but tasty enough to make us eat and avoid the high-altitude "bonk."

LUCKILY, WE HAD NO NEED TO CLIMB VERTICAL ICE THIS YEAR, BECAUSE VICTOR FOUND a notch at the western rim of the crater that led to a ramp giving easy access to the surface of the glacier quite close to its highest point, which was the best place to drill.

But obstacles continued to arise. Only a handful of porters had the strength to reach the summit on a regular basis (most being Julias's brothers and stepbrothers), and they were now being asked to perform more extraordinary feats than before, with little more in the way of food or equipment. An ordinary day now required them to climb five or more thousand feet with an awkward load up the rickety and occasionally hair-raising terrain of the Western Breach. A couple of rusty pairs of crampons were left on the sand by the snow ramp leading to the top of the glacier; so they would stop there, drop their loads, sit down, and lash just one to a sneaker or beaten-up street shoe before proceeding up. Many risked snow blindness, as they had no sunglasses—I lent mine whenever I was asked—and since they had no sleeping bags, those who ended up spending the night in the crater, which they did only out of sheer necessity, would pile into the same tent and warm one another with body heat. Meanwhile, a meteorology station Keith had set up in the crater was producing readings in the low twenties at night, and it was getting colder.

At Phil's exorbitant rate, none of which was going to improve the porters' lot, Lonnie could afford no more, but he knew better than anyone how much work we had to do, so he pressured Key's for more help over the radio every morning. Phil had hired a pair of British business consultants, John and Jean Butcher, to help him manage things on his end, but they kept pointing the finger at Julias and even talked of firing him. Lonnie prevented that, not so much as a vote of confidence in Julias, but because he didn't want to "switch horses in mid-stream." It was a bad scene all around.

As we waited for the minimum of equipment we needed to begin drilling, we acclimatized, assembled the drilling dome up top, and arranged camp in the crater at the glacier's edge. A reflective cover was lashed over the top of the dome, because here—in contrast to Dasuopu and Guliya—there was a need to cool the drill and the new core segments in the intense heat of midday. It is difficult to extract a core from a warm drill, and there is also an increased chance for the driller's worst nightmare: the drill getting stuck in the hole.

Finally, on our fifth day on the glacier, an air of excitement and anticipation filled the dome as we fished our first segment of ice core up from a "practice" hole. Victor was at center stage and definitely feeling the pressure, as this was the first real test of a drill he had designed. It had already failed once on Sajama and again on Dasuopu, but he had methodically improved it each time, and Lonnie wanted to give it a fair shake here. (Victor would work out most of the bugs by the end of this trip, and Lonnie would go on to use his drill successfully on every deep-drilling expedition he's taken since—including Alaska's Bona-Churchill in the spring of 2002, where they retrieved the longest mountain ice core ever: 623 meters, or 2,044 feet.)

In the center of the circular dome stood a pole with a pulley at the top and a winch at the bottom, similar to the rig I had first seen on Sajama. The cylindrical drill hung from the pulley on a strong cable, by which it was lowered and raised from the hole and which also supplied it with electrical power. When the drill managed to fill itself with core or stopped for some other reason (usually the latter), Victor would grab the cable with both hands, lean back, and yank on it in order to break the set of the drill in the ice. He hurled himself at this task with an intensity that evoked a hushed and respectful silence from the rest of us, wearing a frown and a furled brow, his gaze fixed, his eyes bloodshot, his face unshaven. Topped off by the bombardier's cap, he reminded me slightly of Klaus Kinski in the film *Aguirre: The Wrath of God.*

We were all rooting for Victor and for the drill, but there was some trouble at first. The drill was designed to retrieve more than a meter of core each time it went down, but it kept binding every twenty or thirty centimeters. And although Bruce was the most knowledgeable driller there and quickly fathomed a solution, he was in an awkward position because Victor saw him as competition. Victor had insisted, for instance, on using a gas-powered generator, even though it produced less power than the solar panels engineered by Bruce that we had also brought along—and the drill did seem underpowered. Nevertheless, Bruce managed very diplomatically to make two suggestions that got the drill coring about a meter per drop by the end of the day. Even so, there was not much laughing, although some excitement was generated by the discovery of a layer of dust just below the surface, which Lonnie guessed to have been deposited during the dry La Niña of the previous winter.

Once Victor's lunging had dislodged the drill, he would switch on the winch and pull it up, then swing it with Vladimir's help onto a stand, where they would begin extracting the core. In the interim, Lonnie would lie on the ice and drop a temperature probe down the hole. This revealed that this glacier was not far from turning to water: near the surface, it was less then a degree below the melting point; six meters down, which was as far as they got that day, it was only about three degrees colder.

Things began to look up the next day, as the drill hummed and we got twenty-five meters of "real" ice core. The segments were extracted by the same two-stage process—designed to minimize contamination of the ice—that I had seen on Sajama. This time Lonnie and Mary were logging and packing away the cores, and as Lonnie gathered in each new segment, he looked at times like an expectant father and at others like a poker player raking in his winnings.

The ice became clearer the deeper we went—an effect Lonnie attributed to climate change: as they drilled back in time (rolling the movie reel backward, as it were) the summit crater slowly filled with ice, restricting the amount of nearby ash available to ride to the icy surface on the wind.

When the drill broke down the next morning, Mary was reminded of a note she'd written in her diary at Dasuopu: "No good day goes unpunished."

And as the sun set on that frustrating day, Bruce and I sat with our backs against our packs, while he told me the story of his career. Regarding our present impasse, he observed, "One thing a driller always has to ask is forgiveness from the mountain for taking the ice. If you don't, it won't part with it very easily."

"And has that been done here?"

"Yes. By me at least. But maybe the reason things aren't going well for Victor is that he hasn't. It's not that the mountain will kill you or anything, but it will try to screw things up."

"So that's why it worked so well on Sajama?" I asked, thinking of the llama sacrifice.

"I think so. But Sajama still didn't like it. It raged some. Huascarán didn't like it either. It'll test you. And you don't lie to the mountain; it'll find out. It's been around a long time. . . . I'm a canoeist, I came from rivers, and you don't lie to the river either; if you do, you'll die."

Indeed, Kilimanjaro began to rage openly that night, with a vicious wind that Keith measured at thirty knots steady with higher gusts. Aside from the usual irritations—flapping tents keeping us awake all night, difficult walking, lost belongings—it peppered everything with the fine ash on which we were living, which not only filled the tents but even filtered through the closed zippers of bags lying inside—and coated the lungs. I was constantly blowing dark brown dots onto my handkerchief and shielding my eyes, of course. No wonder the ice was so dusty.

Somewhat sheltered in the flimsy cook tent the next morning, Lonnie wished Mary a happy birthday, and she requested forty meters of ice core as a present. But the mountain had different ideas: the wind continued to rage, and the drill continued to hiccup. Only twelve more meters by late afternoon, when Victor hit an impenetrable layer of dirt at a depth of fifty meters, less than half the depth we had sounded the previous year, in a different location, admittedly. At first Lonnie's imagination ran away with him, and he guessed this might be a layer of volcanic ash, which might enable him to make the first estimate ever of the number of years since Kibo's most recent eruption.

Victor proceeded to ram the drill into the bottom of the hole with more intensity than before, if that were possible. And the first attempt was worthwhile, as it brought up a fragment of core that had been left at the bottom on the previous try. But they finally decided they had actually reached bedrock—or "ash pit," more precisely—and the drilling of this core was complete. The ramming must have knocked something out of alignment, though, for Victor and Vladimir simply could not extract the screw after they pulled the drill up for the last time. They tried everything:

hammers, screwdrivers, antifreeze in the gap between the screw and the housing; it took hours. It was dark, and everyone else had long since descended to the cook tent by the time they finally succeeded.

The struggle also continued on the balloon front. JR and Phil had still not obtained permission from the park bureaucracies; with the clock ticking, JR had decided to leave the negotiations up to Phil, pack the balloon for the porters to bring up later, and climb to the crater himself in order to begin acclimating. Their latest scheme was for Phil to meet personally with the president of Tanzania.

As I descended from the notch—Victor and Vladimir still wrestling with the drill on the summit—I spied a weary figure, accompanied by porters and chased by a buffeting wind, staggering toward our camp with the aid of two ski poles. It turned out to be JR. He'd never climbed a mountain anywhere near this high before, yet he'd made it up in only four days, and he turned out to be quite an agreeable and easygoing camp mate.

But we *had* retrieved our first core to bedrock, after all, and that may explain Lonnie's high spirits in the cook tent that night. When we'd finished a surprisingly good dinner, he asked me to fish around in the corner for a cardboard box labeled PROFESSOR LONNIE THOMPSON—INSTANT COFFEE that had been carried up in a recent resupply. Inside was a cake for Mary, complete with birthday candles. We lit the candles and sang her the song and fought the wind to our respective tents, feeling very uplifted. Raging mountain, be damned.

THEN THINGS BEGAN TO MOVE QUICKLY, FOR THE NECESSITY OF CARRYING THE CORES reared its head; so I volunteered to join the porters on the first carry to the freezer down at Simba Farm, to help work out the bugs in that process. Fighting my way up the snow ramp in the still-raging wind at about noon the next day, I met Keith on his way down. The screw had stuck in the housing again, and he was headed to the cook tent to retrieve a Primus stove in the hope that heat might do the trick. I found a disgruntled group sprawling in and around the dome: Bruce lay outside on a core box in the turbulent lee of the structure, feigning sleep. Inside, Lonnie was reading a magazine, Mary was sitting like a sphinx, Vladimir was writing in his journal, and Victor was brooding in silence. Having concluded, I guess, that all the scientists would be needed for drilling, Lonnie decided to take me up on my offer.

EIGHT PORTERS, JULIAS, AND I LEFT SHORTLY AFTER LUNCH THE NEXT DAY. IT WOULD have been nice to have left earlier, but the porters—who still had no sleeping bags nor adequate warm clothing—had taken longer than we to thaw out and had then been required to struggle through the wind to the dome to pick up the core boxes, which they carried down on their heads. (They had no use for the pack frames Lonnie had brought along. I've seen many a Kilimanjaro porter trudging up the hill with a perfectly functional backpack on his head.)

Dropping off the edge of the crater, we finally escaped the wind—but entered the treacherous terrain of the Western Breach. I felt guilty wearing the latest in outdoor clothing and a comfortable daypack, with ski poles, water and food, and Julias's enormous half brother Stephen to carry my camping gear, while the porters had only the fraying clothes on their backs and large, heavy boxes teetering on their skulls. The way they steered was to tilt their top-heavy loads in the direction they wanted to go, then skitter along to catch up. They slipped frequently on the loose gravel, landing hard on their asses and elbows. With the loads balanced on their heads, they couldn't look down to see where to place their feet, so on difficult sections each had to slide the box onto his shoulder, crook his neck, and cradle the box painfully between head and arm. It was every man for himself. We soon left agonizingly slow "Old Mike" behind. Like the tortoise and the hare, however, he later passed us as we rested in a cloud—appearing from the mist above and disappearing into the mist below—but then missed a shortcut bypassing Arrow Camp and arrived behind us in the end.

The porters cursed their loads and their cold hands and constantly asked me for food and water, but I doled both out sparingly and never let anyone put his lips to my water bottle, for they almost all had runny noses and coughs. Mercifully, halfway down the Breach we began to cross rivulets of meltwater and, just above Lava Tower, a full-fledged stream. At the Tower at 4:30 P.M., the porters finally had a meal of *ugali* with a vegetable sauce, which they scooped by hand from a communal pot. (*Ugali,* known as *millipop,* or *millimilli,* in South Africa, is made from maize and looks something like cream of wheat. The porters believe it to be the best "energy food" for their kind of work.)

And true to the continuing chaos of this situation, only three reinforcements were waiting at the Tower, so the strongest of the original eight were forced to soldier on—Julias's brothers, mainly: Geoffrey, George, Deo, and Stephen.

The last of us reached Shira Plateau at 7:30 P.M., just after dark. We piled the boxes in and onto the roofs of two 4×4s, and Stephen, Julias, and I rode with them across the plateau. There was no room for Stephen inside, so he lay with the boxes in the dark on the roof of one of the trucks as we lurched down—no change evident in his gentle, smiling demeanor.

The farm compound was protected by an impressive fence, and the gate was manned by a sleeping guard with a gun. (The Dutch manager told me later that his *askaris* were under orders to shoot anyone who tried to break in.) We passed inside without much trouble and made our way to the freezer mounted on a tractor trailer that Lonnie had arranged for Phil to buy a few months earlier. The freezer was attended twenty-four hours a day by an unusually talkative refrigeration technician, who drove us so crazy in the short time it took to stow the ice that he must have been driving the poor electrician camped with him utterly mad.

It was after midnight when Julias, Stephen, and I finally set my tent up in a field outside the farm perimeter. Stephen, who was acting as personal porter for his half brother, dutifully laid out Julias's sleeping bag and pad and stretched out between us on the bare floor: no bag for himself, only the spare clothing we lent him. We woke to the farm's million-dollar view of Mount Meru.

FEELING OBLIGED TO CONTACT KEY'S THE NEXT MORNING, WE DROVE OUT TOWARD THE main road looking for a phone—and who should we meet driving in the opposite direction but just the people we were looking for: Phil and the Butchers, who had decided to venture forth for a firsthand look at the freezer and the arrangements they had made for transporting the ice from the plateau. As an example of the regal manner in which Phil ran his business, he had not entrusted our driver with enough money to buy his own gas, and we were now running low, so it was lucky we met him: he paid for a fill-up. We also talked briefly about logistics, but Phil, who is the only Tanzanian I have ever met who is always in a hurry, did not take much part; he soon dragged the Butchers off and charged up to Shira.

When we met them partway across the plateau on their way down, some animosity flared between Julias and Jean Butcher, while Phil's only contribution was to ask impatiently if he and the Butchers could please leave. He assured me that he had just talked to the porters and everyone was happy.

This statement was belied by the degrading spectacle that greeted us at the trailhead: Phil had distributed the equivalent of about three U.S. dollars as a bonus among the destitute porters; one had ended up with more than his share, and this had led to a fistfight. Julias broke it up and confiscated the money, promising to distribute it equitably later, and they all swiftly focused on me to demand more pay—even Julias. Although I agreed with them, I suggested that griping wouldn't work: they should concentrate on the job; Lonnie was a fair man, and I believed he would compensate them as well as he could. The tension dissolved quickly, and we settled in to the evening meal, which converted me instantly to the virtues of *ugali*.

Then Julias entertained us with stories around the campfire. I had noticed that the porters tended to stay up late at night, howling with laughter, and that Julias's deep and resonant voice was usually the most prominent. I had figured he was just one of the guys, the party animal type; but I now learned that the porters saw him as the "elder" of their motley tribe, and one of the elder's roles is to entertain. There was something primordial about it. They whiled away their sleepless, high-altitude nights in the communal swapping of tales, while we white folks fidgeted alone in our tents, trying to read or whatever, by the light of our headlamps, irritated by their racket, when in fact we wouldn't have been able to sleep anyway.

Julias's nickname means "two men" in Chagga, meaning he has the strength of two, and, indeed, he was as imposing physically (though not quite as gigantic) as his half brother Stephen. I learned that the Minja family hailed from Marangu and that the original Chief Mareale used to invite their great-grandfather to attend his meetings with the Germans simply to awe them with his fearsome appearance.

I couldn't understand Julias's stories—they were all in Chagga—but the hilarity was so infectious that I laughed as loud as the rest. He tended to hold the floor. Others would chime in occasionally—usually to tease him, it seemed—and he would respond with some quip that would produce new gales of laughter. As we howled and the fire sent streams of sparks into the starry sky, new porters silently appeared in our small ring of light. One, by the name of Omar, had walked fifteen miles and seven thousand vertical feet from the town of Machame that day, in flip-flops and with only a plastic see-through raincoat for mountain gear, yet he stayed up baying at the moon as late as the rest of us.

I remember pausing to take in the scene at one point and thinking to myself, "It doesn't get any better than this."

But as Mary says, "No good day goes unpunished."

I WAS DAUNTED THE NEXT MORNING AT THE PROSPECT OF CLIMBING SEVEN THOUSAND vertical feet in a day, but the porters were so nonchalant that we didn't leave until nine A.M. I husbanded my energy—and my food and water—and things went well until early afternoon, when I happened to be working my way up a steep slope above Lava Tower as a storm blew in. It began to snow, and a cloud settled in around me, limiting the visibility to no more than twenty feet and muffling all sound. I thought I found the junction for the shortcut bypassing Arrow Camp, but I couldn't see or hear anyone, and there were no footprints. So I climbed by intuition for at least an hour, not sure I was on any path. Eventually I caught and passed a sweet-faced porter named Bariki, to whom I gave food. Neither of us was sure where we were, but we kept plodding upward.

This was not a major mountain storm even by the standards of my tiny home mountains in New England, but it was the biggest I have experienced on Kilimanjaro. It dropped a few inches of wet snow and then lifted to reveal magnificent views of freshened, white terrain—and to assure Bariki and me that we were on the right path. Through the clear air I now perceived another porter one or two hundred feet above me, climbing through the rickety gendarmes that guard the top of the Western Breach, carrying only an empty core box, no pack. A short time later, I sat on a rock to rest, eat, and drink. The porter noticed me from above and motioned for some water, but I waved him off, figuring we'd find an opportunity when I eventually caught up to him.

About a minute later, I heard a small cascade of stones, a grunt, and the bouncing of a box. Then came the sickening sound of a man bouncing down the hill, grunting on every impact. I looked up and saw an empty core box lying alone on the promontory where I had just seen the thirsty porter. I heard him bounce for about three hundred feet and come to a stop somewhere below me to my right, out of sight, then yell feebly a few times and fall silent.

You always wonder what to do first in these situations. There's a danger of panicking, getting hurt yourself, and only adding to the difficulties. I put down my pack, took a few deep breaths to calm myself, walked out to an exposed spur, and yelled for help. I could see a line of porters trudging with bowed heads below me, but none seemed to hear; their heads remained bowed.

The victim had fallen down a steep gully filled with loose scree just to my east. Picking my way delicately around the rickety ridge we were climbing, I perceived his body. It was bent in an unnatural V with his stomach pointing downhill. His legs were dead straight, pointing back up the hill behind him, and there was much too sharp a crook at the base of his spine. His back was obviously broken; in fact, I thought he might be dead. But he let out a few faint groans as I traversed over to him on extremely loose rocks. Reaching him, I straightened his body as gently as I could and laid flat him on his back. Blood was splattered on the rocks around his head and ran in streaks down his face, but he showed no signs of shock.

I noticed that he had no gloves or hat and was wearing plain street shoes with smooth soles. Julias would later inform me that he had never climbed higher than Lava Tower before.

Bariki was the first person to find us, and he emitted a soft whimper when he realized it was his brother, Wilfred; but at least Wilfred had someone to talk to now. The bright sun was also keeping him warm enough for the time being. When Julias and his brothers arrived, I asked Geoffrey, who was thin and athletic and about six feet five, to run downhill for help; when Stephen appeared a while later carrying an enormous load, I told him to drop it and run uphill for help from the crater.

By a miracle, some trekker had dropped a thick, fully functional air mattress, probably from the exact spot Wilfred had fallen, and it had lodged in a level scoop in the scree just twenty feet below us. We fetched the mattress and managed to get Wilfred onto it, then half carried, half dragged him down to the scoop.

The sun fell quickly behind the rim. Wilfred would be spending the night up here.

Scanning the teetering and dripping cliffs above us, Julias then realized we were in a dangerous place: with the falling of night, the dripping water would freeze and expand and the danger of rockfall would increase. We had to move Wilfred, but we didn't have enough people to risk moving a man with a broken back, so we decided to wait for reinforcements. I scrambled back to the pile of my personal gear, which Stephen had dropped, and began setting up our only tent in an alcove on the trail. As it followed a ridge, it was less exposed to rockfall.

As I finished setting up the tent, I heard yelling from above: Stephen and six new porters. We now had enough people to risk moving Wilfred. About a dozen porters were soon carrying him across the loose scree, as I crawled sideways on all fours below them, placing my hands on the heels of their shoes to keep them from sliding downhill.

Managing to zip Wilfred into my sleeping bag, we placed him in my tent at eighteen thousand feet and left Bariki and Julias's brother Deo, who was Wilfred's close friend, to watch him for the night. Eight of us then climbed the last thousand feet of the Western Breach in the dark, with Stephen picking up the rear to make sure no one fell behind—and me keeping an eye on the exhausted Omar, the porter in the see-through raincoat and flip-flops who had climbed all the way from Machame the previous day. He would lean against a wall and gasp every few minutes, then summon the strength to stumble on. We climbed in silence.

In the crater, the first tent I passed was Vladimir's. He and the rest of the scientists had not heard about the accident, for they had been working on the glacier when Stephen had come up for help. When I told Vladimir what had happened he quickly handed me the same kind of sleeping bag that Bruce had offered me on the top of Sajama three years before.

THE NEXT MORNING I DESCENDED INTO THE WESTERN BREACH WITH JULIAS AND Stephen, but stopped on a spur and left them to descend alone when I picked out the cluster of twenty or thirty porters carrying Wilfred down. Geoffrey had run all the way to a camp just above Shira Plateau, where he had found a few park rangers who had radioed Moshi, then picked up a stretcher and carried it back up alone, reaching Wilfred again at eleven P.M. (thus nearly climbing "Kili" twice that day). It took the porters ten hours to carry Wilfred to the trailhead, from which he was shuttled to the Moshi hospital—a good one, thanks to tourism. Over the next few days, we would learn by radio that he had broken an arm, a leg, a hip, and his back; he would remain paralyzed from the waist down and be confined to a wheelchair. Phil Ndesamburo had no insurance for his employees, of course, but some time after the expedition, Lonnie would send Wilfred a personal check for one thousand dollars to help him buy a wheelchair.

IT SEEMED THAT THE MOUNTAIN WAS NOW SATISFIED. THE WIND HAD STOPPED, AND THE drill had begun humming. The drilling team had recovered a second core, near the first, on the day Julias, Stephen, and I had wandered around in the lowlands, and would complete a third on the day of Wilfred's evacuation. The bulk of the drilling was complete; the expedition had been transformed into an endgame in just three critical days.

Over the next week we moved the drill first to the Furtwängler for one short, watery core and then to the southern ice field, only two hundred yards from Uhuru's summit, for a final two.

The last site was the most spectacular. Within view of the summit marker, on a smooth white surface that dropped away in a gentle arc, south toward Moshi, we felt as though we were on the edge of a small planet. Mawenzi loomed darkly to the east across the saddle. A sharp temperature inversion brought the clouds and haze to an abrupt halt a few thousand feet below us, so I was almost tempted to walk out onto the flat white surface of the clouds, which stretched to the horizon under a flawless blue sky.

Five days after the accident, the axe finally fell on the balloon. Phil had never met with the president. JR bid farewell and sauntered off toward the Breach as good-natured as ever, and our primary focus shifted to the carrying of the ice. (Sometime later, wasting time in the cook tent, a few of us determined that the balloon didn't really make sense anyway: it would take just as many porters to carry it up as it would to carry its load of ice down, and climbing is harder than descending.)

We were now brittle from the altitude, and the accident had been a blow to our collective stomach—adding to our edginess as Lonnie briefed us on Wilfred's condition every morning after his radio call. Beyond his concern for Wilfred, he was terrified at the prospect of another lawsuit. The Shawn Wight trial still loomed.

Since we had been at nineteen thousand feet for almost two weeks, some of the long-term effects were also setting in. But this group knew how to handle it; they retreated into their shells for the most part. Mary told me later that she was hit by a combination of anxiety and depression familiar to her from Guliya, Huascarán, and Dasuopu (not Dunde, though, where she had experienced her magical introduction to this game): "I get a tune in my head that I can't get out," she added. "It plays over and over and over until I'm ready to go insane."

And Lonnie's cough reemerged. He was now hacking away almost every night, and the coughs themselves were erupting in three distinct syllables, one from so deep in his chest that I feared, as I myself lay awake most nights, that he would crack a rib, as he had in the past.

His tenacity was astounding. He would rise early every morning and walk to the notch for his radio call, where he was actually writing grant proposals and keeping up with the latest developments in climatology through the exchange of e-mails with Ellen. (One morning he returned with the news that a satellite analysis in that week's *Science* had claimed that the air at our present altitude had been cooling for the past few decades. "Now there's an example of someone who's been spending too much time at her computer," he griped in the cook tent. "Don't tell me it's cooling. I've been visiting these places for thirty years, and I *know* they're getting warmer. I've seen it with my own eyes.") He was also putting in many hours by the drill, logging samples, and, of course, managing the expedition. To temper the porters' endless complaint, he had produced a sign-up sheet, where they now kept track of the loads they carried. He promised to tip them by the load—as much as he could afford—when he had settled accounts with Phil and knew how much money he had left.

Keith kept to himself, monitoring his "met" stations and taking long walks to place the aerial photography markers. And while we were all getting filthier by the day, he managed to outpace everyone in that category. He would appear in the morning with his shoes untied, his shirt half-untucked, sweatshirt pulled up to reveal half his back, one pantleg hitched on a calf, a sock drooping on a shoe—impervious to the cold. His beard had merged with his chest hair. His strategy seemed to be to wear the same clothes every day (with the idea, I sincerely hope, of burning them at the end of the trip), and his only pants had developed a huge rip in the seat, so we tended to shield our eyes whenever he bent over.

As we rigidified with altitude into shallow caricatures of ourselves, I must admit that even Keith's feisty good humor began to grate. We were all edgy and bored and had told all the stories we knew. There was an air of negativity at mealtimes, much criticizing of people who weren't there.

A few moments after they pulled up the very last segment of the last core, Lonnie asked Victor to teach Vladimir how to run the drill (so Victor would have a backup, just in case), and Victor refused—not the move of a team player. He and Lonnie butted heads for a while, and Lonnie finally fired him—although he would take him back a few weeks later, when they had returned to sea level and their senses.

So that hung over us. I remember Bruce saying at one point that this expedition never really got off the ground; it never had that spark, "that Christmastime feeling," as he put it. Ice core drilling is hard work.

AT THE VERY END, DOUG HARDY AND MATHIAS VUILLE, TWO SCIENTISTS FROM THE UNIversity of Massachusetts whom I had met on Sajama, showed up to erect an automated weather station on the northern ice field. They simply did their job and left, never even bothering to climb Uhuru.

I helped where I could: lugging loads to and from the drill sites, taking GPS measurements to help Keith with his mapping, attending to a scientific errand at the Furtwängler one dawn, and keeping to myself. As our time in that remarkable place drew to a close, the moon waxed full, the wind dropped, and the silence of night and the daytime vistas of sand, sky, ice, and cloud enveloped my small scheming mind in a spacious tranquillity.

When the day came for us to leave, we simply packed up and walked down, reaching Key's and Kenny G by dinnertime. Phil then graciously allowed us one good night's sleep before exacting his last pounds of flesh, my own bill working out to more money per day to eat abominable food in the squalid cook tent than to stay in his hotel. As I wandered back to my room, somewhat dazed, I bumped into Keith and asked him where Lonnie was.

"Up in his room recovering from the bill," was the response.

The porters alone had cost him thirty-seven thousand dollars. He tipped them as well as he could, but not as much as he might have liked—nor as they might have expected.

ON OUR LAST TWO MORNINGS IN AFRICA, JR FINALLY PUT THE BALLOON TO USE BY GIVing us rides, which turned into town-wide celebrations. Perceiving us in the sky, kids on the way to school dropped their books and chased us through the streets and across fields. Whenever JR let the balloon cool and sink toward earth, the children's joy would turn to fear and they would turn and run; but as we rose again, they would turn back and resume the chase. There must have been a thousand people waiting to welcome us when we landed in the middle of town the first day.

Lonnie made sure the ice made our flight to Amsterdam, but personal gear was less important. Mine wasn't even off the mountain yet. It arrived in a DHL box six weeks later.

·21·

THE STORY IN THE SNOWS

> The human race is like an awkward adolescent whose political and social mechanisms are not keeping up with his physical growth.
>
> ROBERT KAPLAN

The mountain's main message floated into view without any help from the balloon. In February 2001, almost exactly a year after our return, Lonnie told an audience at the annual meeting of the American Academy for the Advancement of Science—perhaps the premier annual convocation of the international scientific community—that the snows of Kilimanjaro are likely to disappear within twenty years and "little can be done to save them."

This made the front page of the *New York Times* the next day and headlines around the world for about two weeks. All in all, it attracted more media attention than any story on a non–sports-related topic ever to have emerged from Ohio State. The poignant image of the mountain certainly gave it lift, but Lonnie's timing wasn't bad either. About a month earlier, the IPCC had begun rolling out its Third Assessment Report, representing the definitive statement by mainstream science of its current understanding of global warming; and on the very day Lonnie spoke, the IPCC released a volume on the expected impacts to society.

They had stopped waffling. In their first volume, *The Scientific Basis,* they had stated, "There is new and stronger evidence that most of the warming observed over the last 50 years is attributable to human activities." Adding that "the increase in temperature in the 20th century is likely to have been the largest of any century during the past 1,000 years," they went on to predict an average temperature increase of between 2.5 and an unnerving 10 degrees Fahrenheit by the end of the twenty-first century, that sea levels would rise between three inches and three feet, and that extreme weather events—heat waves, heavy rainstorms, cyclones, and summer droughts—would increase in frequency.

In the volume released on the day of Lonnie's talk, the panel pointed out not only that many of the earth's physical and biological systems had already been affected, but that humans were beginning to feel the pain as well. Annual economic losses from catastrophic events had increased tenfold from the 1950s to the 1990s, for example, from about four billion dollars per year to forty billion dollars per year in adjusted 1999 dollars.

The impacts will probably be minor if the warming in this century comes in at the low end of the spectrum, the report continued: most of us would probably be able to adapt. But a change of nine or ten degrees would almost certainly cause widespread catastrophe. Available freshwater would plummet, taking crop yields along with it. The prevalence of malaria, cholera, and other vector-borne diseases would increase. Increased flooding would pose a deadly risk to tens of millions of residents of low-lying areas—Bangladesh and the small Pacific island states being most vulnerable there. And the hunch Lonnie had voiced in 1999 in the bar at Key's was also confirmed: "Those with the least resources [particularly the inhabitants of Africa, Asia, and Latin America] have the least capacity to adapt and are the most vulnerable."

There would be some positive changes, such as an increase in the global timber supply and higher crop yields in some parts of the midlatitudes, but, as Harvard's James McCarthy, the lead author of the impacts report, summed it up, "Most of the earth's people will be on the losing side."

Propelled by Lonnie's clear words at that pivotal time, the melting snow on Kilimanjaro has evolved into one of the central icons of global warming. A week after his talk, the lead editorial in the *New York Times,* "A Global Warning to Mr. Bush," joined the messages from the mountain and the IPCC and called upon the new president to lead. In those halcyon early days, some environmentalists actually believed that as a friend of industry, George W. Bush stood a better chance of persuading Congress to support the Kyoto Protocol, or something like it, than his predecessor had. Indeed, Bush *had* made a campaign promise to regulate carbon dioxide from the greatest emitters, which are electrical power plants.

As the flurry of interest began to fade, however, only a month after

Lonnie's talk and two months into his first term, the president reneged on his promise. Ever since—and despite the advice of his own government's top climatologists—he and his aides have frequently made the astounding claim that the science is not clear on this issue and that more research must be done before concrete steps to limit carbon dioxide emissions are justified.

Shortly after Bush's reversal, I spoke with Michael Oppenheimer, who then worked at Environmental Defense and has since moved to Princeton. A co-author of the IPCC volume on the scientific basis, Oppenheimer has been involved in the greenhouse issue since its very beginnings. Perhaps I caught him on a bad day. He sounded exhausted if not downright depressed:

"We live in a very difficult time right now, because of the attitude of the Bush administration," he said. "We will determine in the coming months, really, whether the last ten years of trying to build a diplomatic solution was all wasted time; and if so, that will set the world back in trying to solve the problem. It will mean a guarantee of significant, additional warming before we bring the thing under control.

"The world is warming. I think there's going to be more and more warming no matter what we do. But I think the future remains in our hands to the extent of being able to determine whether we have a moderate warming that most people and even some ecosystems can adapt to, or whether we have a warming that is so large that society and certainly most ecosystems simply can't successfully adapt. I think that decision is yet to be made.

"The diplomatic process was very difficult. There will be a lot of distrust no matter who the U.S. president is four years from now, and it will take a long time to rebuild. There will be a decade lost, I think. So in these times when the political world is discouraging, the thing we have to fall back on, those of us who worry about the environment, is the solid, strong voice of the scientific community—people like Lonnie who tell it like it is and who understand both what solid science is and how to make it accessible to policy makers and to average people. There is no replacement for that kind of quality."

Now that four more years have passed, some believe the threshold to an inevitable, destructive warming has been crossed.

Meanwhile, other industrial nations have taken the first small step toward action. In order for the Kyoto Protocol to take effect, it had to be ratified by a large enough portion of the industrialized world to account for 55 percent of its total annual emissions. The withdrawal of the United States, which accounts for 25 percent alone, set the bar at an effective 73 percent; but the threshold was reached, nevertheless, in November 2004, a few days after Bush's reelection, when President Vladimir Putin signed the ratification documents for Russia. The only industrialized countries besides the United States not to sign were Australia (a major coal producer), Liechtenstein, and Monaco.

IN FACT, IT HAD NOT BEEN ALL THAT DIFFICULT FOR LONNIE TO FIGURE OUT HOW MUCH time the snows of Kilimanjaro have left. As he told *Science*, "All we did was connect the dots. The local people could have told you it's melting just as well as I can." (He was probably thinking of Julias Minja.)

Actually, Lonnie had produced one of the "dots" himself, with the estimate of the area of Kibo's ice fields that he had made from the aerial photographs taken during our expedition. The idea of using historical and contemporary maps to track the retreat of the snows had been conceived in 1980 by the Swiss scientist Bruno Messerli, whose own analysis began with the map Hans Meyer had drawn by hand on his final visit to the mountain in 1898. Messerli's work was extended in 1997 by Lonnie's old friend and sometime coworker Stefan Hastenrath, who doubted the accuracy of Meyer's map and decided not to use it, but had a new dot to add by that time. Hastenrath showed that the mountain had lost three-quarters of its snow cover between 1912 and 1989, but he didn't discuss the future. In the first year of the new millennium, Lonnie simply plotted Hastenrath's numbers together with the new one he had obtained himself and showed that the area covered by snow has been dropping linearly with time—losing about 1 percent of its 1912 area every year—and that the line that best fits the numbers crosses zero in 2019. In the primary paper on the Kilimanjaro ice cores, which appeared in *Science* in 2002, the Thompsons wrote, "the ice on Kilimanjaro will likely disappear between 2015 and 2020."

But let us not discount the intuition of Julias Minja. On our reconnaissance in 1999, he told me the snows would be gone in about ten years.

WE HAD ALSO GONE TO THE TROUBLE OF DRILLING SIX ICE CORES, OF COURSE, AND they tell a more detailed and, perhaps, more sobering story.

Four days before Lonnie informed the world of the imminent disappearance of the snows, Doug Hardy had returned from a visit to the northern ice field, where he had retrieved the first year of data from his weather station. Upon reaching the top, he had discovered that the station, which looks something like a radio tower, had fallen over, for the hard ice in which he and Mathias had planted it the year before had melted away.

Doug had also checked the accumulation stakes we had left the previous year and found that *accumulation* was something of a misnomer: the surface had dropped more than three feet. Put another way, the ice caps on Kilimanjaro seem to be "losing time" not from the bottom, as on Quelccaya, but from the top.

This made the job of dating the ice cores more difficult than usual, because it meant Lonnie and Ellen couldn't simply count back years from the present. (It *was,* as Lonnie had suspected, a complicated record. They had more trouble sorting out this chronology than they'd had with any other.) Luckily, the ice deposited in the 1950s had not yet been stripped away, so they found the chlorine-36 spike from the IVY Mike bomb test and used that as a starting point. There was also enough biological material to permit carbon dating farther down, and they matched the isotope profiles with one from a speleothem in the Soreq Cave in Israel.* And when the chronological puzzle had finally been pieced together, it confirmed the bold prediction Lonnie had made in January 2000 as we were driving through the dark to Key's: the snows of Kilimanjaro were born about 11,700 years ago.

HE HAD BASED THAT HUNCH ON HIS ASYNCHRONY THEORY, OF COURSE; IT IS ONLY A COincidence that this climate record was born at roughly the same time as human civilization and not far from its cradle. But this coincidence happens to give the snows a long enough memory to tell us a story about the sweep of climate and human history.

The record begins in about the middle of the cold Younger Dryas (which the Thompsons might call the "deglaciation climatic reversal"). The isotopes in our third core from the northern ice field, which has turned out to reach back the furthest in time, show that the Horn of Africa was recovering from a cold spell as the ice cap was born. You will remember that the end of that cold spell marks the beginning of the present Holocene epoch and that the lean conditions associated with the onset of the Younger Dryas had prompted a few tribes of hunters and gatherers living by the headwaters of the Tigris and Euphrates rivers, about two thousand miles north of Kilimanjaro, to begin raising crops and keeping animals for the very first time. As the Chicago archeologist Robert Braidwood, who is credited with the discovery of agriculture's beginnings (though not their link to climate change), said in 1963, "Somewhere in

*A speleothem is a mineral deposit that has formed in a cave by the action of water—stalactites and stalagmites being the most obvious examples. The Thompsons matched Kilimanjaro's isotope profiles to the speleothem in the same way that they had matched those from Huascarán and the Portuguese seabed core.

one of perhaps a dozen places in the Middle East about 12,000 years ago, some man made a remarkable observation: he observed that a common weed which he had doubtless collected for eating was growing where he had previously spilled seeds.

"Once man was able to remain in one spot, he was able to start thinking about matters other than gathering food. He was able to begin thinking about his new relationship to other men, new relationships to his immediate surroundings and to those forces in nature which played such a large part in his existence."

It was a profound invention. The rate of social change accelerated dramatically at that point.

The period from eleven thousand to about four thousand years ago is known as the African Humid Period. The same precession-driven increase in summer insolation that had just given birth to the snows of Kilimanjaro moved north to the Sahara Desert, centered on the Tropic of Cancer, which had been dry for the previous ten thousand years. An increase in rain, carried by a strengthened monsoon and the prevailing westerlies of winter, transformed the earth's largest desert into something like the Garden of Eden from about 6500 to about 4500 B.C. Lake Chad, at the southern edge of the present Sahara, expanded to twenty-five times its present size. Grasses grew, lakes formed, and large African game animals (not to mention people) took up residence there. The carbon dating of bone hooks, harpoons, and piles of fish bones found in the remains of neolithic villages places the height of human settlement in the Sahara from about 4500 to 3000 B.C. "The possibility of settling near the clay pans [shallow lakes on clay-rich earth] favored the introduction and expansion of a rich ceramics industry," wrote the French archeologist Nicole Petite-Maire and her colleagues. "The variety and quality of pottery decoration and of stone and bone adornments, suggest spare time for refinement, incompatible with a hard struggle for life in a hostile environment."

This age of plenty was not entirely free of excitement, however. Somewhere around 6400 B.C., most of the earth experienced what has been called a "climate shock": a stretch of cold, dry weather that lasted about two hundred years. In the long Kilimanjaro ice core this event shows up mainly as a spike in fluorine ions, which the Thompsons see as a sign that nearby lake levels dropped. The wind would have whipped up the fluorine from salt flats at the edges of shrinking "soda lakes," such as Lake Manyara, only three hours' drive west of the mountain, and added it to the dust in the air. Methane levels in Greenland's GRIP ice cores also dip for two hundred years at this time, indicating that the tropics dried and probably cooled (as Guliya showed that they also had during the coldest stretches of the Wisconsinan glacial stage).

Since the 6400 B.C. climate shock also happens to have coincided with the last major draining of Lake Agassiz and her sister, Lake Ojibway, from the surface of the Laurentide ice sheet, primarily into Hudson Bay, certain researchers have been quick to suggest that this shock may have been caused by a shutting off of Wally Broecker's great conveyor. Recently, however, a few of those same researchers have pulled the rug out from under that explanation by pointing out that the flood of freshwater does not appear to have slowed the conveyor down all that much. (Incidentally, the recent work also shows that the draining of Lakes Agassiz and Ojibway would have flooded the North Atlantic at a rate ten times faster than melting glaciers could possibly do today, and that if *that* couldn't shut the conveyor down, it is very unlikely that today's warming could—or, therefore, that there is much chance at all that the present warming will cause an ice age.) Indeed, the Thompsons believe that the methane dip in Greenland and the fluorine spike at Kilimanjaro together "strongly suggest that [the 6400 B.C. climate shock] was driven by abrupt and large hydrologic changes in the tropics, particularly in Africa"—the old "tropics versus poles" debate again.

Jared Diamond, author of *Guns, Germs and Steel,* points out that agriculture spread primarily east and west after its invention, along latitudinal lines, so that apprentices of the new technology would have had no need to "adapt to new day lengths, climates, and diseases." By 6400 B.C., tilled fields could be found in a "huge swath of Eurasia, from Pakistan to the Balkans," but had only recently reached northern sections of the Nile River Valley. It would take a few more millennia for agriculture to extend farther south into Africa.

Archeological sites in the Levant and Mesopotamia (the wide, flat alluvial plain of the Tigris-Euphrates) indicate that the 6400 B.C. climate shock led to profound changes in many primitive societies. Some sedentary farmers resumed the nomadic life of herdsmen. Others "habitat-tracked": they moved to new agricultural sites where water remained plentiful, such as the floodplain at the confluence of the Tigris and Euphrates in what is now southeast Iraq, from which the two great rivers drain as one into the Persian Gulf.

According to a number of climate records, including the Kilimanjaro ice cores, the African Humid Period ended with a bang in about 3000 B.C. The transition to what has now been five thousand years of African drought commenced with a blast of extreme drought and cold that lasted for about one hundred years. Agricultural technology had evolved to the point that the Late Uruk (Sumerian) civilization on the floodplain of the Tigris-Euphrates now employed an irrigation system fed by canals. The Uruk civilization collapsed swiftly right around 3000 B.C., and archeologists are now beginning to realize that drought probably had something to do with it. (Four thousand years later and half a world away, another canal-dependent civilization, the Tiwanaku, would succumb for similar reasons.)

Meanwhile, about a thousand years earlier, a bald spot had appeared in the center of the Sahara. As it grew, the relaxed neolithic fishermen and potters, some of whom now kept goats or cattle, were forced to follow the retreating edge of the savannah. By the end of the African Humid Period, they had been displaced to the banks of large rivers, such as the Niger, the Senegal, and the Nile, or to mountain highlands or the ocean's edge, where there was enough seasonal rain to water crops and replenish wells.

Nile fishing villages evolved into small towns, trade intensified, cultures cross-fertilized, and "extractive" petty states emerged: hierarchical societies in which kings, bureaucrats, soldiers, full-time craftspeople, and the like subsisted by the taxation of those who produced food. In 3100 B.C., the first predynastic state arose in Egypt; and its first true nation-state, the Old Kingdom, ruled by pharaohs, began to flourish in about 2600 B.C. Monumental pyramids were built on the hills near the capitol of Memphis to tower above the tilled fields on the slender floodplain of the Nile. The stepped pyramid was built at Zoser. The great pyramids were built at Giza.

The world's first cities had arisen a few centuries earlier, in Mesopotamia, where the first imperial monarch, Sargon of Akkad, eventually established a great empire, reaching west as far as the Mediterranean and north as far as the Black Sea, over a sixty-year reign that began in about 2350 B.C. The city of Akkad was located in southern Mesopotamia, at a strategic spot on the plane between the Tigris and Euphrates, where they veer to within about thirty miles of each other—not far from the spot where Baghdad would be built about three thousand years later.

Mainly by conquest, Sargon and his successors dominated Sumer to the south and established a network of city-states on the high, dry plains to the north, where, writes Yale archeologist Harvey Weiss, "regional settlement was redistributed to maximize regional efficiency and massive city walls were constructed to control labor." Irrigation was not a viable option in the north; it was all dry farming. Each of the northern cities was charged with managing and gathering revenue from the large agricultural region surrounding it and with transporting revenue south to the imperial center in Akkad.

An ancient cuneiform text, *The Curse of Akkad,* reads,

Ships brought the goods of Sumer upstream to Akkad, . . .
Elam and Subir carried goods to her with pack-asses,
all the provincial governors, temple administrators, and land registrars at the edge of the plains
regularly supplied the monthly and New Year offerings there.

Until, in about 2200 B.C., after less than two centuries of glory, the Akkadian empire suddenly collapsed. Your encyclopedia may tell you that Akkad was overrun by tribal Gutians from the Zagros Mountains in what is now western Iran and that Mesopotamia descended into chaos for more than a century thereafter.

But Egypt's Old Kingdom collapsed at precisely the same time, and a century of chaos ensued there, too . . .

IN 1993, AFTER FIFTEEN YEARS OF EXCAVATION AT TELL LEILAN, AN URBAN CENTER ON the Subir plain mentioned in the above poem, Harvey Weiss announced that he had uncovered a layer of fine, dry windblown dust just above the last signs of Akkadian occupation and proceeded to touch off a tempest in the archeological teapot by suggesting that mighty Akkad had fallen not to lowly mountain tribesmen but to the social upheaval brought about by a remarkable three-century drought.

This notion was treated as a curiosity or local anomaly at best, until 1998, when Lamont paleoclimatologists Heidi Cullen and Peter deMenocal found a similar dust layer in a seabed core from a site about 1,400 miles away, in the Gulf of Oman. (The Shamal, a hot northwest wind, blows Mesopotamia's dust toward the gulf in summer.) Both the climatologists' and Weiss's layers had been fixed in time by carbon dating, which has an inaccuracy of about 150 years for that epoch, so that method alone was not sufficient to match the layers conclusively. It so happens, however, that Weiss had also found a layer of volcanic tephra, evidence of an eruption, about 140 years below his dust layer. Tephra is an ideal stratigraphic marker, because each eruption produces a unique chemical signature. Cullen and deMenocal found a layer with the same signature, spaced by just the right amount of time, in their deep-sea core. Thus Cullen, as a climatologist, lent additional credibility to Weiss's hypothesis when she told *Science*, "There's something going on, a shift of atmospheric circulation patterns over a fairly large region."

That has turned out to have been an understatement.

THIRTY-TWO AND A HALF METERS DOWN IN THE THIRD AND LONGEST ICE CORE FROM Kilimanjaro's northern ice field we found a black layer of dirt more than an inch thick, interrupting very clear ice above and below it. This extraordinary feature proceeded to intrigue and perplex Lonnie for more than a year. On the mountain, he was guessing (as he had when Victor had first rammed the drill against what had turned out to be the ash pit) that the layer had been produced by an eruption of Kibo itself. About six months later, chemical analysis had revealed the presence of salt—which Lonnie had not yet identified with shrinking soda lakes—and he imagined that the dark layer had come from an eruption under the sea. The timing was beginning to come into focus at that point, and his favorite candidate was the so-called Minoan eruption of Santorini: in 1650 B.C., Greek towns on that Aegean island were destroyed by a nearby eruption as well as the tsunami it generated, which also inundated the north coast of Crete. Some believe this calamity gave rise to the myth of Atlantis, the magical land that sank into the sea.

But when they finally nailed down the chronology, Lonnie and Ellen realized that Kilimanjaro's dust layer lines up precisely with the layer Harvey Weiss found at Tell Leilan. The Thompsons found, furthermore, that the two other cores from the northern ice field, both of which were taken from spots closer to the crater than the long core—that is, nearer the edge of the ice—start just as the dust event ends. This told the Thompsons that the glacier must have retreated past the first two drilling spots during Weiss's drought and that "the major dust layer . . . was deposited during extremely dry conditions either before or during the retreat. . . ." Presumably, the ice would have begun to advance again after the drought ended.

Other records show that the lakes of central Africa retreated at this time. The speleothem from the Soreq Cave shows that precipitation in Israel dropped by 20 to 30 percent for about four hundred years. A core retrieved from the bed of Lake Van in eastern Turkey, in which annual sediment layers—so-called varves—are so distinct that they can be counted like tree rings, displays a "sharp five-fold dust spike at [about] 2290 B.C., continuing until [about] 2000 B.C." There is evidence for a simultaneous drought in Greece, Italy, and Spain. Spikes show up in pollen records from West Asia and in seabed sediments from the Arabian Gulf off the coasts of Pakistan and India. Lakes also dropped in Ethiopia, the all-important headwaters of the Nile.

And there are signs in the New World. There's a dust spike at 2100 B.C., give or take two centuries, in a sediment core from Elk Lake, Minnesota, and its magnitude indicates that there was three times as much dust in the air back then than there ever was during the 1930s Dust Bowl. The widths of tree rings from Indian Gardens, Nevada, decrease sharply from 2278 to 2056 B.C. Droughts occurred in Mexico. And, finally, Lonnie's Huascarán ice cores display an enormous dust event in about 2200 B.C., lasting more than a century, which would appear to mark the greatest drought Peru has experienced in the last seventeen thousand years. Since the timing of this event is inaccurate by about two hundred years, he and Ellen have not had much to say about it yet, but as Harvey Weiss has continued to build his case that whatever felled Tell Leilan probably reached around the globe, they have begun to look more closely. They're testing for a possible link to Old World records by looking at the chemistry of the dust.

Furthermore, every record with fine enough resolution shows that this global eruption of dust and drought came to a head very rapidly, reaching its peak within two or three decades at most.

Probably the most interesting thing about it is that it is completely unexplained (although it has been linked, inevitably, to changes in the North Atlantic). Early on, Weiss halfheartedly suggested that it might have resulted from an asteroid impact: "Hundreds of years after the event," he wrote, "a cuneiform collection of 'prodigies,' omen predictions of the collapse of Akkad, preserved the record that 'many stars were falling from the sky.' Closer, perhaps as early as 2100 B.C., the author of the *Curse of Akkad* alluded to 'flaming potsherds raining from the sky.'" But no one has found an impact crater that might serve as the smoking gun as yet.

WHATEVER IT WAS THAT HAPPENED IN 2200 B.C. SEEMS TO HAVE WREAKED AS much havoc upon human society as any single event of the Holocene—excluding, perhaps, the havoc we wreaked upon ourselves with World Wars I and II.

In the scenario favored by Weiss, the farmers on the alluvial plain upstream of Akkad responded to this drought in the same way that their ancestors had four thousand years earlier, and that the "Okies" would respond to the Dust Bowl four thousand years later: they picked up and moved. Some took up pastoralism, others searched for fields nearer reliable water, hoping to keep farming. The latter strategy was unsuccessful for the most part, for the new agricultural technology, coupled with the previous time of plenty, had allowed the population to grow, and most of Mesopotamia's agricultural niches were already taken. Since tenant farmers abandoned their fields and stopped paying taxes, the administrative cities on the plain were abandoned in turn. The Assyrian King List, pressed into cuneiform tablets, tells of "seventeen kings living in tents" in northern Mesopotamia at this time, a sign that control passed swiftly from urban administrators to nomadic warlords.

The result was a diaspora of environmental refugees, all moving south toward the imperial center of Akkad, on the water-rich floodplain of the

Tigris-Euphrates. Weiss believes the capital was overwhelmed by a wave of destitute immigrants, who inundated the city just as the extractive economy that supported it collapsed as well. The ancient, northern "Okies" were known to their imperial rulers as Amorites; as evidence that Akkad attempted to stem the human tide by force, Weiss points to the "Repeller of the Amorites" wall, which was built to the north of the city between 2054 and 2030 B.C. It stretched 110 miles: from the bank of the Euphrates on the west to bank of the Tigris on the east.

So, opportunistic raiding by the Gutians may have been the straw that broke the camel's back, but the ground had been laid by climate change.

Similar scenarios played out all through the Middle East. The level of the Dead Sea dropped more than three hundred feet: "The well-scoured archeological landscapes of Palestine between the twenty-third and nineteenth centuries B.C. show that walled towns declined and were replaced by unwalled villages, cave occupations, and campsites," writes Weiss. The second city of Troy, in Anatolia, either collapsed, was abandoned, or was destroyed. Populations crashed in mainland Greece and on the island of Crete—where, after the drought ended, the nascent political center at Knossos built its famous cistern and system of aqueducts, perhaps as insurance against another.

Fifteen hundred miles to the east of Akkad, in the watershed of the Indus River, the Sarasvati, "foremost mother, foremost of rivers, foremost of goddesses," dried up and retreated into myth; and in what anthropologists know as the "Indus Collapse," the urban centers of the Harappan civilization were abandoned—as their taxpayers, too, migrated to wetter land to look after themselves.

As with all great changes, a small minority thrived. In Central Asia, forests and arable land were gradually replaced by steppe, which is more amenable to sheep and goat herding, so sedentary villagers loaded their belongings onto horse-driven carts or chariots, gathered up their herds, and took to the road. The population actually increased in a large swath of what is now mainly Kazakhstan, north of the Black, Caspian, and Aral seas. And although the southern regions of present-day Turkmenistan and Uzbekistan turned mainly to desert, in certain scattered oases, the people learned irrigation and prospered.

Finally, in Peru, the 2200 B.C. climate shock seems to have initiated a swing of the cultural seesaw: population centers shifted from the coast to the Andean highlands.

AS EARLY AS 1971, THE EGYPTOLOGIST BARBARA BELL HAD GUESSED THAT THE COLLAPSE of the Old Kingdom, which coincided precisely with Akkad's, may have been caused by a persistent drop in Nile flood levels. (The evidence for a simultaneous drop in Ethiopian lake levels dovetails well.) Furthermore, this collapse seems to have occurred at a peak in Egypt's power: Pepi II, the last pharaoh of the Old Kingdom (who, incidentally, seems to hold the record for the longest kingly reign of all time: ninety years, from 2275 to 2185 B.C.), extended Egyptian hegemony from Nubia, in southern or Upper Egypt, to Phoenicia, on the eastern shores of the Mediterranean. But the First Intermediate Period, which came next, is described variously as an "obscure period" or a "dark age." The Greek historian Manetho, of the third century B.C., may have been embellishing when he alluded to a procession of "seventy kings in seventy days" during this period, but there is little doubt that it was a chaotic time. The central government in Memphis lost control of Middle and Upper Egypt, and local princes or chieftains were left to shepherd their people through a severe famine that is mentioned repeatedly in hieroglyphic texts. One of those minor rulers, who was unfortunate enough to live in southernmost Upper Egypt, wrote, "All of Upper Egypt was dying of hunger, to such a degree that everyone had come to eating his children, but I managed that no one died of hunger in his home. . . . The entire country has become starved like a starved grasshopper, with people going to the north and to the south (in search of grain)."

Some historians and archeologists believe that the diaspora not only from Upper Egypt but also from the Levant and Mesopotamia eventually found its way to the Nile River Delta, and that the resulting social upheaval upended Egypt's imperial center just as a similar scenario played out in Akkad.

One should tread lightly in reading the Bible as history, of course, but the argument can be made that this was the time of the Hebrew patriarchs: Abraham, Isaac, and Jacob. (Joseph was born a few generations later, according to the Scriptures. William Quinn used historical sources to place him in 1700 B.C., after the drought.)

According to Genesis 12:10, Abraham had dwelt for only a few years by the Dead Sea in the land of Canaan when,

> . . . there was a famine in the land: and Abram went down into Egypt to sojourn there; for the famine was grievous in the land.

There is much talk of famine and the wrath of God in the Book of Genesis, and this story of migration, for reasons of famine, of a Semite from the east to the rich Nile Delta dovetails well with historical accounts of Asiatics flooding Egypt in its First Dark Age.

• • •

"THE DISAPPEARANCE OF KILIMANJARO'S ICE FIELDS . . . WILL BE UNPRECEDENTED FOR the Holocene," the Thompsons write. "This will be even more remarkable given that the [northern ice field] persisted through a severe [approximately] 300-year drought that so disrupted the course of human endeavors that it is detectable from the historical and archaeological records throughout many areas of the world." Thus, in disappearing, the snows of Kilimanjaro are recording a change that is unprecedented in the twelve-thousand-year history of human civilization.

In the book *Environmental Disaster and the Archeology of Human Response,* which has provided a few of the examples I have used here, Michael Moseley, perhaps the foremost archeologist of South America, encourages his colleagues to "embolden themselves and investigate natural disasters as significant subjects of contemporary relevance and not merely as quaint curios of antiquity." He and his colleagues have provided copious evidence that the most recent global environmental disaster, in early biblical times, caused quite a few civilizations to cross their particular "environmental thresholds" (a phrase coined by Moseley's student Alan Kolata). Simply put, their agricultural, economic, and social habits put them on a collision course with a climate change that was in their case unforeseen.

The IPCC report on the societal impacts of global warming, released on the day Lonnie informed the world that the snows of Kilimanjaro will soon disappear, seems to be an assessment of our own proximity to an environmental threshold. By 2050, the human population is expected to grow by a third—to nine or ten billion people. In spite of our technological gains, most of us still produce food not much differently from the way the Mesopotamians did four thousand years ago: by subsistence agriculture; and as the earth fills with people, migration to greener pastures becomes less and less of an option. We seem to be headed for a bottleneck. There is every chance that we will surpass the high-water mark of misery we have established through war in the century just past. Again, we will be bringing it upon ourselves, only this time by less direct means.

Jim McCarthy, the lead author of the impacts report, seemed to be unaware of the archeological work when I spoke to him (after all, he *had* been rather busy lately), but it struck me that his description of the current situation would probably cause the students of Akkad and Tiwanaku to nod in recognition.

"There's a problem with the convenient shorthand of 'global warming,' " he said. "If your local climate were to change by only a degree or two, you'd probably think, 'Who would notice?' or 'Why would anyone care?'

"The big story is the extremes. The big story is that we are increasingly seeing these extremes reach limits that are right up against the system boundaries that we've designed into our agriculture, our socioeconomic systems, and our infrastructure, to make them safe. They are no longer safe.

"If you had looked at the Chicago heat wave in 1995—the summer heat wave that killed something like seven hundred people—and you had done a careful analysis of where that could occur with even more devastating consequences, you might have thought of France; you might have thought of Paris. And then you might have thought, 'Well, what if it happens in June?' and you would have said, 'Yeah, that would be bad . . . but, *God, what if it happens in August?'*—because of the bi-generational holiday: parents and children go on holiday and leave their elders behind, and all the health care workers go with them. If you've ever been in Paris in August . . . I mean, it's an eerie place!"

Fifteen thousand people died in Paris in August of 2003. And although it did turn out to be the hottest summer on record in Europe—and, globally, 2003 came in as the third-hottest year on record (meaning that four of the hottest years on record have now occurred in the last six years)—it was not the average temperature that killed them. It was the collision of an innocent cultural tradition with a new weather extreme—one of many such collisions that will take place this century, I fear.

In describing what he sees in store, McCarthy employs a phrase similar to those used by archeologists to describe the four- and eight-thousand-year-old catastrophes they have uncovered: "People are beginning to use terms like 'climate disruption' or 'climate shock.' The consequences of these terrible hot periods for human health, for agriculture; this is what will ultimately take its toll."

AND THE SNOWS OF KILIMANJARO ARE NOT THE ONLY ONES THAT ARE DISAPPEARING. About a year after our deep drilling expedition, I got a call from my friend Alison Osius, the editor at the climbing magazine who had long ago given Máire the tip that had sent me off on the adventure that has become the writing of this book. At a recent editorial meeting, several of Alison's colleagues had related stories from their climbing friends, just back from their own adventures, of the snow and ice disappearing from dreamlike routes on some of the most beautiful and famous mountains in the world. They wanted someone to write a story about it.

Alison gave me a few names. They gave me others. After one climber told his story of the changes he'd seen in less than ten years on Everest's "little sister," Pumori, he said, "You ought to talk to so-and-so; he has an amazing story about Alaska." Before getting to Alaska, the next climber had to tell me about Huascarán . . . then Alaska, then the Mexican volcanoes, then Chimborazo, then the New Zealand Alps. He gave me the phone number of one of his climbing partners, who told a story about Patagonia and another about Chamonix and another about Colorado . . .

I had known Lonnie for almost four years and had joined him on three expeditions by that time, but I would say that it was in those conversations with fellow climbers—whose relationship with mountains is simple love—that the reality of global warming first really hit me. They expressed some hint of anger at the heedlessness of our political leaders, but little self-righteousness. Their primary emotion was sadness, and their most common lament was that they would not be able to share the rapture of the high white wilderness with their children. As the snows disappear, not only do the routes we climb become more difficult and dangerous, but the climbing itself becomes less aesthetic. And, to most of us, the mountain world loses some of its luster. A peak looks better with a clean crest of snow on top.

The most shocking stories came from the greatest range of them all. I learned that Ama Dablam, a sublime white spire on the approach to Mount Everest that is often described as the most beautiful peak in the world, had been climbed without crampons recently. A glacier that was very close to the first camp used by Sir Edmund Hillary and Tenzing Norgay when they made the first ascent of Everest in 1953 has retreated about three miles since their success, and changes to the snow and ice higher up have greatly altered the scenes of some of the most legendary climbing sagas. Even the snow on the mountain's very summit has thinned enough to make it about four feet shorter than it was in Hillary and Norgay's day.

The Worldwatch Institute reports that the eastern Himalaya have lost one-fifth of their glaciers—two thousand in all—over the last century, and that the rate is accelerating. According to the Swiss-based International Commission on Snow and Ice, "Glaciers in the Himalayas are receding faster than in any other part of the world and, if the present rate continues, the likelihood of them disappearing by the year 2035 is very high."

Wise old Shi Yafeng, who reminded Lonnie of Yoda from *Star Wars* even thirty years ago when he first met him, is the lead author of a 2,700-page *Chinese Glacier Inventory,* published in the fall of 2004. Shi and his colleagues (who include Yao Tandong and Lonnie) report that the total area covered by glaciers in China has dropped by more than a thousand square miles since 1960 and, again, that the rate is accelerating. Yao, who led the scientific team, predicts that 64 percent of his country's glaciers will disappear by 2060 and that they *all* stand to disappear by the end of this century.

As usual, the main concern is water, and, in this case, the effects are painfully ironic. Yao estimates that the melting of mountain glaciers in China adds about the equivalent of the annual flow of the Yellow River to the many rivers that originate in China's mountain watersheds, including the Yellow. Although it has led to an increase in catastrophic flooding, the water this adds in the short term has also encouraged development and increased agriculture and population growth, so that almost a quarter of China's billion-plus population now lives in its western areas, below the mountains. Meanwhile, the wasting of the glaciers portends a swift change to drought and desertification not far down the road. The phrase "environmental threshold" springs to mind.

Himalayan glaciers feed three of the earth's greatest and holiest rivers: the Indus, the Brahmaputra, and, holiest of all, the Ganges. Five hundred million people, one-tenth of the earth's population, live on the Gangetic Plain. The river supplies most of their water, and as in the Andes, it is especially critical in the dry season, when the monsoon is gone. The population has swelled in this time of plenty, but the odds are that the Ganges will soon become seasonal.

In Sanskrit, *Himalaya* means "abode of snow," but as crops and people die from lack of water while watching the highest mountains on earth turn from white to black, that name may soon seem grotesquely inappropriate.

And the name of one of Lonnie's stomping grounds in Peru, the Cordillera Blanca, would mean "white mountains," but my climbing friends tell me that they have watched that great range gradually turn black, too, over the past two decades, as the snowline has risen an average of thirty feet per year. Before-and-after photographs of formerly snow-clad peaks abound.

Lonnie's favorite, of course, is the Qori Kalis outlet tongue on Quelccaya, which he has been tracking for about three decades now. He's been going down to check his old friend every summer, lately; and in fact, he, Keith, Vladimir, and a few others retrieved a second set of deep cores at the summit of Quelccaya in the summer of 2003. They always camp in the moraine by the large boulder that Qori Kalis was pushing downhill when Lonnie first saw it with John Mercer in 1974. An eighteen-acre lake now lies between the boulder and the receding glacial margin, a lake that did not exist as recently as 1987.

In 1992, just after he showed photos of Qori Kalis's retreat to Al Gore and his fellow senators in Washington, Lonnie and Henry Brecher took another set of images that demonstrated that the tongue had retreated three times faster over the previous eight years than it had in the twenty years before that. Volume loss, which takes thinning into account, had grown by a factor of seven. More images taken in 1998 showed that the retreat had increased by another factor of three in the intervening five years, to more than five hundred feet per year!

Lonnie estimates that the entire Quelccaya ice cap, which thirty years ago covered twenty-seven square miles and was five hundred feet deep at its eighteen-thousand-foot summit, will disappear completely in his lifetime.

"I expect I'll come back in twenty years and see the marks where our drills punched through to bedrock in '83 and again in 2003," he says. "The world's ice is providing the most visible evidence there is for global warming, and the changes are incredible. To me it's so straightforward."

EVER SINCE 1984, WHEN HE NOTICED THAT QORI KALIS HAD SWITCHED FROM SLOW ADvance to swift retreat in less than one year, he has realized that glaciers work on a threshold principle: when they decide to die, they can die very quickly. He believes that the retreating mountain glaciers he and his colleagues (and my climbing friends) have found all over the world have already passed their thresholds, and he has developed a new interest in the effect this will have on something that is probably more fundamental than the beauty of the mountain environment or tourism around Kilimanjaro: sea level.

The understanding of the relationship between rising temperatures and rising seas has developed rapidly in the four years since the most recent IPCC assessment. While mountain glaciers had been discounted as insignificant, it has now been shown that the pervasive retreat of the glaciers in Alaska and Patagonia alone is probably the greatest single contributor to rising sea levels today.

Lonnie tells me that the U.S. Geological Survey is tracking two thousand Alaskan glaciers right now, and all but thirteen are retreating. He adds that if all the earth's mountain glaciers were to melt, sea level would rise about a foot and a half—and "that's a lot. If you live in Bangladesh, that's a whole lot. The impact of rising sea levels on humans can be tremendous, and it can happen a lot faster than the models predict."

This brings him to what is probably the single most important insight of the past few years: that the vast ice sheets in Greenland and Antarctica seem to work on the same threshold principle as mountain glaciers, and that this looms as the most significant near-term threat from global warming. (Consider the prescience of John Mercer, who first sounded this warning almost fifty years ago.)

None other than Jim Hansen, whose string of accurate predictions stretches back twenty-five years, has been putting his mind to this problem of late. Hansen estimates that even if we were to stop emitting greenhouse gases today, the warming of about one degree that is "in the pipeline" anyway, due to the heat that the ocean's have already stored, will soon bring us not only to the highest temperatures of the Holocene but very close to the high point of the Eemian, the interglacial just before the Wisconsinan, 125,000 years ago, when sea level was between sixteen and twenty feet higher than it is today.

Hansen points to evidence that the polar ice sheets have disintegrated rapidly in the past: various records show that about fourteen thousand years ago, as the Wisconsinan was coming to an end, a pulse of meltwater from the disintegrating polar ice caps sent sea levels up at the rate of three feet every twenty years for five hundred years. Estimating that the earth's average temperature need only rise about two more degrees Fahrenheit—less than the lowest IPCC projection for this century—for us to reach the threshold of ice sheet collapse, Hansen has named the gradual upward drift in our planet's temperature "the global warming time bomb."

BY THE TIME WE HAD LIVED IN KILIMANJARO'S CRATER FOR TWENTY-TWO DAYS, THE moon was full, the air very still, and we had developed a feeling that the northern ice field was almost alive, like some great aquamarine dragon sprawling wounded beside us on the sand. Under the sun's heat it withered and died. At night it was revived painfully by the cold.

By day, water streamed off the top of the glacier and down its steep sidewalls, and pools formed on the sand at the base of the wall by our tents. We used the pools for washing. Sometimes we broke off the long icicles that grew with the dripping and melted them to drink, or left pans or bottles out to catch the thin strings of water that streamed from their tips. In some places, the rivulets wandering along the surface had converged to form streams, which catapulted in waterfalls off the top. There was a constant sound of dripping and gushing water.

In late afternoon, the sun dropped quickly behind the crater wall, and the sounds of flowing and dripping slowed and stopped. Around dusk there was a period of silence, and as the dark and cold deepened, the glacier began to shake off its lethargy and speak. In the middle of the night I was sometimes awakened by a hollow boom or a plastic crack, just in time to hear it echo off Uhuru Peak, a mile away. In the intense clarity of high-altitude sleeplessness I would lie on my back and listen—for hours, it seemed. The sounds were not constant as in daytime. They emerged randomly from random directions, framed by silence: a pop . . . hollow bells . . . tinkling . . . a feathery rumble . . . knuckles tapping on sheet metal . . . a sigh . . . a fragile creak . . . a car hood slamming . . . plastic crinkling . . . stage thunder . . . *whump* . . . a tray of champagne glasses spilling down some stairs . . . *thump* . . . a tap in an iron chamber . . .

Once or twice a night I'd leave my tent to watch it, glowing silver in the moonlight, and it did look alive. The Milky Way swooped overhead; the Big Dipper wheeled above the glacier around the hidden North Star; the Southern Cross stood firm, high over Uhuru.

On our last full day, I walked up to our former drilling site with Vladimir to watch Doug and Mathias put the finishing touches on their weather station. Vladimir soon found something scientific to do, so I wandered off east, alone. We had definitely drilled in the best spot. Walking down a gentle grade, I passed many small lakes, and the surface grew rougher and slushier.

At the glacier's edge, about a mile away, the sun gnawed like a large rodent at the east-facing wall. It had already nibbled out large bays, leaving pools of water on the sand within them. Tortured arms of ice, dwindling to knifelike fins, stretched out east between the bays, their dead-vertical sidewalls facing north and south. Farther away the fins had melted into the sand, leaving smaller fins and erratic needles, some as high as thirty feet, stretched in a line behind. Cut off from the main body of the glacier, they sank slowly into pools of water on the sand.

This is the way a glacier dies.

NOTES

PART I: THE SAJAMA EXPEDITION

9 *like a kid* This image was borrowed from climatologist Richard Alley (2000). In the meantime, there has been some progress: the myriad oscillations may be dwindling down to a fundamental few (Kerr 2004a).

12 *temperature has risen and fallen in concert with atmospheric carbon dioxide* Jouzel et al. 1987; Barnola et al. 1987; Genthon et al. 1987; Petit et al. 1999.

13n *received wisdom* Seager et al. 2002; Kerr 2002b.

16n *Hans Ertl* Yossi Brain provided me with this quote one afternoon in his flat in La Paz, as he showed slides of Sajama and the other mountains of Bolivia.

PART II: EARLY DAYS

32 *popular books* Nash 2002; Keys 2000; Fagan 1999.

33 *competing with the Yukon for the title of the fastest warming place* The glaciologist in question is Dr. Eric Steig, quoted by journalist Larry Rohter (2005). Vaughan et al. (2001) supplied the numbers in the rest of this paragraph, while Petit (2000) tells roughly the same story with less technical detail.

34 *Alaska has warmed by about five degrees* Whitfield 2003.
obvious changes Kerr 2002a; Goldman 2002.
Alaska's white spruce have died Egan 2002; Stevens 1998a; Barber, Juday, and Finney 2000.

35 *EPICA* Augustin et al. 2004; McManus 2004; Walker 2004.

40 *first project* Thompson, Hamilton, and Bull 1973.

41 *second study* Thompson 1974.
coldest of Cold War science Weart 2003b, under "Money for Keeling: Monitoring CO_2 Levels."
estimate the temperature of the overlying atmosphere Dansgaard 1964.

43 *parasol effect* Crutzen et al. 2003.

44 *S. Ichtiaque Rasool* Rasool and Schneider 1971.

44 *ice age might "start next summer"* Stevens 1999, 145.
45 *fanaticism and nationalism* Steele 1998, 196.
46 *'Inesplorado'* Ibid., 200.
neither understood science nor trusted scientists Ibid., 173.
47 *Jay Zwally* Zwally, a physicist by training, worked at NSF for only two years before returning to full-time research at NASA. Research is his true calling, but he remembers his stint at NSF as an exciting time, and he is proud of the role he played in the birth of tropical ice core research.

"It's strange that [the first Quelccaya] study was funded by an Antarctic program," he says, "and it was a bit of a stretch trying to fund a field trip to tropical South America with money earmarked for the polar regions, but Joseph Fletcher, the head of Polar Programs, had an interest in climate. I sold it to him by arguing that it would help us interpret the Antarctic cores."

Zwally now works on a satellite-based laser-altimetry project, monitoring changes in the shape and volume of the Greenland and Antarctic ice sheets, and their effect upon sea level (Leary 2002; Zwally et al. 2002).

48n *Cedo Maranguníc* Maranguníc and Shipton made at least three significant first ascents and remote traverses in the far south: over the austral summer of 1960 and 1961 they accomplished the first north–south traverse of the south Patagonian ice cap; the following year, they were first to reach the summit of Mount Darwin, the highest point on Tierra del Fuego; and later they made the first traverse of the north Patagonian ice cap—quite a set of records (Steele 1998).

50n *Ricker published what is still the definitive reference* Ricker 1977.

52 *agriculture and animal husbandry for the very first time* Bar-Yosef 2000, 28; Heun et al. 1997.
certain members of what I will call the North Atlantic School Clapperton 1993, 454–57.

53 Antarctic Journal of the United States Mercer et al. 1975; Thompson and Dansgaard 1975.

55 *his first paper in* Science Thompson, Hastenrath, and Morales Arnao 1979.

56 *AMANDA* Neutrinos are notoriously difficult to detect, because they rarely interact with other forms of matter. (Wolfgang Pauli was almost right when he wrote, upon dreaming the particle up in 1930, "I have done a terrible thing, I have postulated a particle that cannot be detected.") AMANDA, on the South Pole, takes advantage of this fact by using the Earth as a filter against the particles it doesn't want to see: it looks "down"—that is, toward the center of the Earth and, therefore, at the northern sky. AMANDA is designed to detect the eerie blue glow given off by muons born in rare collisions between the neutrinos that streak upward through the diamond-clear Antarctic ice and stationary neutrons or protons in the ice itself. (In other words, the ice is the medium of detection.) Of all the creatures in the wide zoology of particle physics, only the neutrino has any chance of passing all the way through the earth to engage in such collisions. The AMANDA collaboration hopes to use this odd particle to observe the inner workings of cataclysmic events in deep space, such as super novae, gamma ray bursts, and active galactic nuclei. Eventually their instrument—or, actually, its successor, IceCube—will comprise five to eight thousand light detectors, monitoring more than a cubic kilometer of deep Antarctic ice. The best place to learn more is the IceCube Web site: http://icecube.wisc.edu/.

56 *weirdest of the Seven Wonders* Musser 1999.
63 *Lonnie and Ellen were the sole authors* Thompson and Mosley-Thompson 1981.
69 *Huaynaputina* Thompson et al. 1986.
70 *Dansgaard appears as an author* Ibid.
elongated air bubbles They can tell a lot about what has been happening at a particular depth in a glacier by the shape of the air bubbles. If the ice has melted and refrozen they will be stretched vertically, for refreezing generally progresses from top to bottom. You may have noticed bubbles of this shape on lake ice in the wintertime. It's a bad sign in an ice core, because it means that all the layers involved in the melting have been homogenized.

On a healthy polar glacier, which by definition has remained below freezing most if not all the way to its base, the shapes of the bubbles indicate whether or not the layers have been flowing, and in which direction. If the glacier is frozen to its bed, the bottom layer will remain essentially fixed; the upper layers will flow at different rates depending on their distance from the bottom, and in different directions depending on the configuration of the underlying terrain. (The temperature of the ice tends to rise as you go deeper, owing to the pressure of the overburden. If it has reached the melting point at its base, the whole glacier will slide as one piece on a relatively frictionless layer of water, like an ice cube on a slanting tabletop, and it won't deform; the bubbles will remain spherical.) At the top of an ideal dome, the flow will be radially symmetric, and the bubbles will flatten out into disks. At the balance point of a saddle, they'll be shaped like hot dogs or tubes.

PART III: THE WARMING SETS IN

75 *happy hunting ground* Callendar 1961, 2.
77 *published papers* Thompson, Mosley-Thompson, and Morales Arnao 1984; Thompson et al. 1985; Thompson et al. 1986.
79 *Dyurgerov* Dyurgerov and Meier 2000.
would place the transition, similarly, in the mid-seventies IPCC 2001b, 129.
Anthropocene This term seems to have been coined by Paul Crutzen and Eugene Stoermer (2000). The former received the Nobel Prize in chemistry in 1995 for his contributions to the understanding of atmospheric ozone chemistry. Consider also the disturbing conjecture by André Berger and Marie France Loutre (2002) that fossil fuel burning may very well lead to an "irreversible greenhouse effect," which may cause the Anthropocene warm epoch to last as many as fifty thousand years.
greenhouse effect I took portions of this argument, the numbers in particular, from a paper by Ralph Cicerone (2000), who has recently been elected president of the U.S. National Academy of Sciences.
80 *As early as 1824, he suggested* Fourier 1824.
81 *Edme Mariotte* Fleming 1998, 64.
wondrous factory Ibid., 67.
trusting to my legs Eve and Creasey 1945, 341.
82 *not one of the twenty survived to tell its tale* Ibid., 308.
alone amid these scenes of majesty and desolation Eve and Creasey 1945.
84 *I was accosted by a guide* Ibid., 113.
85 *with the Prince Consort in the chair* Ibid., 80.
solar heat possesses Fleming 1998, 66.

85 *held fast in the grip of frost* Ibid., 71.

86 *married at the age of fifty-five* Eve and Creasey 1945, 202.

I measured a teaspoonful of magnesia Ibid., 279–80.

87 *dense twenty-nine-page treatise* Arrhenius 1996. Uppenbrink (1996) estimates how many calculations he made.

took his cues from Tyndall Fleming 1998, 90.

88 *Langley* Langley was an accomplished, self-taught scientist, born in Roxbury, Massachusetts, whose formal education ended in high school. He was appointed secretary of the Smithsonian Institution in 1887 and three years later founded its Astrophysical Observatory, whose first structure housed one of his bolometers. One of his first ideas was to study seasonal changes in the atmosphere's infrared absorption—an effect he had previously observed elsewhere. In his request to Congress for funding, he suggested that these studies, might "enable us to predict the years of good and bad harvests, so far as these depend on natural causes independent of man." Thus, the founding of the Smithsonian Astrophysical Observatory might be seen as the first time the U.S. government invested in greenhouse research.

In the 1890s, Langley was the only reputable scientist to become obsessed with the idea of human flight. He designed two steam-powered "flying-machines," which accomplished the very first unmanned flights by crafts heavier than air. Unfortunately, however, these positive achievements were overshadowed by the spectacular failure of his manned aircraft. Nine days before the Wright Brothers' successful flight at Kitty Hawk, Langley's "Great Aerodrome," lost its rear wings the moment it left its launching catapult on top of a houseboat in the middle of the Potomac, performed a sort of a half gainer, and plunged into the water, nearly drowning its pilot, before a full complement of dignitaries and representatives of the press. And the sad thing is that the contraption probably would have flown: it did when it was reconstructed many years later. It seems that the catapult stripped off its wings as it was taking off. Three years after this much-ridiculed failure, Langley died a broken man. Langley Air Force Base, the nation's oldest, and NASA's Langley Research Center are named after him (Park 1997; Langley 1900).

91 *marks of civilization* Fleming 1998, 16–18.

Thomas Jefferson Jefferson was close to correct. A recent study (Kalnay and Cai 2003; Ball 2003) shows that the replacement of forest by fields does tend to warm the surface locally.

1905 letter Fleming 1998, 89–90.

92 *Snowball Earth* The great Russian climatologist Mikhail Budyko planted the seed of the Snowball Earth idea with his studies of nuclear winter in the early sixties. The phrase was coined in 1992 by Joseph Kirschvink, a geobiologist at the California Institute of Technology. Primary sources: Hoffman et al. 1998; Hoffman and Schrag 2000. An excellent, Web-enhanced perspective: Archer 2003; a review article: Kerr 2000a; a popular book: Walker 2003.

93 *wiped out the dinosaurs about fifty million years ago* For an excellent popular account of this discovery see Alvarez (1998).

94 *temperature of the oceans is now rising* Levitus et al. 2000; Stevens 2000.

95 *paper at a London dinner meeting* Callendar 1938.

regulator of huge capacity Arrhenius 1908, 54.

96 *1939 essay* Callendar 1939.

96 *contentious debate* Regalado 2003b; Revkin 2003.
Hockey Stick Mann, Bradley, and Hughes 1999; IPCC 2001b.
confirmed by sophisticated computer models Crowley 2000; Mann 2000; Mann et al. 2003.
"Can Carbon Dioxide Influence Climate?" Callendar 1949.
A computer simulation would support this prediction Wang et al. 1991.

97 *studies by Ellen Thompson* Mosley-Thompson et al. 1999.
contemporary textbook Blair 1942, 101.
birds had come north Lopez 1986, 160.
The Sea Around Us Carson 1951.

98 *Kipling* The witty meteorologist was one Charles Brooks (Fleming 1998, 108). He was borrowing from Kipling's "In the Neolithic Age":
There are nine and sixty ways of constructing tribal lays,
And every single one of them is right!

99 *final exam* Lewis Lapham quoted Fuller in an essay in *Harper's* (2002).
Gilbert Plass Biographical sketch of Plass: Fleming 1998, 121–22.

100 *90 percent* Weart 1997a, 321.
sponsored by the Office of Naval Research Ibid., 332.
error was discovered in the spectroscopy studies Ibid., 333–34.
water vapor in the atmosphere falls off very rapidly Weart 1997b, 36.

101 *ENIAC* Weik 1961.
in on the development of the ENIAC from the start Ibid.
implosion lenses Rhodes 1986, 575.
climatological warfare Weart 1997a, 335.
hat trick Weart 2003a, 57; Weart 2003b, under "General Circulation Models of the Atmosphere."
Edward Lorenz of MIT Gleick 1987; Rind 1999; Weart 2003a, 2003b.

102 *During a sabbatical* Fleming 1998, 121.
wrote up his ideas Plass 1956a, 1956b, 1959.
several orders of magnitude larger Both quotes in this paragraph come from Plass 1956b, 149.

103 *seventeen times its preindustrial level* Ibid.
seen to be a very serious problem Plass 1956a, 385.

104 *Hans Suess* Biographical details: Suess papers.

105 *Werner Heisenberg* In his 1993 book, Thomas Powers argued that Heisenberg purposely sabotaged the Nazi nuclear program. The book formed the basis for the recent Broadway hit *Copenhagen,* which centered upon a controversial meeting between Heisenberg and Niels Bohr, at which, according to Powers, Heisenberg expressed moral qualms and consciously aided the enemy by subtly divulging the existence of his program to Bohr, who then passed the knowledge on to the Allies, which prompted them to develop the bomb themselves. Richard Rhodes, who is perhaps the foremost historian of the bomb, had previously (1986) offered the obvious interpretation that Heisenberg was simply trying to recruit Bohr, which was indeed the ostensible purpose of the meeting. Rhodes's interpretation was confirmed more or less definitively in 2002, when the Bohr family released a letter that Bohr had drafted but never sent to Heisenberg, demonstrating that Bohr himself believed Heisenberg to be enthusiastic about his bomb work (Glanz 2002).
Suess is the only one to have confessed Powers 1993, 158.

105 *invited Suess to join* Suess papers.

106 *Suess Effect* Suess 1953.

rough estimate of the characteristic time The estimate and the subsequent quotes come from Suess (1953).

second, more careful, radiocarbon study Houtermans, Suess, and Munk 1967.

Roger Revelle Biographical details: Revelle biography.

whatever anyone at Scripps does Malone, Goldberg, and Munk 1998, 302.

107 *private papers began to include calculations* Weart 1997a, 342.

wastes introduced into the upper layer Revelle 1988, quoted by Weart (2003b, under "Roger Revelle's Discovery").

108 *Craig once described* This and the quote in the following paragraph come from Lawren 1989.

109 *three back-to-back papers* Arnold and Anderson 1957; Craig 1957; Revelle and Suess 1957.

Environmental Defense Environmental Defense (2003). This fine organization has innocently bought into the erroneous myth that is found everywhere in the literature: Gale Christianson (1999, 155) writes, "Thus far [Revelle and Suess] were siding with Callendar." William K. Stevens (1999, 138) concurs: "This finding [that the oceans were not taking up as much carbon dioxide as had previously been thought] was reported by two highly respected scientists, Roger Revelle and Hans E. Suess." And according to Michael Oppenheimer and Robert Boyle (1990, 36), "[T]hey reported that the ocean had *not* absorbed as much carbon dioxide as previously assumed."

Revelle said later that he had not been particularly concerned about global warming Weiner 1990, 58; Weart 1997a, 347. Even Revelle's imagery was old: Plass had spoken of a global experiment years earlier. However, none of this prevented Revelle, later in life, from promoting the revisionist view of his early contributions. He even once claimed to be the "Granddaddy of Global Warming" (Fleming 1998, 122).

110 *not too happy about the whole thing* Weart 1997a, 345.

Revelle, evidently, was telling them all Weart 1997b, 38–39.

finding a late manuscript Weart 1997a, 347 (n82).

111 *observe geophysical phenomena* National Academy 2004.

earliest research programs The specific purpose of Brown's Caltech laboratory was to develop radioactive methods similar to carbon dating, based on other, longer-lived isotopes, for determining the ages of rocks. In the mid-fifties, his protégée Claire Patterson used these techniques to make the first modern estimate of the age of the Earth, 4.5 billion years—still reckoned to be true.

informal discussion with some students and Keeling's literature search: Weiner 1990, 16.

112 *ten times more sensitive than it needed to be* Weiner 1990, 19.

behind the eight ball The main sources for the details of Keeling's personal and scientific life at this time are Weiner's book (1990, 19–25), Keeling (1978), a telephone interview I conducted with Dr. Keeling on October 2, 2001, and two of his papers (1958, 1961) on the measurements he took by hand in 1955 and 1956 in what he later called "rural and marine" locations all over the country. The tables in those papers indicate exactly where he was at specific times on specific dates, that he stayed up all night a number of times, and that he was quite peripatetic. He took his glass flasks everywhere from Washington State to

Big Sur to Organ Pipe Cactus National Monument, Arizona, to Otter Creek Township, Indiana, to Assateague Island off the coast of Virginia—the last around the day of his fateful meeting with Harry Wexler.

112 *His outstanding characteristic* Weiner 1990, 31–32.

113 *Rossby suggested that his institute prepare the ground* Keeling 1978.

114 *Keeling packed thirty flasks into his car* Weiner 1990, 24–25.

115 *Wexler asked a number of questions in rapid-fire* Keeling 1978.
a few ten-thousand-dollar infrared analyzers Keeling 1978; Weiner 1990, 32.

116 *table Keeling published later* Keeling 1961.
once Revelle had Keeling under his thumb The story of Revelle's perhaps overbearing behavior and the setting-up of the station at Mauna Loa is told by both Keeling (1978) and Weiner (1990, 32–33).

117 *hopelessly erratic* Keeling 1978.

118 *witnessing for the first time nature's borrowing* Keeling 1978.
as soon as we had two Marches Stevens 1999, 141.
announced in a peer-reviewed journal Keeling 1960.

119 *air at the South Pole* Imagine walking out into the perpetually dark, six-month polar winter in temperatures that often dropped below −100°F, to fill one of those flasks. You would have worn a full suit of navy-issue ECW (Extreme Cold Weather) gear, and at temperatures that low, you would have made absolutely sure that you had covered every square millimeter of your skin, including your face. You would probably have rigged up some sort of tube or modified snorkel in order to breathe through your face mask without fogging your ski goggles. The scarlet and blue Aurora Australis would frequently have been shimmering overhead.

For consistency, only one member of the South Pole field team was authorized to take air samples, and he or she received two days of training at Scripps before heading south. As Keeling and subsequent collaborators later wrote (Keeling, Adams et al. 1976, 553), "Although this procedure is simple to execute, special precautions must be consistently observed to avoid contaminating a high proportion of the samples. The sample taker, to minimize contamination from his own breath, was instructed to sample only when the wind was at least 5 knots. After first breathing normally near the site for some moments, he exhales, then inhales slightly, and finally without exhaling again, walks 10 steps into the wind, where he takes the sample [by opening the stopcock of the flask]. He should have a clear idea of the wind direction and be certain that no local source of CO_2, even another human being, is upwind."

120 *more than a decade's worth of data from the South Pole and Mauna Loa* Back-to-back papers in *Tellus*: Keeling, Bacastow et al. 1976; Keeling, Adams et al. 1976.
Fossil Fuel Age Harrison Brown, Keeling's mentor at Caltech, seems to have coined this phrase about fifty years ago, in the book he was writing in Jamaica as Keeling was struggling alone, back home, to develop his carbon dioxide method (Brown 1954, 168).

121 Graph of carbon dioxide emissions The numbers behind the graph and in the text come from Marland et al. 2003.

122 *we stand to double the preindustrial level of carbon dioxide sometime in the second half of the present century* IPCC 2001b, 527; Hansen 2003.
A gallon of gasoline weighs about seven pounds Six of the seven pounds in a

gallon of gasoline is carbon. Upon burning, each molecule of that carbon combines with two heavier oxygen molecules from the local air. The molecular weight of carbon is about twelve; oxygen's about sixteen; carbon dioxide's forty-four. Thus, six pounds of carbon produce twenty-two pounds of carbon dioxide.

This also explains why emissions estimates by different organizations sometimes differ by a factor of 3.7 (22 ÷ 6): some report carbon dioxide, as I do; some report carbon alone, which obviously weighs less.

122 *one hundred tons of ancient plant matter* *Science* 2003b.

Industrial Eruption Weiner 1990, 41.

123 *Keeling Curve* Keeling and Whorf 2004.

126 *as Roger Revelle would put it* The quote comes from Oppenheimer (1998).

Nature *in 1978* Mercer 1978. Mercer had actually raised the specter of Antarctic ice sheet disintegration to the public ten years earlier (Mercer 1968), but it was not until his report in *Nature* that his warning was widely heard.

Scientific American Schneider 1998; Hall 1998.

interest again in 2002 Kaiser 2002; Rignot and Thomas 2002.

seen as optimistic by many of today's experts Hansen 2003, 2004; O'Neill and Oppenheimer 2002; Oppenheimer 1998; Zwally et al. 2002; De Angelis and Skvarca 2003; Shepherd et al. 2003; Kaiser 2003; Schiermeier 2004.

new science advisor, Roger Revelle In a biographical remembrance produced by the National Academy of Sciences (Malone, Goldberg, and Munk 1998), Revelle is described as "Homeric." His fatal flaw seems to have been a tendency toward high-handedness and autocracy, which caused such resentment among his colleagues during his early years at Scripps that senior staff members stepped in to prevent him from assuming the directorship the first time he was nominated. During the IGY, while he was keeping Dave Keeling on rather a short leash, Revelle was exhibiting similar aggressiveness in the pursuit of what he later called the greatest accomplishment of his life: the founding of the University of California at San Diego. He succeeded in founding it in 1960, but he created such bad blood with statewide board of regents in the process that they chose someone else to be the university's first chancellor, a position everyone assumed would go to Revelle. He left California the following year to become Udall's science advisor—the first such advisor and at a critical time. He did much good work over the next few decades, mostly in policy, but the regents' decision was a terrible blow, forcing him to leave his home. He became a sort of wandering scientific statesman after that.

you ought to back her Revelle biography.

127 *Johnson's Science Advisory Panel* Revelle biography.

Joint Numerical Weather Prediction Unit Weart 2003b, under "General Circulation Models." JNWP has gone through two name changes over the years. It is now known as the Geophysical Fluid Dynamics Laboratory.

128 *inability of "simple" radiation models* Möller 1963.

first global warming study Manabe and Wetherald 1967.

incorporated water vapor feedback into a more realistic, three-dimensional GCM Manabe and Wetherald 1975.

129 *Reid Bryson* Weart 2003a, 90–91.

129n *warming of the Indian Ocean* Giannini, Saravanan, and Chang 2003; Kerr 2003b.

129 *"Darkness"*

I had a dream, which was not all a dream.
The bright sun was extinguish'd, and the stars
Did wander darkling in the eternal space,
Rayless, and pathless, and the icy earth
Swung blind and blackening in the moonless air . . .

130 *soon after 1980* Schneider and Mass 1975.

first strange indications of flickering Dansgaard et al. 1971; Weart 2003b, under "Rapid Climate Change"; Weart 2003a, 78.

Wallace Broecker, from Columbia University's Lamont-Doherty Earth Observatory, also predicted a warming Broecker 1975.

Broecker based his thinking not on computers (which he distrusts to this day) Stevens 1998b.

131 *we would soon "experience a substantial warming."* Broecker 1999a.

James Hansen entered the picture Hansen's biographical details: H. W. Wilson 1996.

Mars Cicerone 2000.

dry ice Tillman.

cautionary aspect to this tale Information about the early photographs from the Viking space probe and about the evolution of climates on Venus and Mars comes from Wang et al. (1976, 689). Moore (2004), Kerr (2004b), and Leary (2004) tell the story of the definitive proof, produced in early 2004 by the two NASA rovers and the European Space Agency's Mars Express orbiter, that large bodies of liquid water once existed on the surface of Mars.

132 *greenhouse potency of most freons* Blasing and Jones 2003.

By 1976, he and Yung Wang et al. 1976.

results from their first true GCM Hansen 1996.

announcement in the midst of the hottest July Weart 2003a, 115.

133 *that range has held up amazingly well* Kerr 2004c.

Only in 2005 did a serious study call it into question Stainforth et al. 2005; Kerr 2005.

change the very face of the planet This particular characterization of the Charney panel's conclusions was offered by Hansen (1996).

133n *Manabe's estimate of two degrees Celsius as the low bound and Hansen's of four degrees Celsius at the top bound* Hansen 1996; Kerr 2004c.

133 *radar screen* Oppenheimer and Boyle 1990, 38–39.

134 *Hansen's group revealed these results in* Science Hansen et al. 1981.

135 *approaching the warmth of the Mesozoic* Hansen 1996.

his funding would be terminated Ibid.

152 *paper they published in* Science Thompson, Mosley-Thompson et al. 1989.

154 *reason for concern, but not panic* Weart 2003b, under "Government: The View from Washington."

substantial increases in global warming Ibid.

155 *a rise of global mean temperature . . . greater than any in man's history* Agrawala 1997, 2.

born in politics Ibid., 8.

156 *case for limiting greenhouse emissions was at least as strong* Ibid., 6.

Shardul Agrawala Dr. Agrawala's interest in the policy aspects of climate change has led him on quite an odyssey over the past decade. He earned his doctorate at Princeton's Woodrow Wilson School of Public and International Affairs in

1999, then put in a stint at Columbia's International Research Institute for Climate Prediction. He now works at the Environment Directorate of the Organization for Economic Cooperation and Development in Paris. This quote comes from page 7 of a discussion paper he prepared in 1997, while taking a leave from Princeton at Harvard's Kennedy School of Government.

156 *peer-review process* "For example," writes Agrawala (1997, 11), "draft chapters of the 1995 Working Group II Second Assessment report went through two full scale reviews: the first involving anywhere from twenty to sixty expert reviewers per chapter (a total of 700 experts from 58 countries were involved), and the second involving all IPCC member governments and the experts who had sent their reviews in the first round."

line-by-line approval Agrawala 1997, 12.

157 *worst growing season since the Dust Bowl* Weiner 1990, 87–90.

accepted by the Journal of Geophysical Research Hansen et al. 1988.

Jeopardy Hansen 1996, 181.

158 *It's time to stop waffling* Ibid., 181.

159 *"Hansen vs. the World"* Kerr 1989; Broecker et al. 1989.

160 *three exquisite back-to-back papers* Jouzel et al. 1987; Barnola et al. 1987; Genthon et al. 1987.

a deeper look at the physics Lorius et al. 1990.

PART IV: THE ESSENCE OF LIFE

165 *Technology is a blessing* Hardin 1993, 101.

1988 report in Nature Thompson et al. 1988.

167 *entirely man-made* Kolata 1993, 104.

168 *openings in the body of Pacha Mama* Aveni 2000, 66–67.

169 *unfit for agriculture* Kolata 1996, 265.

170 *landmark paper* Denevan 1970.

economic well-being was synonymous with agriculture Ortloff and Kolata 1993.

surrounded by raised fields and linked by roads and elevated causeways Binford et al. 1997, 243.

171 *studied virtually every aspect of Tiwanaku civilization in mind-boggling detail* His study of the raised bed is described in Kolata (1993, 183–205).

172 *try it themselves* Kolata 1996, chap. 6.

173 *(Malthusian) response to abundant surplus* Kolata 2001, 176.

174 *statistical study* Ortloff and Kolata 1993.

175 *Sometime around A.D. 1150* Kolata 1996, 179ff.

177 *Kolata and Ortloff later showed* Ortloff and Kolata 1993.

concurrent warming in his ice cores from Huascarán Thompson 1995.

Medieval Warm Period The notion of a possible Medieval Warm Period was first proposed by the British climatologist Hubert Lamb (1965). Bradley, Hughes, and Diaz (2003) weigh the evidence for and against.

Popular books For instance, Stevens 1999, 44.

newspaper articles For instance, Brooke 2001.

scientists themselves For instance, Broecker 2001. (Bradley et al. [2001] rebut Broecker's argument.)

anecdotes are almost all false In a sidebar in a story on climate change that appeared on the cover of *Chemical & Engineering News,* Bette Hileman (2003) debunks the myth of the Medieval Warm Period. I am indebted to Michael Mann, who is quoted in the sidebar, for communicating further details via e-mail, particularly those concerning wine production in medieval England.

178 *up very slightly from around A.D. 800 to 1400* Mann and Jones 2003.

paper in the journal Climate Research Soon and Baliunas 2003. These two published another, essentially identical paper in collaboration with three somewhat less academically inclined global warming contrarians in the obscure and clearly partisan journal *Energy and Environment* at about the same time (Soon et al. 2003).

Regarding the politics: Regalado 2003b; Revkin 2003; AGU 2003; *Science* 2003a.

methods employed in the paper were shown to be faulty Mann et al. 2003.

179 *Some died in A.D. 1100* Stine 2001.

bristlecone pines Hughes and Funkhouser 1998; Hughes, Funkhouser, and Fenbiao 2002, 17.

Patagonia Stine 1994.

180 *impressive study of tree-ring records* This story of the Anasazi, including the quote, is taken from a fascinating article by Linda Cordell (2001).

most colossal urban and agricultural infrastructure in the entire world Stine 2001.

181 *Los Angeles quietly bought 320,000 acres of land in the Owens Valley* Broder 2004a.

snowpack acts as a natural reservoir The discussion of retreating western snowpack in this and the following few paragraphs is based on a news article in *Science* written by Robert Service (2004).

182 *the latter would seem, from the newspapers, to be on the increase already* As I was putting the finishing touches on this manuscript, *Nature* published an ingenious study in which a team from New Mexico and Arizona examined the charcoal residues of ancient forest fires that had washed to valley bottoms in two study areas in the northern Rocky Mountains, one in Idaho, the other in Wyoming's Yellowstone National Park (Pierce, Meyer, and Jull 2004; Whitlock 2004). The study demonstrates that over the past eight thousand years, smaller, less damaging fires have tended to occur when it was cold, while severe, stand-clearing fires have occurred more often during warm periods (not all that surprising, I guess). The inference would be that the severe fires of recent years (the infamous Yellowstone blaze around the time of Jim Hansen's 1988 testimony to Congress, for instance) may have been encouraged by the warmth and aridity that has settled upon the West with the onset of global warming—and, of course, that we can expect more in the future.

period since 1999 is now officially the driest Most of the points raised in the following discussion of the Colorado watershed were discovered in Johnson and Murphy (2004).

183 *federal government found it necessary to step in* After much wrangling and much prodding from the Interior Department, Californians finally agreed, in October 2003, on how to share their allotment from the Colorado River (Jehl 2002e; AP 2003).

no "excess" at all Murphy 2003.

Daniel McCool Quoted by Johnson and Murphy (2004).

184 *down card in the poker game* Leslie 2000, 50.

185 *JG Boswell Company* Broder 2004b.

fifteen thousand tons of water to produce one ton of cotton Gleick 2000.

United States subsidizes cotton farmers For example: Becker 2004a.

185 *World Trade Organization* The WTO drama, at least as it played out in 2004, is described in Becker 2004a, 2004b, 2004c; Becker and Benson 2004; and Bloomberg 2004.
six Chicago counties Egan 2001.
Atlanta Jehl 2002b, 2003b.
the Pee Dee (and the other eastern disputes) Jehl 2003a.
Ogallala Aquifer Jehl 2002d.

186 *divert the White River* Jehl 2002b.
government guarantees rice farmers more than double *New York Times* 2002.
T. Boone Pickens (and Enron) Egan 2001.
irrigation project in the headwaters of the Euphrates Jehl 2002c.
wrongheaded Peter Gleick (2003) presents a superb discussion of the pitfalls and costs of the traditional "dam and aqueduct" solution to the water problem, as well as the promising transition to smaller, decentralized, community-scale solutions that is now under way.
a billion lack safe drinking water This and the disease-related death statistic come from Gleick (2003).
report issued in 2002 UNEP 2002.

187 *more attention to blood than to water* This quote (Leslie 2000, 37) comes from the magazine article that has provided the seed for Leslie's book in progress.

188 *most irrigation-based civilizations fail* Postel 1999.
called the richest in the New World Moseley 1992, 180.

189 *mine its base hydraulically* Shimada 1994, 14.
crown emits squirming rays The details of the Decapitator's getup come from Shimada (1994, 232).

192 *forty million deaths in India and China* Kininmonth 1999.
Sir Gilbert Walker Mark Cane told me some of Sir Gilbert's story when I interviewed him in July 2000. He put more into the paper he was writing at the time (Cane 2000).

Sir Gilbert turns out to have been yet another climatologist with a side interest in mountaineering. As I was searching for Guy Callendar's obituary notice in the *Quarterly Journal of the Royal Meteorological Society* one day, I stumbled across the following in the obituary for Dr. Tom Longstaff, one of the greatest mountaineer-explorers of the Victorian era, who also happened to be *the* senior fellow of the Meteorological Society at the time of his death: "Longstaff was one of the last of those accomplished Victorians, those men whose instinct was to climb out of the crowd and to whom we owe so much. Mountaineering and an interest in weather as an element in natural history not uncommonly went together; among that generation it is not generally known that the late Sir Gilbert Walker made the first direct ascent of Scawfell Pinnacle [in Britain's Lake District] in 1897 when C. T. R. Wilson was fresh from Ben Nevis" (Manley 1965).

Charles Thomson Rees Wilson's love of the mountains also led to a useful discovery (University of Cambridge Web site). Wilson was a research physicist from Cambridge University and an avid climber, who worked at the summit observatory on Scotland's Ben Nevis for a short while in 1894. Upon returning to the lowlands, he wrote (sounding something like John Tyndall) that "the wonderful optical phenomena shown when the sun shone on the clouds surrounding the hill-top, and especially the coloured rings surrounding the sun [coronas] or surrounding the shadow cast by the hill-top or observer on mist or cloud [glories], greatly excited my interest and made me wish to imitate them in the laboratory." Back in Cambridge, he found a way to simulate cloud formation in glass chambers and soon discovered that X-rays (known as "uranium rays" in his day) caused tiny dotted lines of water droplets to form in his simulated clouds. After Ernest Rutherford's early elucidation of atomic structure in 1912, Wilson realized that he could use his "cloud chamber" to track the paths of radioactive particles. The cloud chamber went on to become one of the most important experimental tools in particle physics and, in 1927, earned Wilson a Nobel Prize.

Wilson's biographer described him as "gentle and serene, indifferent to prestige and honour, a man whose work had come from an intense love of the natural world, and a delight in its beauties."

192 *towering landmark in the understanding of ENSO* Cane 2000.

193 *eighteen inches higher around Indonesia* National Academy of Sciences.

194 *An intensifying Walker Circulation* International Research Institute for Climate Prediction.

a dog chasing its tail This phrase—and most of the intuitive description of El Niño found here—comes from an interview with Mark Cane that I conducted in July 2000. I am responsible for all the mistakes, of course.

196 *sea gave the topsoil back* Moseley 1992, 211.

remarkable relative to the entire 1,500-year glacial data set Shimada et al. 1991.

197 *Viracocha* Moseley 1992, 211.

flourishing of novel religious forms Cordell 2001, 188.

the gods have failed Ibid., 190.

198 *tiny and lie at the edge of the city, behind high walls* Bawden 1996, 288.

used at least once as a fortification Ibid., 314.

199 *large secular bureaucracy* Shimada 1994, 179.

200 *dealing a new round of blows to the bases of the huacas* Ibid., 242.

raised beds soldiered through Moseley 1992, 228–29.

Naymlap The quote comes from Donnan (1990, 243–45), who adapted it from a translation by Means (1931, 51–53). Balboa's original manuscript, *Miscelánia antártica,* is in the Universidad Nacional Mayor de San Marcos. There are copies in the New York Public Library and the University Library of the University of Texas, Austin.

202 *Chimú then expanded back to the north* Shimada 1990, 371.

Allison Paulsen Paulsen 1976.

203 *two more cultures gave last gasps in chorus with the Moche* Shimada et al. 1991, 253–54.

Nazca Lines The Nazca Lines, on the coastal Pampa Colorada, or "Colored Plain," south of Lima, were rediscovered by airplane pilots in the 1920s. Nearly a thousand gigantic figures appear to be etched on nearly two hundred square miles of the desert floor—dozens of stylized line drawings of plants and animals: a 360-foot monkey with a long spiraling tail, a 210-foot killer whale, a 443-foot condor, a 935-foot pelican, a hummingbird, a spider, flowers, trees; there were geometric designs as well: huge spirals, triangles, trapezoids, sets of parallel lines, more than a hundred ray systems consisting of as many as twenty-five perfectly straight lines up to two and a half miles long that emanate from small man-made hills. A few solitary straight lines, some as long as thirty miles, rise from nowhere and go nowhere. Like the man-made huacas of the Moche and

Tiwanaku, the Nazca Lines seem to have been related to water: they are usually aligned with underground channels and other water sources (Aveni 2000).

204 *The Thompsons believe an El Niño dynamic may explain this* Thompson 1992, 1995.

205 *Methuselah Walk* Hughes, Funkhouser, and Fenbiao 2002; Hughes and Funkhouser 1998.

PART V: MORE PIECES FOR THE PUZZLE

210n *1912 was an El Niño year* Solomon and Stearns 1999.

211 *4-by-13-meter snow pit* Thompson, Mulvaney, and Peel 1989.

214 *world's two highest satellite weather stations* I found this fact in a fine profile of Lonnie that appeared in the *Los Angeles Times* (Abramson 1992). First sentence: "At a glance, there is nothing about Lonnie Thompson that evokes Indiana Jones."

1993 paper Thompson et al. 1993.

peninsula had started warming in the 1930s Thompson et al. 1994.

215 *other simulations* Wang et al. 1991.

216 *Mike's fireball alone would have engulfed Manhattan* Rhodes 1995, 510.

217 *testimony Lonnie delivered in February 1992* Thanks to the U.S. Government Printing Office there is a complete transcript of the hearing (U.S. Senate 1992).

1990s will replace the 1980s as the hottest decade on record U.S. Senate 1992, 1, 4.

219 *contrarians* Some will never be convinced. Consider a letter Lonnie received in 1994 (spelling and punctuation intact):

Dear Dr. Thompson;

I had been told that you are an world-famous expert on the study and interpretation of ice-cores. However *Ice Cores and the Age of the Earth* by Dr. Larry Vardiman clearly demostrates that you are nothing more than an evolutionist intent spreading Humanistic Falsehoods in order to defame and deny the **TRUTH** about the **WORD** of our **LORD** as given to us in the **HOLY BIBLE**. Dr. Vardiman clearly exposes and denounces the fraud perpetrated knowingly on True CHRISTIANS by bogus scientific research conducted by you and other Anticreationists intent on defaming and denying our **LORD** in the name of evolutionism.

Be warned!!!! Just as Rev. Tim Lahaye and Dr. Dobson are actively working with Congress to defund and abolish the US Geological Survey for the blasphemous Anticreationist and AntiChristian pornography that it publishes as alledged geological and paleontological research, good Christians will force state legislatures to defund AntiChristian "scientists" and the universities that support them. However, *Ice Cores and the Age of the Earth* by Dr. Larry Vardiman will convinced an intelligent person like you of your errors and cause you to join true scientists in their support of the **WORD** of our **LORD**. For *John* 3:14–18 clearly demostrates that JESUS CHRIST is truly the SON OF GOD who was sacrificed such that we would not perish, but rather have eternal life.

Yours in Christ;

R.H.M.

An Lamb of Christ

The evidence is very clear U.S. Senate 1992, 15.

There was clear evidence for warming in the tropics at that time, but slightly less evidence for warming in the polar regions. Back to the time of Arrhenius, equilibrium calculations had predicted that the poles would heat up more than anywhere else when everything had settled out, and in the fourteen years since the hearing, temperatures in the Arctic and on the Antarctic Peninsula have indeed skyrocketed. Thus this particular change may have begun in the tropics and spread to the poles—as Lonnie has been arguing for some time that most changes will.

219 *smoking gun* Ibid., 48.

220 *These programs were not undertaken to look for evidence for global warming* Ibid., 13.

221 *Lao Tzu* Bernbaum 2001, 86.
Shangri-La Ibid., 3.

222 *Dalai Lama* Ibid., 10.

235 *Rising temperatures were forcing Lonnie to go to higher and more dangerous locations* Most of the detailed information in this paragraph comes from Davis et al. (1995).

244 tour de force *characteristic of the man* Griffin 1999.

245 *feat of high-altitude endurance* Amazingly, the fifty-three days Lonnie and Vladimir spent in the Garganta Col did not constitute a high-altitude endurance record even on Huascarán itself. Nicolas Jaeger, who had been a member of the first French team to climb Everest in 1978, lived for sixty-six days on Huascarán's south summit in 1979. As a doctor, he mixed business with pleasure (or pain) by using himself as the subject in a study of high-altitude physiology. He even took his own blood samples.

Jaeger's record was then eclipsed or tied—it's hard to say which, since it involved a slightly higher mountain for a slightly shorter time—by a Spaniard named Fernando Garrido, who spent sixty-two days on the 22,831-foot summit of Aconcagua (627 feet higher than Huascarán) over the austral summer of 1985/86. Garrido claims that he did it solely for the "personal experience," and he seems to have had one. He calls it "the strongest and most intense experience" of his life. He lost all of his toe- and fingernails and nearly forty pounds.

Later, on Dasuopu, a few members of Lonnie's team would tie or, arguably, surpass these two climbers' feats.

250 *SU81-18* Bard et al. 1987; Bard et al. 2000.

252n *species pump* Haffer 1969; Schneider 1996.

253 *He instituted the greatest funding increase of all time* Sarewitz and Pielke (2000) argue convincingly that Bush's "research as policy" strategy was very effective at muting activism by scientists about the greenhouse.

254 *wild times in the North Atlantic* The story of these exciting discoveries, along with their "thermohaline explanation," is found in Kerr (1993a, 1993b) and Broecker (1995a). Broecker and Hemming (2001) give a more recent perspective.

255 *Gerard Bond* Bond and Lotti 1995.
two elegant studies Bond et al. 1997; Bond et al. 2001.

256 *the Younger Dryas began in Greenland precisely 12,880 years ago* Lonnie told me GISP2's exact dates for the Younger Dryas in an e-mail dated April 9, 2000. Kerr (1993a, 890) corroborates Lonnie's numbers.

257 *The basic idea came to me in 1984* The quote comes from Broecker (1997c, 4). Kerr provides additional insight (1993a).

258 *Achilles' Heel of Our Climate System* Broecker 1997b.

262 *meeting in Venice* This PAGES-CLIVAR Workshop on Paleoclimate Variability and Prediction took place from November 16 to 20, 1994. Lonnie gave the keynote address: "The low-latitude ice core record (including paleomonsoons)."

262 *Huascarán results in* Science Thompson et al. 1995
tenacity of a lone scientist Broecker 1995b.

264 *bronze age man* For instance, Shouse 2001.

266 *In 1993, at the annual December meeting* Kerr 1994.
Rick Fairbanks Fairbanks and his friends (Guilderson, Fairbanks, and Rubenstone 1994) based their inferences on two so-called paleothermometers: in a theme that is probably familiar by now, the oxygen isotope ratio in the calcium carbonates in a coral reef reflects the temperature of the water in which the reef is immersed. Additionally, strontium, which is chemically similar to calcium, sometimes replaces it in the carbonates, and the rate of this substitution also depends on water temperature.
Martin Stute Stute examined noble gases dissolved in water that had been sequestered in the earth long ago (Stute et al. 1995). As *Science* reporter Richard Kerr put it (1995), "The colder the water, the more neon, argon, krypton, and xenon dissolve in it before it sinks out of contact with the atmosphere."
Schrag Schrag, Hampt, and Murray 1996. Broecker (1996) offers a comment.

267 *Lonnie and Willi Dansgaard had revealed* Thompson and Dansgaard 1975.
Dansgaard had achieved legendary status in 1964 Dansgaard 1964.
primitive mathematical model Broecker 1997a.

268 *Raymond Pierrehumbert* Pierrehumbert 1999.

270 *other scientists' studies* They reviewed the work of Reginald Newell from MIT.
published this thinking in 2000 Thompson, Mosley-Thompson, and Henderson 2000.

271 *their definitive statement* Thompson et al. 2003.
Hockey Stick Mann, Bradley, and Hughes 1999; IPCC 2001b; also see p. 96.
plant uncovered by Quelccaya's retreating Qori Kalis Lonnie told me about the plant shortly after he found it, and it subsequently became the subject of a science column in the *Wall Street Journal* (Regalado 2004).

272 *origin of the term climate* Budyko 1974, xii.
NATO Advanced Research Workshop Thompson 1996.

273 *definitive report on Guliya* Thompson et al. 1997.
said as much in a press release OSU 1992.

274 *EPICA* Augustin et al. 2004; Walker 2004; McManus 2004.

275 *record reaching back 420,000 years* Petit et al. 1999; Stauffer 1999.
Lake Vostok Gavaghan 2001.
people who should know better still believe Vostok has a longer memory than Guliya Reporters for *Science* magazine, for instance (Stone 1999).

277 *age of the ice is almost secondary* OSU 1997.

278 *John Imbrie and his daughter Katherine have told the story* Imbrie and Imbrie 1979. The initial report of the discovery appeared in *Science* (Hays, Imbrie, and Shackleton 1976).
Croll is the hero of the story Croll's biographical details: Imbrie and Imbrie 1979, chap. 6.

279 *recent speculation by Wally Broecker* I am speaking here of Wally's alleged "bipolar seesaw" (Broecker 1998), which comes up again on page 309.
Milutin Milankovitch Milankovitch's biographical details: Imbrie and Imbrie 1979, chap. 8.
revelation similar to Lonnie's Ibid., 101.
the task of the exact natural sciences Ibid., 173.

280 *Johannes Kepler* I boned up on orbital variations with the help of the Imbries (1979, chap. 5) and Ray Bradley's excellent textbook (Bradley 1999, 35–46).

282 *accurate date for the very end of the Pleistocene* Imbrie and Imbrie 1979, 129.

283 *examination of the world's fossil reefs* Broecker 1965; Imbrie and Imbrie 1979, 143.
terraced reefs in Barbados Broecker et al. 1968; Imbrie and Imbrie 1979, 145.
one-hundred-thousand-year beat Broecker and van Donk 1970; Imbrie and Imbrie 1979, chap. 14.
swells in eccentricity line up well with the Czechoslovakian soil changes Imbrie and Imbrie 1979, 158–59.

284n *Broecker's great conveyor is another excellent example* Wunsch (2002) explains exactly how specious the thinking behind the mightily named thermohaline circulation actually is.

284 *Shackleton spent ten years* Imbrie and Imbrie 1979, 163–64.

285 *Todd Sowers* Bender, Sowers, and Labeyrie 1994.
Its original goal Imbrie and Imbrie 1979, 161–62.

287 *orbital changes as an input and the climate record as an output* Hays, Imbrie, and Shackleton 1976, 1122.
It is concluded Hays, Imbrie, and Shackleton 1976, 1131. Story of discovery: Imbrie and Imbrie 1979, chap 15. Also useful: Bradley 1999, 280–83.
Imbrie teamed up with his son Imbrie and Imbrie 1980.
Shackleton undertook a combined analysis Shackleton 2000; Kerr 2000b.

288 *highest levels in that span* Petit et al. 1999; Stauffer 1999.

289 *Temperature Tracks Carbon Dioxide at Vostok* Ice core data in the graph: Petit et al. 1999. The spike in CO_2 in the twentieth century is basically the Keeling Curve (p. 123; Keeling and Whorf 2004).

290 *a call came in from Al Gore's office* Lafferty 1994.
Johannes Oerlemans Oerlemans 1994.

291 *Western Fuels Association* Gelbspan 1997, 36.

292 *ICE campaign* Gelbspan 1998.
according to the Union of Concerned Scientists Union of Concerned Scientists 2003.

294 *paper in* Science Thompson et al. 1998.
Hard to explain those contradictory bits of evidence with anything except a temperature change The complexity of this issue is exemplified, however, by a recent paper on which Lonnie appears as an author, written primarily by his friend Ray Bradley from the University of Massachusetts (Bradley, Vuille et al. 2003). This study demonstrates that the isotopes on Sajama track sea surface temperatures in the equatorial Pacific, but Bradley explains the correspondence by resorting to the amount effect. Lonnie disagrees with Ray's explanation and points to the bottom line: the simple fact is that Sajama's isotopes track temperature.

296 *A recent study* Baker, Seltzer et al. 2001; Baker, Rigsby et al. 2001. The second of these papers uses Pierrehumbert's questionable analysis to argue that Sajama's isotopes are more a measure of precipitation than temperature, since the southern Altiplano was wetter during the glacial than it is now. That the isotopes clearly measure temperature as well is demonstrated in the Thompsons' definitive statement on the amount effect, mentioned on page 271 (Thompson et al. 2003). That they have also tracked Pacific sea surface temperatures is demonstrated in the paper by Ray Bradley, Lonnie, and their colleagues, mentioned in the previous note (Bradley, Vuille et al. 2003).
"Will Our Ride into the Greenhouse Future Be a Smooth One?" Broecker 1997c.

296 *Achilles' Heel* Broecker 1997b.
vivid press release Columbia 1997.
Iceland would become one large ice cap Broecker 1999b.

297 *difficult to comprehend the misery* Broecker 1997c.
Science *itself even jumped on the bandwagon* Kerr 1998b.
"The Great Climate Flip-Flop" Calvin 1998. This article has now been incorporated into a book (Calvin 2002).

298 *poking an angry beast* Broecker 1999b.
sun-driven 1,500-year beat Gerard Bond and collaborators had demonstrated this in the two elegant studies, previously mentioned (p. 255), which explained the 1,500-year spacing between Wisconsinan Dansgaard-Oeschger events (Bond et al. 1997; Bond et al. 2001).
stir the pot in social and media circles Regalado 2003a.
it may have been a false alarm Broecker 2000.
Pentagon released a study Schwartz and Randall 2003. Associated news articles: Townsend and Harris 2004; Hertsgaard 2004; Revkin 2004d.

299 Fortune *magazine* Stipp 2004.
greatly out of touch with current scientific thinking Weaver and Hillaire-Marcel (2004) provide an excellent overview of the plentiful evidence refuting the notion that global warming is at all likely to bring on an ice age.
previously ignored—and some say altered—others from the National Academy of Sciences and the Environmental Protection Agency George W. Bush's propensity for bending science to suit his political agenda has been well documented. Revkin (2004a) provides an overview. In January 2003, *Science* found it necessary to publish an editorial on this disturbing practice (Kennedy 2003).

299n *"The Coming Anarchy"* Kaplan (1994) borrowed the quote about the stretch limo from Thomas Homer-Dixon, head of the Peace and Conflict Studies Program at the University of Toronto.

300 *letter to* Science *dismissing the report* Broecker 2004.
a popcorn movie that's actually a little subversive Revkin 2004d.
(perhaps tongue-in-cheek) survey Balmford et al. 2004.

302 *as Mark Cane put it in an elegant perspective* Cane 1998.
Sleeping Dragon Wakes Pierrehumbert 2000.
His background is applied mathematics and meteorology Most of the details of Cane's background come from a fine profile by William K. Stevens (1996) that appeared in the *New York Times*.

304 *precipitated the decline of American civilization* Stevens 1996.
wisely followed the dictum of Wee Willie Keeler Cane and Evans 2000.
core from Indonesia's Makassar Strait Visser, Thunell, and Stott 2003; Dunbar 2003; Kerr 2003a.

305 *began to relax in Peru about five thousand years before it did in Greenland* Seltzer et al. 2002.
he writes elsewhere Broecker 1997c.

306 *It is well established that El Niño years are warmer* Cane 1998.
controlled by the twenty-two-thousand-year precession cycle Clement, Seager, and Cane 2000; Clement and Cane 1999; Clement, Hall, and Broccoli 2004.
Shackleton showed Shackleton et al. 1984.

307 *Cane and Molnar* Cane and Molnar 2001; Wright 2001.
a role in human evolution deMenocal 1995; Potts 1996; Feibel 1997.

308 *"What Is the Thermohaline Circulation?"* Wunsch 2002.
Matthew Maury Maury's pioneering book, *The Physical Geography of the Sea and Its Meteorology,* published in 1855, was reprinted many times and translated into three languages. The quote about the "Emerald Isle of the Sea" opens the paper by Seager et al. (2002). Literary flourishes from the arts, history, or the world of sports are often found in Mark Cane's papers, and since he is an author on the Seager paper, he may have supplied this one.
fine study by a group led by Richard Seager Seager et al. 2002.

309 *Wally Broecker accepts the conclusions (and apologizes for his "previous sins")* Kerr 2002b.
Science proceeds one funeral at a time This nugget was offered up by archeologist Harvey Weiss, who appears later in this story.
GLAMAP Mix 2003.
Southern Ocean School For example, Sowers and Bender 1995; Henderson and Slowey 2000; Schrag 2000.
bipolar seesaw Broecker 1998.

310 *crosshairs focused on both poles* For instance, Blunier et al. 1998; Steig et al. 1998; Steig et al. 2000; Morgan et al. 2002; Knorr and Lohmann 2003; Stocker 2003.
it is hard to tell which takes the lead Steig and Alley 2002.
"Of Ice and Elephants" Schrag 2000.
which end is the head? Interesting supporting views: White and Steig 1998; Steig 2001.

311 *textbooks in the mid-1990s* Clapperton 1993, 454–57.
solved the problem named for Lonnie's mentor Thompson, Mosley-Thompson, and Henderson 2000.

313 *in his diary* This particular quote from Shawn Wight's diary comes from the memory of Mary Davis, who later read it. Other references to and quotes from the diary come from an article by Julianne Basinger (2001).

321 *records for high-altitude endurance* These feats certainly rival and arguably exceed those of Fernando Garrido on Aconcagua and Nicolas Jaeger on Huascarán, since those summits are both lower and significantly warmer than Dasuopu's Camp IV. (See source note on page 411: *feat of high-altitude endurance.*)
satisfactory condition Post 1997.

323 *paper based on Shawn's thesis work* Cole-Dai et al. 2000.

PART VI: KILIMANJARO

325 *more than merely "travel"* Meyer 1891, xiii.

327 *comprise more than one thousand cubic miles of rock and dirt* Downie et al. 1956.

330 *three hundred thousand people in the Dodoma region faced starvation within five months EastAfrican* 1999.

331 *The recent El Niño* The descriptions and quotes about the devastation in East Africa found in this and the following paragraph were taken from a report by Harvard Medical School's Center for Health and the Global Environment (1999). The consequences of climate change for public health in the largest sense, that is, the health of people everywhere (obviously a huge subject), is addressed in a paper by Paul Epstein (1999), which is "enhanced" with online links on the *Science* Web site. See also Linthicum et al. 1999.

333 The Imperial Gazetteer of India Quinn 1992, 125.

333 *Nilometers* Ibid., 137.
334 *famine foreseen by Joseph set in about 1708 B.C.* Ibid.
335 *twenty-three days south by sail from Ras Hafun* Reader 1982, 1.
336 *Fernandez de Encisco* Meyer 1891, 5.
kalima *means "mountain"* Krapf [1860] 1968, 255. (From Rebmann's journal, which Krapf quotes extensively.)
To the Maasai it is Ngaje Ngai Tilman 1938.
the coy sources of the White Nile Burton [1860] 1961, vol. 1, 5.
336n *God allows great things to arise out of small and insignificant beginnings* Krapf [1860] 1968, 485.
337 *people who have ascended the mountain have been slain by spirits* Ibid., 192.
The mountains of Jagga gradually rose more distinctly to my sight Reader 1982, 8.
338 *I deny altogether the existence of snow on Mount Kilimanjaro* Ibid., 9.
Let Geography perish Bridges 1968, 61.
von der Decken Meyer 1891, 9; Reader 1982, 15–16.
easier to believe in the misrepresentations of man Reader 1982, 16.
a certain degree of unenviable notoriety Meyer 1891, 10.
Charles New Ibid., 11; Reader 1982, 17–18.
339 *Mandara received the missionary New* Johnston 1885.
Johnston was a tiny man Vauxhall Society.
British secret agent Reader 1982, 22.
My map of Africa lies in Europe Kimambo and Temu 1969, 101.
Hans Meyer Meyer told the story of his African explorations in his magnificent book *Across East African Glaciers* (1891). His first two expeditions were covered in the preface (pp. vi–xi).
340 *His company included an eight-man Somali bodyguard* Meyer 1891, 45–52.
condign punishment Ibid., 52.
341 *great glacier, which issues from a stupendous fissure* Ibid., 253.
342 *enormous cleft, through which the ice that at this point covers the bottom of the crater* Ibid., 155.
346 *We had reached a height of 14,200 feet* Ibid., 141.
349 *this meant a reduction of about 13 percent* Bolin 1998; May 1997.
354 *calculated insolation curves* Berger 1978a, 1978b.
355 *Salar de Uyuni grew to maximum size just as the ice cap was born on Sajama* Baker, Rigsby et al. 2001.
snows of Kilimanjaro had been born about eleven thousand years ago A few weeks later, lounging in a tent in Kilimanjaro's summit crater, we talked about another implication of asynchrony:

"So," I said to Lonnie, "you're saying all the tropical glaciers on the planet are going to disappear in the next fifty years—or it looks that way—and that humans are causing this. But on the other hand, you're telling me that natural cycles have caused them to appear and disappear in the past, with your asynchrony theory and the precession cycle and all that."

"Yes," he replied, "and that's why you have to look at *where* the glaciers are disappearing, because they *have* disappeared in the past—through a natural process. What sets the twentieth century apart is that they're disappearing at every latitude. See, with the precession cycle, they're disappearing at one latitude but growing at another: the cycle feeds through. But now the cycle is not feeding through. They're disappearing on Sajama, Quelccaya, Huascarán, here on Kilimanjaro, the Himalayas; doesn't matter where you are. None of them are growing."

In that sense the present time seems to be unique in at least the last 125,000 years, since the last interglacial, the Eemian.

373 *satellite analysis* Gaffen et al. 2000.

376 *snows of Kilimanjaro are likely to disappear within twenty years* OSU 2001.

front page of the New York Times Revkin 2001.

volume on the expected impacts IPCC 2001a.

377 *Kilimanjaro has evolved into one of the central icons of global warming* As such it has attracted undue, and in some cases downright weird, attention since Lonnie made his prediction in 2001. A London-based Zimbabwean scientist has even proposed the dubious, Christo-like solution of draping huge reflective tarps down the vertical walls of the summit icefields in order to forestall further melting (Mason 2003; Morton 2003); and, of course, the global warming contrarians went on the attack immediately.

Confusion has been sown in the more sober scientific community as well. In 2002, Jeanne Altmann, a behavioral ecologist at Princeton who has spent decades studying the baboon communities living in Kenya's Amboseli Basin, a few miles north of the mountain, reported that daily measurements taken at her field stations over the final quarter of the twentieth century show a temperature rise of twelve degrees Fahrenheit: far too much to be attributed solely to global warming (Altmann et al. 2002). Noting that not only the snow but also the forests on Kilimanjaro's flanks have retreated in the time she's been working there, Altmann attributed the warming to local land-use changes caused by humans, who have cleared forest to make way for farmland, and to "damage from an increasing, and increasingly resident, elephant population." *Nature* then took the word of this expert on baboons, admittedly, but not climate (Mason 2003); and in an editorial that was clearly based on the *Nature* story, the *New York Times* (2003) began questioning the connection between global warming and the retreating snows.

Temperature is a difficult thing to measure, especially in the heat of day, and there is a sharp rise about halfway through Altmann's measurements, which seems to coincide with a relocation of her field station. But she also measured daily temperature minima, which are less problematic, and still found an increase of three degrees. On the other hand, most other weather stations in the region have recorded little or no change over the last few decades.

This was confusing enough; then to add to it, members of the Massachusetts team that set up the weather station on the summit while we were up there took part in a theoretical study (Kaser et al. 2004) that suggested that temperatures have held steady on Kilimanjaro ever since Hans Meyer climbed it in 1889 and that the main reason for the retreat of the snows has been a change to persistently arid conditions that commenced abruptly in about 1880. According to this proposal, it has been too cold near the summit for the snow to have been melting; it must actually have been subliming—turning directly from ice to gaseous water vapor—and although sublimation is a very slow process, the mountain has been receiving too little snow for the past century to offset even that.

Pat Michaels, the vigilant editor of the *World Climate Report* (founded by Western Fuels, as previously mentioned), found no use for Altmann's work, because he tends to steer clear of any indication that the earth might be warming; but he seized upon the sublimation study in his monthly screed, the *World Climate Alert* (Greening Earth Society 2004). "Despite countless reports blaming global warming," he began, "it turns out Mt. Kilimanjaro's glaciers are retreating because of a climate shift that occurred more than 120 years ago, long before humans could have caused it." He then bragged that he had notified Andrew Revkin, the main global warming correspondent for the *New York Times,* of the new report and was waiting to "see what happens."

Revkin took the bait (2004b). Under the headline, "Climate Debate Gets Its Icon: Mt. Kilimanjaro," he wrote, "[N]ow the pendulum has swung. This month, the mountain was taken up as a symbol of eco-alarmism by a cluster of scientists and anti-regulation groups" (although I count just one point in this cluster: the publishers of the *Alert*). While he did provide room for the report's lead author, Austrian glaciologist Georg Kaser; Ray Bradley and Doug Hardy from the University of Massachusetts, who joined Kaser in the study; and Lonnie Thompson to rebut Michaels's claims, the net result was that Revkin played into Michaels's hands: his article cast a shade of doubt not only on the connection between Kilimanjaro's dwindling snow and global warming, but on the reality of global warming itself.

This is an excellent example of the way the news media, in its earnest but misguided belief that it is being "objective" by presenting "both sides of the story," has assisted Michaels and his friends in their active misrepresentation of greenhouse science.

Ray Bradley says, "Revkin's headline should have been, 'Kilimanjaro's Glaciers Are Evaporating Faster Than Ever.' That would have been an interesting perspective on it. Clearly, nobody's arguing that they're not disappearing. It's a question of why they're disappearing, and I don't think any of us are arguing, either, that it's not related to global warming."

Ray would be seen as the senior author of the sublimation paper. He holds the rank of Distinguished Professor at the University of Massachusetts and is director of its Climate System Research Center. He is also the author of a respected textbook on paleoclimatology and is often asked to write perspectives in the leading journals.

He remembers that when he climbed the mountain with Doug Hardy and one of Kaser's students in 2002, "They were taking the devil's advocate position on whether Kilimanjaro's snow was disappearing due to global warming, and I said, 'Look. There's no way on earth that every glacier in the world outside of Scandinavia* is disappearing as a result of global warming; and here's this glacier in the tropics that is somehow disappearing, but it's not related to global warming. Give me a break.'"

In Ray's opinion, Kaser, who actually wrote the paper, "wasn't very successful, and I didn't pay enough attention in trying to steer it to a clearer presentation. I thought that paper would basically sink like a stone in some obscure little journal, but the whole thing was blown out of proportion—and it didn't help that Hillary Clinton and John McCain had said, just the week before,

*In a recent phone conversation, Lonnie informed me that even those few glaciers in Scandinavia that had been advancing have now begun to retreat.

'Look, you can say all you like about global warming, but one thing you can't argue about, and that's that Kilimanjaro is melting!' That just raised the stakes for these climate skeptic people."

Indeed, Senator McCain, who had recently co-sponsored a bill aimed at curbing greenhouse emissions, had displayed before-and-after photographs of the mountain in a Senate debate.

The only new *data* in Kaser's study are statements that the summit weather station had measured an average yearly temperature of nineteen degrees Fahrenheit over the two years it had been operating and that it had never recorded a temperature above freezing. Reflecting upon the weird shape of the ice fields—which had also surprised Lonnie and Keith Mountain when they first saw them—and reviewing local climate records that do seem to indicate a change in aridity in about 1880, Kaser speculates reasonably that sublimation must be playing a role.

But Doug Hardy claims that they wrote the article "as a discussion piece for scientists. There are all kinds of questions as to whether [Revkin] should have even engaged [Pat Michaels] in that discussion."

The fact is that it would be difficult if not impossible to prove a direct and unarguable link between any particular change and global warming per se. (Remember that in his Senate testimony Jim Hansen even refused to link the hot summer of 1988 to global warming.) The logic actually flows in the other direction: it is the overwhelming accumulation of evidence from so many locations that proves that the earth is getting warmer. Incidentally, Hardy and Kaser floated a proposal aimed at studying the connection between the changes on Kilimanjaro and global warming, but it was turned down—for just this reason, most likely. (Lonnie did support Doug's successful proposal simply to keep tracking the weather, however.)

Ray, Doug, and Lonnie all believe, by the way (as Guy Callendar did more than sixty years ago), that humans had put quite enough carbon dioxide into the atmosphere by 1880 to have affected climate noticeably. As I have pointed out (p. 96), the hockey stick–shaped graph (Mann, Bradley, and Hughes 1999) that was the centerpiece of the most recent IPCC assessment (2001b) seems to support this conclusion. It shows that the average temperature of the northern hemisphere began to spike near the end of the 1800s, after having dropped for almost a thousand years.

And, of course, we saw dripping and flowing water everywhere on the upper reaches of the mountain both times we climbed it—even on the surface of the glacier under the very feet of Doug and his colleague Mathias as they set up their weather station.

Lonnie does not discount sublimation, but he still believes melting is the main factor in the present retreat. He points out that only in the upper sixty-five centimeters of the cores from the northern ice field did he find elongated air bubbles, indicating that only the upper ice has melted and refrozen, and adds, "What is happening there now has not happened at any time in the past eleven thousand years. There is not a time in the history of that ice field that these conditions have existed. The bubbles would have been preserved, and we would see them in the stratigraphy."

377 *"A Global Warning to Mr. Bush"* *New York Times* 2001.

378 *the president reneged* Jehl and Revkin 2001.

378 *astounding claim that the science is not clear on this issue* In an essay in *Science* in December 2004, science historian Naomi Oreskes, from the University of California at San Diego, cogently explained just how clearly scientists have spoken on this issue. "IPCC is not alone in its conclusions," she wrote. "In recent years, all major scientific bodies in the United States whose members' expertise bears directly on the matter have issued similar statements." She cited formal position statements by the National Academy of Sciences, the American Meteorological Society, the American Geophysical Union, and the American Association for the Advancement of Science.

To test whether these organizations might have "downplay[ed] legitimate dissenting opinions," Oreskes analyzed the abstracts of more than nine hundred papers on the subject of global climate change that had been published in refereed scientific journals between 1993 and 2003. Of these, she found, 75 percent either endorsed the consensus position, dealt with the evaluation of impacts, or proposed steps for mitigation; 25 percent "dealt with methods or paleoclimate, taking no position on current anthropogenic climate change"; and "remarkably, none of the papers disagreed with the consensus position."

"Politicians, economists, journalists, and others may have the impression of confusion, disagreement, or discord among climate scientists," she concluded, "but that impression is incorrect."

some believe the threshold to an inevitable, destructive warming has been crossed This statement is based on a report issued jointly in January 2005 by the Institute for Public Policy Research in Britain, the Center for American Progress in the United States, and the Australia Institute (AP 2005; Institute for Public Policy Research 2005).

379 *All we did was connect the dots* Krajick 2002.

Messerli Messerli 1980.

Hastenrath Hastenrath and Greischar 1997.

primary paper on the Kilimanjaro ice cores Thompson, Mosley-Thompson et al. 2002.

380 *raising crops and keeping animals for the very first time* Bar-Yosef 2000; Heun et al. 1997.

380–81 *Somewhere in one of perhaps a dozen places* Lavietes 2003.

381 *Nicole Petite-Maire and her colleagues* Petite-Maire, Beufort, and Page 1994.

climate shock Weiss and Bradley 2001. (This is a succinct and fascinating paper.)

382 *shock may have been caused by a shutting off of Wally Broecker's great conveyor* Barber et al. 1999; Clarke et al. 2003.

flood of freshwater does not appear to have slowed the conveyor down Weaver and Hillaire-Marcel 2004.

Thompsons believe Thompson, Mosley-Thompson et al. 2002.

Jared Diamond Both quotes come from Diamond (1997b).

profound changes in many primitive societies Weiss 2001.

383 *Uruk civilization collapsed swiftly right around 3000 B.C.* Weiss and Bradley 2001.

displaced to the banks of large rivers Petite-Maire, Beufort, and Page 1994.

"extractive" petty states emerged Hassan 1994.

regional settlement was redistributed Weiss 2001, 85.

384 *Ships brought the goods* Ibid.

Tell Leilan Weiss et al. 1993.

384 *Cullen and deMenocal* Cullen et al. 2000.
There's something going on Kerr 1998a.

385 *dust layer . . . was deposited during extremely dry conditions* Thompson, Mosley-Thompson et al. 2002.
Lake Van Weiss (2001) presents the widespread evidence for the 2200 B.C. dust event that is reviewed in this and the next paragraph.

386 *greatest drought Peru has experienced in the last seventeen thousand years* Kerr 1998a.
linked, inevitably, to changes in the North Atlantic deMenocal 2001, 670; deMenocal et al. 2000.
many stars were falling from the sky Weiss 1994, 719.

387 *well-scoured archeological landscapes* Weiss 2001, 89.
Indus Collapse Possehl 1994, 63. This author does not believe that drought caused this particular society to collapse. Weiss, however, counters that argument convincingly (1994, 2001).
Central Asia Hiebert 1994, 51.
Peru Kerr 1998a.
Barbara Bell Bell 1971.

388 *starved like a starved grasshopper* Bell 1971, 9.
Asiatics flooding Egypt in its First Dark Age Hassan 1994.

389 *disappearance of Kilimanjaro's ice fields* Thompson, Mosley-Thompson et al. 2002.
quaint curios of antiquity Moseley 2001.

391 *glacier that was very close to the first camp used by Sir Edmund Hillary* Danielson 2002.
snow on the mountain's very summit has thinned enough to make it about four feet shorter Watts 2004.
Worldwatch Institute Mastny 2000.
Glaciers in the Himalayas are receding faster Ramachandran 2001.
Chinese Glacier Inventory *People's Daily* 2004; Chinese Academy of Sciences 2004.
Yao, who led the scientific team, predicts People's Daily 2004; Watts 2004.

392 *Ganges* Ramachandran 2001.
Lonnie and Henry Brecher took another set of images Brecher and Thompson 1993; Thompson 2000; and subsequent discussions with Lonnie Thompson.

393 *understanding of the relationship between rising temperatures and rising seas has developed rapidly* Miller and Douglas 2004; Munk 2002; Meier and Wahr 2002.
Alaska Arendt et al. 2002; Meier and Dyurgerov 2002.
Patagonia Rignot, Rivera, and Casassa 2003.

394 *global warming time bomb* Hansen 2003, 2004. (Also: O'Neill and Oppenheimer 2002; Oppenheimer 1998; Zwally et al. 2002; De Angelis and Skvarca 2003; Shepherd et al. 2003; Kaiser 2003; Schiermeier 2004.)

REFERENCES

Abramson, Rudy. 1992. Drilling for frozen secrets. *Los Angeles Times,* 2 January.

Agrawala, Shardul. 1997. Explaining the evolution of the IPCC structure and process. ENRP Discussion Paper E-97-05, Kennedy School of Government, Harvard University.

Alley, Richard B. 2000. *The Two-Mile Time Machine: Ice-Cores, Abrupt Climate Change, and Our Future.* Princeton, N.J.: Princeton University Press.

Altmann, J., S. C. Alberts, S. A. Altmann, and S. B. Roy. 2002. Dramatic change in local climate patterns in the Amboseli Basin, Kenya. *African Journal of Ecology* 40:248–51.

Alvarez, Walter. 1998. *T. Rex and the Crater of Doom.* New York: Vintage/Random House.

American Geophysical Union (AGU). 2003. Leading climate scientists reaffirm view that late 20th century warming was unusual and resulted from human activity. Press Release. Washington, D.C.: American Geophysical Union, 7 July. http://www.agu.org/sci_soc/prrl/prrl0319.html.

Archer, David. 2003. Who threw that snowball? *Science* 302:791–92.

Arendt, Anthony A., Keith A. Echelmeyer, William D. Harrison, Craig S. Lingle, and Virginia B. Valentine. 2002. Rapid wastage of Alaska glaciers and their contribution to rising sea level. *Science* 297:382–86.

Arnold, James R., and Ernest C. Anderson. 1957. The distribution of carbon-14 in nature. *Tellus* 9:28–32.

Arrhenius, Svante. 1896. On the influence of carbonic acid in the air upon the temperature of the ground. *London, Edinburgh, and Dublin Philosophical Magazine and Journal of Science* 41:237–76.

———. 1908. *Worlds in the Making: The Evolution of the Universe.* Translated by H. Borns. London: Harper.

Associated Press (AP). 2003. Last California district approves pact on Colorado River water, 4 October.

———. 2005. Report: Global warming at critical point, 24 January.

Augustin, Laurent, Carlo Barbante, Piers R. F. Barnes, Jean-Marc Barnola, Matthias Bigler, Emiliano Castellano, Olivier Cattani, Jerome Chappellaz, Dorthe Dahl-Jensen, Barbara Delmonte, Gabrielle Dreyfus, Gael Durand, Sonia Falourd, Hubertus Fischer, Jacqueline Flückiger, Margareta E. Hansson, Philippe Huybrechts, Gérard Jugie, Sigfus J. Johnsen, Jean Jouzel, Patrik Kaufmann, Josef Kipfstuhl, Fabrice Lambert, Vladimir Y. Lipenkov, Geneviève C. Littot, Antonio Longinelli, Reginald Lorrain, Valter Maggi, Valerie Masson-Delmotte, Heinz Miller, Robert Mulvaney, Johannes Oerlemans, Hans Oerter, Giuseppe Orombelli, Frederic Parrenin, David A. Peel, Jean-Robert Petit, Dominique Raynaud, Catherine Ritz, Urs Ruth, Jakob Schwander, Urs Siegenthaler, Roland Souchez, Bernhard Stauffer, Jorgen Peder Steffensen, Barbara Stenni, Thomas F. Stocker, Ignazio E. Tabacco, Roberto Udisti, Roderik S. W. van de Wal, Michiel van den Broeke, Jerome Weiss, Frank Wilhelms, Jan-Gunnar Winther, Eric W. Wolff, and Mario Zucchelli. 2004. Eight glacial cycles from an Antarctic ice core. *Nature* 429:623–28.

Aveni, Anthony F. 2000. *Between the Lines: The Mystery of the Giant Ground Drawings of Ancient Nasca, Peru.* Austin: University of Texas Press.

Baker, Paul A., Catherine A. Rigsby, Geoffrey O. Seltzer, Sherilyn C. Fritz, Tim K. Lowenstein, Niklas P. Bacher, and Carlos Veliz. 2001. Tropical climate changes at millennial and orbital timescales on the Bolivian Altiplano. *Nature* 409:698–701.

Baker, Paul A., Geoffrey O. Seltzer, Sherilyn C. Fritz, Robert B. Dunbar, Matthew J. Grove, Pedro M. Tapia, Scott L. Cross, Harold D. Rowe, and James P. Broda. 2001. The history of South American tropical precipitation for the past 25,000 years. *Science* 291:640–43.

Ball, Philip. 2003. Cities and fields make the world seem warmer. *Nature Science Update,* 29 May. http://www.nature.com/nsu/030527/030527-6.html.

Balmford, Andrew, Andrea Manica, Lesley Airey, Linda Birkin, Amy Oliver, and Judith Schleicher. 2004. Hollywood, climate change, and the public. *Science* 305:1713b.

Barber, D. C., A. Dyke, C. Hillaire-Marcel, A. E. Jennings, J. T. Andrews, M. W. Kerwin, G. Bilodeau, R. McNeely, J. Southon, M. D. Morehead, and J.-M. Gagnon. 1999. Forcing of the cold event of 8,200 years ago by catastrophic drainage of Laurentide lakes. *Nature* 400:344–48.

Barber, Valerie A., Glenn Patrick Juday, and Bruce P. Finney. 2000. Reduced growth of Alaskan white spruce in the twentieth century from temperature-induced drought stress. *Nature* 405:668–73.

Bard, Edouard, Maurice Arnold, Pierre Maurice, Josette Duprat, Jean Moyes, and Jean-Claude Duplessy. 1987. Retreat velocity of the North-Atlantic polar front during the last deglaciation determined by C-14 accelerator mass-spectrometry. *Nature* 328:791–94.

Bard, Edouard, Frauke Rostek, Jean-Louis Turon, and Sandra Gendreau. 2000. Hydrological impact of Heinrich events in the subtropical northeast Atlantic. *Science* 289:1321–24.

Barnola, J. M., D. Raynaud, Y. S. Korotkevich, and C. Lorius. 1987. Vostok ice core provides 160,000-year record of atmospheric CO_2. *Nature* 329:408–14.

Bar-Yosef, Ofer. 2000. The impact of radiocarbon dating on Old World archaeology: Past achievements and future expectations. *Radiocarbon* 42:23–39.

Basinger, Julianne. 2001. Research at what cost? *Chronicle of Higher Education,* 27 July.

Bawden, Garth. 1996. *The Moche.* Cambridge: Blackwell.

Becker, Elizabeth. 2004a. Acting conciliatory, U.S. seeks to revive global trade talks. *New York Times,* 12 January.

———. 2004b. Lawmakers voice doom and gloom on W.T.O. ruling. *New York Times,* 28 April.

———. 2004c. W.T.O. rules against U.S. on cotton subsidies. *New York Times,* 27 April.

Becker, Elizabeth, and Todd Benson. 2004. Brazil's road to victory over U.S. cotton. *New York Times,* 4 May.

Bell, Barbara. 1971. The Dark Ages in ancient history I. The first Dark Age in Egypt. *American Journal of Archaeology* 75:1–26.

Belluck, Pam, and Andrew C. Revkin. 2001. A mittenless autumn, for better and worse. *New York Times,* 23 December.

Bender, Michael L., T. Sowers, and L. Labeyrie. 1994. The Dole effect and its variations during the last 130,000 years as measured in the Vostok ice core. *Global Biogeochemical Cycles* 8:363–76.

Berger, André. 1978a. Long-term variations of caloric insolation resulting from the Earth's orbital elements. *Quaternary Research* 9:139–67.

———. 1978b. Long-term variations of daily insolation and Quaternary climatic changes. *Journal of Atmospheric Science* 35:2362–67.

Berger, André, and Marie France Loutre. 2002. An exceptionally long interglacial ahead? *Science* 297:1287–88.

Bernbaum, Edwin. 2001. *The Way to Shambhala: A Search for the Mythical Kingdom Beyond the Himalayas.* Boston: Shambhala.

Binford, Michael W., Alan L. Kolata, Mark Brenner, John W. Janusek, Matthew T. Seddon, Mark Abbot, and Jason H. Curtis. 1997. Climate variation and the rise and fall of an Andean civilization. *Quaternary Research* 47:235–48.

Blair, Thomas A. 1942. *Climatology, general and regional.* New York: Prentice-Hall.

Blasing, T. J., and Sonja Jones. 2003. Current greenhouse gas concentrations. Carbon Dioxide Information Analysis Center. Oak Ridge National Laboratory. Oak Ridge: U.S. Department of Energy, November. http://cdiac.esd.ornl.gov/pns/current_ghg.html.

Bloomberg News. 2004. Brazil wins rulings on 2 trade issues. *New York Times,* 9 September.

Blunier, Thomas, J. Chappellaz, J. Schwander, A. Dällenbach, B. Stauffer, T. F. Stocker, D. Raynaud, J. Jouzel, H. B. Clausen, C. U. Hammer, and S. J. Johnsen. 1998. Asynchrony of Antarctic and Greenland climate change during the last glacial period. *Nature* 394:739–43.

Bolin, Bert. 1998. The Kyoto negotiations on climate change: A science perspective. *Science* 279:330–31.

Bond, Gerard C., Bernd Kromer, Juerg Beer, Raimund Muscheler, Michael N. Evans, William Showers, Sharon Hoffmann, Rusty Lotti-Bond, Irka Hajdas, and Georges Bonani. 2001. Persistent solar influence on North Atlantic climate during the Holocene. *Science* 294:2130–36.

Bond, Gerard C., and Rusty Lotti. 1995. Iceberg discharges into the North Atlantic on millennial timescales during the last glaciation. *Science* 267:1005–10.

Bond, Gerard C., William Showers, Maziet Cheseby, Rusty Lotti, Peter Almasi, Peter deMenocal, Paul Priore, Heidi Cullen, Irka Hajdas, and Georges Bonani. 1997. A pervasive millennial-scale cycle in North Atlantic Holocene and glacial climates. *Science* 278:1257–66.

Bradley, Raymond S. 1999. *Paleoclimatology: Reconstructing Climates of the Quaternary.* 2d ed. San Diego: Academic Press.

Bradley, Raymond S., Keith R. Briffa, Thomas J. Crowley, Malcolm K. Hughes, Philip

D. Jones, and Michael E. Mann. 2001. The scope of medieval warming. (Letter) *Science* 292:2011b–12b.

Bradley, Raymond S., Malcolm K. Hughes, and Henry F. Diaz. 2003. Climate in medieval time.

Bradley, Raymond S., Mathias Vuille, Douglas Hardy, and Lonnie G. Thompson. 2003. Low latitude ice cores record Pacific sea surface temperatures. *Geophysical Research Letters* 30:1174–77.

Brecher, Henry H., and Lonnie G. Thompson. 1993. Measurement of the retreat of Qori Kalis in the tropical Andes of Peru by terrestrial photogrammetry. *Photogrammetric Engineering and Remote Sensing* 59:1017–22.

Bridges, Roy C. 1968. Introd. to *Travels, Researches, and Missionary Labors During an Eighteen Years' Residence in Eastern Africa Together with Journeys to Jagga, Usambara, Ukambani, Shoa, Abessinia, and Khartoum And a Coasting Voyage from Mombaz to Cape Delgado,* by Johann Ludwig Krapf. 2d ed. London: Frank Cass.

Broder, John M. 2004a. Los Angeles mayor seeks to freeze valley growth. *New York Times,* 8 August.

———. 2004b. Spun and unspun tales of a California cotton king. *New York Times,* 8 January.

Broecker, Wallace S. 1965. Isotope geochemistry and the Pleistocene climatic record. In *The Quarternary of the United States*. Edited by H. E. Wright, Jr., and D. G. Frey, 737–53. Princeton, N.J.: Princeton University Press.

———. 1975. Climatic change: Are we on the brink of a pronounced global warming? *Science* 189:460–64.

———. 1995a. Chaotic climate. *Scientific American,* November, 62–68.

———. 1995b. Cooling the tropics. *Nature* 376:212–13.

———. 1996. Glacial climate in the tropics. *Science* 272:1902–4.

———. 1997a. Mountain glaciers: Recorders of atmospheric water vapor content? *Global Biogeochemical Cycles* 11:589–97.

———. 1997b. Thermohaline circulation, The Achilles' Heel of our climate system: Will manmade CO_2 upset the current balance? *Science* 278:1582–88.

———. 1997c. Will our ride into the green-house future be a smooth one? *GSA Today* 5:1–7.

———. 1998. Paleocean circulation during the last deglaciation: A bipolar seesaw? *Paleoceanography* 13:119–21.

———. 1999a. Climate change prediction. *Science* 283:175f.

———. 1999b. What if the conveyor were to shut down? Reflections on a possible outcome of the great global experiment. *GSA Today* 9:1–7.

———. 2000. Dust: Climate's Rosetta Stone. Manuscript. Submitted to *Key Reporter,* newsletter of the Phi Beta Kappa Society. Kindly provided to the author by Professor Broecker.

———. 2001. Was the medieval warm period global? *Science* 291:1497–99.

———. 2004. Future global warming scenarios. *Science* 304:388b.

Broecker, Wallace S., and Sidney Hemming. 2001. Climate swings come into focus. *Science* 294:2308–9.

Broecker, Wallace S., Michael E. Schlesinger, James Risbey, and Richard A. Kerr. 1989. Hansen and the greenhouse effect. (Letter) *Science* 245:451–52.

Broecker, Wallace S., David L. Thurber, John Goddard, Teh-Lung Ku, R. K. Matthews, and Kenneth J. Mesolella. 1968. Milankovitch hypothesis supported by precise dating of coral reefs and deep-sea sediments. *Science* 159:297–300.

Broecker, Wallace S., and J. van Donk. 1970. Insolation changes, ice volumes, and the ^{18}O record in deep-sea cores. *Reviews of Geophysics and Space Physics* 8:169–97.
Brooke, James. 2001. Story of Viking colonies' icy 'Pompeii' unfolds from ancient Greenland farm, *New York Times,* 8 May.
Brown, Harrison. 1954. *The Challenge of Man's Future*. New York: Viking.
Budyko, Mikhail I. 1974. *Climate and Life*. Translated by David H. Miller. New York, London: Academic Press.
Burton, Richard F., Sir. [1860] 1961. *The Lake Regions of Central Africa*. Repr., with an introd. by Alan Moorehead. New York: Horizon.
Callendar, Guy S. 1938. The artificial production of carbon dioxide and its influence on temperature. *Quarterly Journal of the Royal Meteorological Society* 64:223–40.
———. 1939. The composition of the atmosphere through the ages. *Meteorological Magazine* 74:33–39.
———. 1940. Variations of the amount of carbon dioxide in different air currents. *Quarterly Journal of the Royal Meteorological Society* 66:395–400.
———. 1941. Infra-red absorption of carbon dioxide, with special reference to atmospheric radiation. *Quarterly Journal of the Royal Meteorological Society* 67:263–75.
———. 1949. Can carbon dioxide influence climate? *Weather* 4:310–14.
———. 1957. The effect of fuel combustion on the amount of carbon dioxide in the atmosphere. *Tellus* 9:421–22.
———. 1958a. On the amount of carbon dioxide in the atmosphere. *Tellus* 10:243–48.
———. 1958b. On the present climatic fluctuation. *Meteorological Magazine* 87:204–7.
———. 1961. Temperature fluctuations and trends over the Earth. *Quarterly Journal of the Royal Meteorological Society* 87:1–11.
Calvin, William H. 1998. The great climate flip-flop. *Atlantic Monthly,* January, 47–64.
———. 2002. *A Brain for All Seasons: Human Evolution and Abrupt Climate Change*. Chicago: University of Chicago Press.
Cane, Mark A. 1998. A role for the tropical Pacific. *Science* 282:59–61.
———. 2000. Understanding and predicting the world's climate system. In *Applications of Seasonal Climate Forecasting in Agricultural and Natural Ecosystems: The Australian Experience*. Edited by G. Hammer, 29–50. Netherlands: Kluwer.
Cane, Mark A., and Michael Evans. 2000. Do the tropics rule? *Science* 290:1107–8.
Cane, Mark A., and Peter Molnar. 2001. Closing of the Indonesian seaway as a precursor to East African aridification around 3–4 million years ago. *Nature* 411:157–62.
Carson, Rachel. 1951. *The Sea Around Us*. New York: Oxford University Press.
Chinese Academy of Sciences. 2004. CAS researchers completed the work on Chinese Glacier Inventory. CAS news item. 17 September. http://english.cas.ac.cn/eng2003/news/detailnewsb.asp?InfoNo=25162.
Christianson, Gale E. 1999. *Greenhouse: The 200-year Story of Global Warming*. New York: Walker.
Cicerone, Ralph J. 2000. Human forcing of climate change: Easing up on the gas pedal. *Proceedings of the National Academy of Sciences of the United States of America* 97:10304–6.
Clapperton, C. 1993. *Quaternary Geology and Geomorphology of South America*. Amsterdam: Elsevier.
Clarke, Garry, David Leverington, James Teller, and Arthur Dyke. 2003. Superlakes, megafloods and abrupt climate change. *Science* 301:922–23.
Clement, Amy C., and Mark A. Cane. 1999. A role for the tropical Pacific: Coupled

ocean-atmosphere system on Milankovitch and millennial timescales. Part I: A modeling study of tropical Pacific variability. In *Mechanisms of Global Climate Change at Millennial Timescales*. Edited by P. U. Clark, R. S. Webb, and L. D. Keigwin, 363–71. Washington, D.C.: American Geophysical Union.

Clement, Amy C., A. Hall, and A. J. Broccoli. 2004. The importance of precessional signals in the tropical climate. *Climate Dynamics* 22:327–41.

Clement, Amy C., R. Seager, and M. A. Cane. 2000. Suppression of El Niño during the mid-Holocene by changes in the Earth's orbit. *Paleoceanography* 15:731–37.

Cole-Dai, Jihong, Ellen Mosley-Thompson, Shawn P. Wight, and Lonnie G. Thompson. 2000. A 4,100-year record of explosive volcanism from an East Antarctica ice core. *Journal of Geophysical Research* 105, no. D19:24, 431–41.

Columbia University. Columbia Earth Institute. 1997. Leading climate change expert warns of possible collapse of Earth's ocean systems. Press release, 27 November.

Cordell, Linda. 2001. Aftermath of chaos in the pueblo southwest. In *Environmental Disaster and the Archeology of Human Response*. Edited by Garth Bawden and Richard M. Reycraft. Albuquerque: University of New Mexico, Maxwell Museum of Anthropology.

Craig, Harmon. 1957. The natural distribution of radiocarbon and the exchange time of carbon dioxide between atmosphere and the sea. *Tellus* 9:1–17.

Crowley, Thomas J. 2000. Causes of climate change over the past 1,000 years. *Science* 289:270–77.

Crutzen, Paul J., and Eugene F. Stoermer. 2000. The "Anthropocene." *IGBP Newsletter* (Stockholm) 41:17–18. http://diotima.mpch-mainz.mpg.de/~air/anthropocene/.

Crutzen, Paul J., V. Ramanathan, Theodore L. Anderson, Robert J. Charlson, Stephen E. Schwartz, Reto Knutti, Olivier Boucher, Henning Rodhe, and Jost Heintzenberg. 2003. The parasol effect on climate. *Science* 302:1679–81.

Cullen, Heidi M., Peter B. deMenocal, S. Hemming, G. Hemming, F. H. Brown, T. Guilderson, and F. Siroco. 2000. Climate change and the collapse of the Akkadian empire: Evidence from the deep sea. *Geology* 28:379–82.

Danielson, Stentor. 2002. Everest melting? High signs of climate change. *National Geographic News,* 5 June. http://news.nationalgeographic.com/news/2002/06/0605_020604_everestclimate.html.

Dansgaard, Willi. 1964. Stable isotopes in precipitation. *Tellus* 16:436–47.

Dansgaard, Willi, S. J. Johnsen, H. B. Clausen, and C. C. J. Langway. 1971. Climatic record revealed by the Camp Century ice core. In *The Late Cenozoic Glacial Ages*. Edited by Karl K. Turekian, 37–56. New Haven: Yale University Press.

Davis, Mary E., Lonnie G. Thompson, Ellen Mosley-Thompson, Ping-Nan Lin, Victor N. Mikhalenko, and Jihong Dai. 1995. Recent ice core climate records from the Cordillera Blanca, Peru. *Annals of Glaciology* 21:225–30.

De Angelis, Hernán, and Pedro Skvarca. 2003. Glacier surge after ice shelf collapse. *Science* 299:1560–62.

deMenocal, Peter B. 1995. Plio-Pleistocene African climate. *Science* 270:53–59.

———. 2001. Cultural responses to climate change during the late Holocene. *Science* 292:667–73.

deMenocal, Peter B., Joseph Ortiz, Tom Guilderson, and Michael Sarnthein. 2000. Coherent high- and low-latitude climate variability during the Holocene warm period. *Science* 288:2198–202.

Denevan, William M. 1970. Aboriginal drained-field cultivation in the Americas. *Science* 169:647–54.

Diamond, Jared. 1997a. *Guns, Germs, and Steel: The Fates of Human Societies*. New York: Norton.

———. 1997b. Location, location, location: The first farmers. *Science* 278:1243–44.

Donnan, Christopher B. 1990. An assessment of the validity of the Naymlap dynasty. In *The Northern Dynasties: Kingship and Statecraft in Chimor: A Symposium at Dumbarton Oaks, 12th and 13th October 1985*. Edited by Michael E. Moseley and Alana Cordy-Collins, 243–74. Washington, D.C.: Dumbarton Oaks Research Library and Collection.

Downie, Charles D., W. Humphries, W. H. Wilcockson, and P. Wilkinson. 1956. Geology of Kilimanjaro. *Nature* 178:828–30.

Dunbar, Robert B. 2003. Leads, lags and the tropics. *Nature* 421:121–22.

Dyurgerov, Mark B., and Mark F. Meier. 2000. Twentieth-century climate change: Evidence from small glaciers. *Proceedings of the National Academy of Sciences of the United States of America* 97:1406–11.

EastAfrican (Kenya). 1999. Tanzania hardest hit as famine stalks E. Africa, 12 January.

Egan, Timothy. 2001. Near vast bodies of water, land lies parched. *New York Times,* 12 August.

———. 2002. As trees die, some cite the climate, *New York Times,* 25 June.

Environmental Defense. 2003. Global warming: The history of an international scientific consensus. New York: Environmental Defense, January. http://www.environmentaldefense.org/documents/381_FactSheet_globalwarming_timeline.pdf.

Epstein, Paul R. 1999. Climate and health. *Science* 285:347–48.

Eve, A. S., and C. H. Creasey. 1945. *The Life and Work of John Tyndall*. London: Macmillan.

Fagan, Brian. 1999. *Floods, Famines, and Emperors: El Niño and the Fate of Civilization*. New York: Basic Books.

Feibel, Craig S. 1997. Debating the environmental factors in hominid evolution. *GSA Today* 7:1–7.

Fleming, James Rodger. 1998. *Historical Perspectives on Climate Change*. New York: Oxford University Press.

Fourier, Joseph. 1824. Remarques générals sur les températures du globe terrestre et des espaces planétaires. *Ann. chim. phys.* (Paris) 27:136–67. (English translation: Burgess, Ebeneser. 1837. *American Journal of Science* 32:1–20.)

Gaffen, Dian J., Benjamin D. Santer, James S. Boyle, John R. Christy, Nicholas E. Graham, and Rebecca J. Ross. 2000. Multidecadal changes in the vertical temperature structure of the tropical troposphere. *Science* 287:1242–45.

Gavaghan, Helen. 2001. Researchers plan probe into Antarctic lakes. *Nature* 414:573.

Gelbspan, Ross. 1997. *The Heat Is On: The High-Stakes Battle over Earth's Threatened Climate*. Reading, Mass.: Addison-Wesley.

———. 1998. A good climate for investment. *Atlantic Monthly,* June.

Genthon, Christophe, J. M. Barnola, D. Raynaud, C. Lorius, J. Jouzel, N. I. Barkov, Y. S. Korotkevich, and V. M. Kotyakov. 1987. Vostok ice core: Climatic response to CO_2 and orbital forcing changes over the last climatic cycle. *Nature* 329:414–18.

Giannini, Alessandra, R. Saravanan, and P. Chang. 2003. Oceanic forcing of Sahel rainfall on interannual to interdecadal time scales. *Science* 302:1027–30.

Glanz, James. 2002. New twist on physicist's role in Nazi bomb. *New York Times,* 7 February, 1.

Gleick, James. 1987. *Chaos: Making a New Science*. New York: Viking.

Gleick, Peter H. 2000. *The World's Water 2000–2001: The Biennial Report on Freshwater Resources.* Washington, D.C.: Island.

———. 2003. Global freshwater resources: Soft-path solutions for the 21st century. *Science* 302:1524–28.

Goldman, Erica. 2002. Even in the high Arctic, nothing is permanent. *Science* 297:1493a–94.

Gore, Al. 1992. *Earth in the Balance: Ecology and the Human Spirit.* Boston: Houghton Mifflin.

Greening Earth Society. 2004. Snow Fooling! *World Climate Alert,* no. 14a (March). http://www.greeningearthsociety.org/wca/2004/wca_14a.html.

Griffin, Lindsay. 1999. High Mountain Info. *High Mountain Sports* (Sheffield, UK), July.

Guilderson, Thomas P., Richard G. Fairbanks, and James L. Rubenstone. 1994. Tropical temperature variations since 20,000 years ago: Modulating interhemispheric climate change. *Science* 263:663–65.

H. W. Wilson Company. 1996. Hansen, James E. in *Current Biography.* New York: H. W. Wilson.

Haffer, Jürgen. 1969. Speciation in Amazonian forest birds. *Science* 165:131–36.

Hall, Alan. 1998. Going, Going—Gone? *Scientific American,* August.

Hansen, James E. 1996. Climatic changes: Understanding global warming. In *One World: The Health and Survival of the Human Species in the 21st Century.* Edited by Robert Lanza, M.D., 173–90. Santa Fe, N.M.: Health Press.

———. 2003. Can we defuse the global warming time bomb? *naturalScience,* 1 August. http://pubs.giss.nasa.gov/docs/2003/2003_Hansen.pdf.

———. 2004. Defusing the global warming time bomb. *Scientific American,* March. http://pubs.giss.nasa.gov/docs/2004/2004_Hansen1.pdf.

Hansen, James E., I. Fung, A. Lacis, D. Rind, S. Lebedeff, R. Ruedy, G. Russell, and P. Stone. 1988. Global climate changes as forecast by Goddard Institute for Space Studies three-dimensional model. *Journal of Geophysical Research* 93:9341–64.

Hansen, James E., D. Johnson, A. Lacis, S. Lebedeff, P. Lee, D. Rind, and G. Russell. 1981. Climate impact of increasing atmospheric carbon dioxide. *Science* 213:957–66.

Hardin, Garrett. 1993. *Living Within Limits: Economies and Population Taboos.* Oxford: Oxford University Press.

Harvard Medical School. The Center for Health and the Global Environment. 1999. Extreme weather events: The health and economic consequences of the 1997/98 El Niño and La Niña. Report, January. http://www.heedmd.org/ensowebsite/disease.html.

Hassan, Fekri A. 1994. Nile floods and political disorder in early Egypt. In *Third Millennium* BC *Climate Change and Old World Collapse.* Edited by H. Nüzhet Dalfes, George Kukla, and Harvey Weiss. NATO ASI Series I. Global Environmental Change. Vol. 49. Berlin, New York: Springer-Verlag.

Hastenrath, Stefan, and Lawrence Greischar. 1997. Glacier recession on Kilimanjaro, East Africa, 1912–89. *Journal of Glaciology* 43:455–59.

Hays, James D., John Imbrie, and Nicholas J. Shackleton. 1976. Variations in the Earth's orbit: Pacemaker of the ice ages. *Science* 194:1121–32.

Henderson, Gideon M., and Niall C. Slowey. 2000. Evidence from U-Th dating against northern hemisphere forcing of the penultimate deglaciation. Nature 404:61–66.

Hertsgaard, Mark. 2004. Weathering the Crisis. *Nation,* 24 February.

Heun, Manfred, Ralf Schäfer-Pregl, Dieter Klawan, Renato Castagna, Monica Accerbi,

Basilio Borghi, and Francesco Salamini. 1997. Site of einkorn wheat domestication identified by DNA fingerprinting. *Science* 278:1312–14.

Hiebert, Fredrik T., 1994. Bronze Age central Eurasian cultures in their steppe and desert environments. In *Third Millennium* BC *Climate Change and Old World Collapse.* Edited by H. Nüzhet Dalfes, George Kukla, and Harvey Weiss. NATO ASI Series I. Global Environmental Change. Vol. 49. Berlin, New York: Springer-Verlag.

Hileman, Bette. 2003. Climate change. *Chemical and Engineering News* 81, no. 50:27–37. http://pubs.acs.org/cen/coverstory/8150/8150climatechange.html.

Hoffman, Paul F., Alan J. Kaufman, Galen P. Halverson, and Daniel P. Schrag. 1998. A Neoproterozoic Snowball Earth. *Science* 281:1342–46.

Hoffman, Paul F., and Daniel P. Schrag. 2000. The Snowball Earth. *Scientific American,* January, 68–75.

Houtermans, J., Hans E. Suess, and Walter Munk. 1967. Effect of industrial fuel combustion on the carbon-14 level of atmospheric CO_2. In *Radioactive Dating and Methods of Low-Level Counting. Proceeedings of a Symposium, Monaco, 1967.* Vienna: International Atomic Energy Agency.

Hughes, Malcolm K., and Gary Funkhouser. 1998. Extremes of moisture availability reconstructed from tree rings for recent millennia in the Great Basin of Western North America. In *The Impacts of Climate Variability on Forests.* Edited by Martin Beniston and John L. Innes, 57–68. Berlin, New York: Springer.

Hughes, Malcolm K., Gary Funkhouser, and Ni Fenbiao. 2002. The ancient Bristlecone pines of Methuselah Walk, California, as a natural archive of past environment. *PAGES News* (Bern) 10, no. 1:16–17.

Imbrie, John, and John Z. Imbrie. 1980. Modeling the climatic response to orbital variations. *Science* 207:943–53.

Imbrie, John, and Katherine Palmer Imbrie. 1979. *Ice Ages: Solving the Mystery.* Short Hills, N. J.: Enslow.

Institute for Public Policy Research (London). 2005. G8-plus group needed to tackle climate change. Press release, 24 January.

Intergovernmental Panel on Climate Change (IPCC). 2001a. *Climate Change 2001: Impacts, Adaptation, and Vulnerability. Contribution of Working Group II to the Third Assessment Report of the Intergovernmental Panel on Climate Change.* Edited by James J. McCarthy, Osvaldo F. Canziani, Neil A. Leary, David J. Dokken, and Kasey S. White. Cambridge: Cambridge University Press. http://www.grida.no/climate/ipcc_tar/wg2/index.htm.

———. 2001b. *Climate Change 2001: The Scientific Basis. Contribution of Working Group I to the Third Assessment Report of the Intergovernmental Panel on Climate Change.* Edited by John T. Houghton, Y. Ding, D. J. Griggs, M. Noguer, P. J. van der Linden, X. Dai, K. Maskell, and C. A. Johnson. Cambridge: Cambridge University Press. http://www.grida.no/climate/ipcc_tar/wg1/index.htm.

International Research Institute for Climate Prediction. Linking Southern Oscillation with El Niño. http://iri.ldeo.columbia.edu/climate/dictionary/enso/stage2.html (accessed in 2000; site now discontinued).

Jehl, Douglas. 2002a. Arkansas rice farmers run dry, and U.S. remedy sets off debate. *New York Times,* 11 November.

———. 2002b. Atlanta's growing thirst creates water war. *New York Times,* 27 May.

———. 2002c. In race to tap the Euphrates, the upper hand is upstream. *New York Times,* 25 August.

———. 2002d. Saving water, U.S. farmers are worried they'll parch. *New York Times,* 28 August.

———. 2002e. U.S. approves water plan in California, but environmental opposition remains. *New York Times,* 31 August.

———. 2003a. A new frontier in water wars emerges in East. *New York Times,* 3 March.

———. 2003b. As cities move to privatize water, Atlanta steps back. *New York Times,* 10 February.

Jehl, Douglas, and Andrew C. Revkin. 2001. Bush, in reversal, won't seek cut in emissions of carbon dioxide. *New York Times,* 14 March.

Johnson, Kirk, and Dean E. Murphy. 2004. Drought settles in, lake shrinks and West's worries grow. *New York Times,* 2 May.

Johnston, Sir Henry Hamilton. 1885. The Kilima-njaro expedition. *Proceedings of the Royal Geographical Society* 7, no. 3, conf. 137.

Jouzel, Jean, C. Lorius, J. R. Petit, C. Genthon, N. I. Barkov, V. M. Kotlyakov, and V. M. Petrov. 1987. Vostok ice core: A continuous isotope temperature record over the last climatic cycle (160,000 years). *Nature* 329:403–8.

Kaiser, Jocelyn. 2002. Breaking up is far too easy. *Science* 297:1494–96.

———. 2003. Warmer ocean could threaten Antarctic ice shelves. *Science* 302:759a.

Kalnay, Eugenia, and Ming Cai. 2003. Impact of urbanization and land-use change on climate. *Nature* 423:528–31.

Kaplan, Robert D. 1994. The coming anarchy. *Atlantic Monthly,* February.

Kaser, Georg, D. Hardy, T. Mölg, R. S. Bradley, and T. M. Hyera. 2004. Modern glacial retreat on Kilimanjaro as evidence of climate change: Observations and facts. *International Journal of Climatology* 24:329–39.

Keeling, Charles David. 1958. The concentration and isotopic abundances of atmospheric carbon dioxide in rural areas. *Geochimica et Cosmochimicha Acta* 13:322–34.

———. 1960. The concentration and isotopic abundances of carbon dioxide in the atmosphere. *Tellus* 12:200–203.

———. 1961. The concentration and isotopic abundances of carbon dioxide in rural and marine air. *Geochimica et Cosmochimicha Acta* 24:277–98.

———. 1978. The influence of Mauna Loa observatory on the development of atmospheric CO_2 research. *In Mauna Loa Observatory: A 20th Anniversary Report.* Edited by John Miller. National Oceanic and Atmospheric Administration special report. Silver Spring: Air Resources Laboratories. http://www.mlo.noaa.gov/HISTORY/Fhistory.htm.

Keeling, Charles David, A. Adams, C. A. Ekdahl, Jr., and P. R. Guehther. 1976. Atmospheric carbon dioxide variations at the South Pole. *Tellus* 28:552–64.

Keeling, Charles David, R. B. Bacastow, A. E. Bainbridge, C. A. Ekdahl, Jr., P. R. Guenther, L. S. Waterman, and J. F. S. Chin. 1976. Atmospheric carbon dioxide variations at Mauna Loa Observatory, Hawaii. *Tellus* 28:538–51.

Keeling, Charles David, and Tim Whorf. 2004. Atmospheric CO_2 records from sites in the SIO air sampling network. In *Trends: A Compendium of Data on Global Change.* Carbon Dioxide Information Analysis Center. Oak Ridge National Laboratory. Oak Ridge: U.S. Department of Energy. http://cdiac.esd.ornl.gov/new/keel_page.html and http://cdiac.esd.ornl.gov/trends/co2/sio-mlo.htm.

Kennedy, Donald. 2003. An epidemic of politics. *Science* 299:625.

Kerr, Richard. 1989. Hansen vs. the world on the greenhouse threat. *Science* 244:1041–43.

———. 1993a. How ice age climate got the shakes. *Science* 260:890–92.

———. 1993b. The whole world had a case of the ice age shivers. *Science* 262:1972–73.

———. 1994. Ancient tropical climates warm San Francisco gathering. *Science* 263: 173–75.

———. 1995. Chilly ice-age tropics could signal climate sensitivity. *Science* 267:961.

———. 1998a. Sea-floor dust shows drought felled Akkadian empire. *Science* 279: 325–26.

———. 1998b. Warming's unpleasant surprise: Shivering in the greenhouse? *Science* 281:156–58.

———. 2000a. An appealing Snowball Earth that's still hard to swallow. *Science* 287:1734–36.

———. 2000b. Ice, mud point to CO_2 role in glacial cycle. *Science* 289:1868.

———. 2001. The tropics return to the climate system. *Science* 292:660–61.

———. 2002a. A warmer Arctic means change for all. *Science* 297:1490–93.

———. 2002b. Mild winters mostly hot air, not Gulf Stream. *Science* 297:2202.

———. 2003a. Tropical Pacific a key to deglaciation. *Science* 299:183–84.

———. 2003b. Warming Indian Ocean wringing moisture from the Sahel. *Science* 302:210a–211.

———. 2004a. A few good climate shifters. *Science* 306:599–601.

———. 2004b. Mars: Opportunity tells a salty tale. *Science* 303:1957b.

———. 2004c. Three degrees of consensus. *Science* 305:932–34.

———. 2005. Climate modelers see scorching future as a real possibility. *Science* 307:497a.

Keys, David. 2000. *Catastrophe: An Investigation into the Origins of the Modern World.* New York: Ballantine.

Kimambo, Isaria N., and Arnold J. Temu, eds. 1969. *A History of Tanzania.* Nairobi: East African Publishing House.

Kininmonth, W. 1999. *The 1997–98 El Niño Event: A Scientific and Technical Retrospective.* Geneva: World Meteorological Organization.

Knorr, Gregor, and Gerrit Lohmann. 2003. Southern Ocean origin for the resumption of Atlantic thermohaline circulation during deglaciation. *Nature* 424:532–33.

Kolata, Alan L. 1993. *The Tiwanaku: Portrait of an Andean Civilization.* Cambridge: Blackwell.

———. 1996. *Valley of the Spirits: A Journey into the Lost Realm of the Aymara.* New York: Wiley.

———. 2001. Environmental thresholds and the 'natural history' of an Andean civilization. In *Environmental Disaster and the Archeology of Human Response.* Edited by Garth Bawden and Richard M. Reycraft, 163–78. Albuquerque: University of New Mexico, Maxwell Museum of Anthropology.

Krajick, Kevin. 2002. Ice man: Lonnie Thompson scales the peaks for science. *Science* 298:518–22.

Krapf, Johann Ludwig. [1860] 1968. *Travels, Researches, and Missionary Labors During an Eighteen Years' Residence in Eastern Africa Together with Journeys to Jagga, Usambara, Ukambani, Shoa, Abessinia, and Khartoum And a Coasting Voyage from Mombaz to Cape Delgado.* 2d ed. with an introd. by Roy C. Bridges. London: Frank Cass.

Lafferty, Michael B. 1994. Melting glaciers: Expert fears loss of climate data. *Columbus Dispatch,* 17 June, Metro section, 1c.

Lamb, Hubert H. 1965. The early medieval warm epoch and its sequel. *Palaeogeography, Palaeoclimatology, Palaeoecology* 1:13–37.

Langley, Samuel Pierpont. 1900. *Annals of the Astrophysical Observatory of the Smithsonian Institution,* vol. 1. Washington: Government Printing Office. http://ads.harvard.edu/books/saoann/.

Lapham, Lewis. 2002. Spoils of war. *Harper's,* March.

Lavietes, Stuart. 2003. Two archaeologists, Robert Braidwood and his wife, Linda Braidwood, Die. *New York Times,* 17 January.

Lawren, Bill. 1989. Harmon Craig: Stalking excellence, leaving controversy in his wake. *The Scientist* 3, no. 8:1.

Leary, Warren E. 2002. Flurry of satellites to monitor earth and examine galaxy. *New York Times,* 9 December.

———. 2004. Scientists Report Evidence of Saltwater Pools on Mars. *New York Times,* 24 March.

Leslie, Jacques. 2000. Running dry: What happens when the world no longer has enough freshwater? *Harper's,* July.

Levitus, Sydney, John I. Antonov, Timothy P. Boyer, and Cathy Stephens. 2000. Warming of the world ocean. *Science* 287:2225–29.

Linthicum, Kenneth J., Assaf Anyamba, Compton J. Tucker, Patrick W. Kelley, Monica F. Myers, and Clarence J. Peters. 1999. Climate and satellite indicators to forecast Rift Valley fever epidemics in Kenya. *Science* 285:397–400.

Lopez, Barry. 1986. *Arctic Dreams: Imagination and Desire in a Northern Landscape.* New York: Scribner's.

Lorius, Claude, J. Jouzel, D. Raynaud, J. Hansen, and H. Le Treut. 1990. The ice core record: Climate sensitivity and future greenhouse warming. *Nature* 347:139–45.

Malone, Thomas F., Edward D. Goldberg, and Walter H. Munk. 1998. Roger Randall Dougan Revelle, March 7, 1909–July 15, 1991. In *Biographical Memoirs,* vol. 75, 288–309. Washington, D.C.: National Academy of Sciences.

Manabe, Syukuro, and Richard T. Wetherald. 1967. Thermal equilibrium of the atmosphere with a given distribution of relative humidity. *Journal of the Atmospheric Sciences* 24:241–59.

———. 1975. The effects of doubling the CO_2 concentration on the climate of a general circulation model. *Journal of the Atmospheric Sciences* 32:3–15.

Manley, G. 1965. Dr. Tom Longstaff. Obituary. *Quarterly Journal of the Royal Meteorological Society* 91:559.

Mann, Michael E. 2000. Lessons for a new millennium. *Science* 289:253–54.

Mann, Michael E., Caspar M. Ammann, Raymond S. Bradley, Keith R. Briffa, Thomas J. Crowley, Malcolm K. Hughes, Philip D. Jones, Michael Oppenheimer, Tim J. Osborn, Jonathan T. Overpeck, Scott Rutherford, Kevin E. Trenberth, and Thomas M. L. Wigley. 2003. On past temperatures and anomalous late 20th century warmth. *Eos* 84:256–58.

Mann, Michael E., Raymond S. Bradley, and Malcolm K. Hughes. 1999. Northern hemisphere temperatures during the past millennium: Inferences, uncertainties, and limitations. *Geophysical Research Letters* 26:759–62.

Mann, Michael E., and Philip D. Jones. 2003. Global surface temperatures over the past two millennia. *Geophysical Research Letters* 30:1820ff.

Marland, Gregg, Tom A. Boden, and Robert J. Andres. 2003. Global, regional and national CO_2 emissions. In *Trends: A Compendium of Data on Global Change.* Carbon Dioxide Information Analysis Center. Oak Ridge National Laboratory. Oak Ridge: U.S. Department of Energy. http://cdiac.esd.ornl.gov/ftp/ndp030/global00.ems.

Mason, Betsy. 2003. African ice under wraps: Secrets locked in Kilimanjaro's ice cap need urgent protection. *Nature Science Update,* 24 November. http://www.nature.com/nsu/031117/031117-8.html.

Mastny, Lisa. 2000. Melting of earth's ice cover reaches new high. Worldwatch News Brief. Worldwatch Institute, Washington, D.C., 6 March. http://www.worldwatch.org/press/news/2000/03/06.

May, Robert M. 1997. Kyoto and beyond. *Science* 278:1691.

McManus, Jerry F. 2004. Palaeoclimate: A great grand-daddy of ice cores. *Nature* 429:611–12.

Means, Philip Ainsworth. 1931. *Ancient Civilizations of the Andes.* New York: C. Scribner's Sons.

Meier, Mark F., and Mark B. Dyurgerov. 2002. Sea level changes: How Alaska affects the world. *Science* 297:350–51.

Meier, Mark F., and J. M. Wahr. 2002. Sea level is rising: Do we know why? *Proceedings of the National Academy of Sciences of the United States of America* 99:6524–26.

Mercer, John H. 1968. Antarctic ice and Sangamon Sea level. *Bulletin of the International Association of Scientific Hydrology* (Wallingford, Oxfordshire, UK) 79:217–25.

———. 1978. West Antarctic ice sheet and CO_2 greenhouse effect: a threat of disaster. *Nature* 271:321–25.

Mercer, John H., Lonnie G. Thompson, Cedomir Maranguníc, and John Ricker. 1975. Peru's Quelccaya ice cap: Glaciological and glacial geological studies, 1974. *Antarctic Journal of the United States* 10:19–24.

Messerli, Bruno. 1980. IAHS-AISH Publication 126:197.

Meyer, Hans. 1891. *Across East African Glaciers, An Account of the First Ascent of Kilimanjaro.* Translated by E. H. S. Calder. London: George Philip and Son.

Miller, Laury, and Bruce C. Douglas. 2004. Mass and volume contributions to twentieth-century global sea level rise. *Nature* 428:406–409.

Mix, Alan C. 2003. Chilled out in the ice-age Atlantic. *Nature* 425:32–33.

Möller, Fritz. 1963. On the influence of changes in the CO_2 concentration in air on the radiation balance of the Earth's surface and on the climate. *Journal of Geophysical Research* 68:3877–86.

Moore, Jeffrey M. 2004. Mars: Blueberry fields for ever. *Nature* 428:711–12.

Morgan, Vin, Marc Delmotte, Tas van Ommen, Jean Jouzel, Jérôme Chappellaz, Suenor Woon, Valérie Masson-Delmotte, and Dominique Raynaud. 2002. Relative timing of deglacial climate events in Antarctica and Greenland. *Science* 297:1862–64.

Morton, Oliver. 2003. The tarps of Kilimanjaro. *New York Times.* Op-ed, 17 November.

Moseley, Michael E. 1992. *The Incas and Their Ancestors: The Archaeology of Peru.* New York: Thames and Hudson.

———. 2001. Confronting Natural Disaster. In *Environmental Disaster and the Archeology of Human Response.* Edited by Garth Bawden and Richard M. Reycraft, 219–23. Albuquerque: University of New Mexico, Maxwell Museum of Anthropology.

Mosley-Thompson, Ellen, John F. Paskievitch, Anthony J. Gow, and Lonnie G. Thompson. 1999. Late 20th century increase in South Pole snow accumulation. *Journal of Geophysical Research (Atmospheres)* 104, no. D4:3977–86.

Munk, Walter. 2002. Twentieth century sea level: An enigma. *Proceedings of the National Academy of Sciences of the United States of America* 99:6550–55.

Murphy, Dean E. 2003. Nevada and part of California may face water shortages. *New York Times,* 12 December.

Musser, George. 1999. Seven wonders of modern astronomy. *Scientific American,* December.

Nash, J. Madeleine. 2002. *El Niño: Unlocking the Secrets of the Master Weather-Maker.* New York: Warner Books.

National Academy of Sciences. El Niño and La Niña: Tracing the dance of ocean and atmosphere. Oceanography's perspective. Washington, D.C.: National Academy of Sciences. http://www7.nationalacademies.org/ opus/elnino_5.html.

———. 2004. The International Geophysical Year. Washington, D.C.: National Academy of Sciences. http://www.nas.edu/history/igy/.

New York Times. 2001. A global warning to Mr. Bush. Editorial, 26 February.

———. 2002. One subsidy too many. Editorial, 12 November.

———. 2003. The shrinking snows of Kilimanjaro. Editorial, 26 November.

Oerlemans, Johannes. 1994. Quantifying global warming from the retreat of glaciers. *Science* 264:243–45.

Ohio State University (OSU). 1992. Chinese ice cores provide climate records of four ice ages. Press release, 30 November. http://www.osu.edu/units/research/archive/icecores.htm.

———. 1997. Researchers date Chinese ice core to 500,000 years. Press release, 19 June. http://www.osu.edu/units/research/archive/guliya.htm.

———. 2001. Ice caps in Africa, tropical South America likely to disappear within fifteen years. Press release, 18 February. http://www.acs.ohiostate.edu/units/research/archive/glacgone.htm.

Oliver, Ronald, and J. D. Fage. 1995. *A Short History of Africa*. 6th ed. London: Penguin.

O'Neill, Brian C., and Michael Oppenheimer. 2002. Dangerous climate impacts and the Kyoto protocol. *Science* 296:1971–72.

Oppenheimer, Michael. 1998. Global warming and the stability of the West Antarctic Ice Sheet. *Nature* 393:325–32.

Oppenheimer, Michael, and Robert Boyle. 1990. *Dead Heat: The Race Against the Greenhouse Effect.* New York: Basic Books.

Oreskes, Naomi. 2004. Beyond the ivory tower: The scientific consensus on climate change. *Science* 306:1686.

Ortloff, Charles R., and Alan L. Kolata. 1993. Climate and collapse: Agro-ecological perspectives on the decline of the Tiwanaku State. *Journal of Archeological Science* 20:195–221.

Park, Edwards. 1997. Langley's Feat—and Folly. *Smithsonian,* November. http://smithsonianmag.com/smithsonian/issues97/nov97/object_nov97.html.

Paulsen, Allison C. 1976. Environment and empire: Climatic factors in prehistoric Andean culture change. *World Archeology* 8:121–32.

People's Daily Online. 2004. Glacier study reveals chilling prediction. http://english.people.com.cn/200409/23/print20040923_158036.html.

Petit, Charles W. 2000. Polar meltdown. *U.S. News and World Report,* 28 February.

Petit, Jean-Robert, J. Jouzel, D. Raynaud, N. I. Barkov, J.-M. Barnola, I. Basile, M. Bender, J. Chappellaz, M. Davis, G. Delaygue, M. Delmotte, V. M. Kotlyakov, M. Legrand, V. Y. Lipenkov, C. Lorius, L. Pépin, C. Ritz, E. Saltzman, and M. Stievenard. 1999. Climate and atmospheric history of the past 420,000 years from the Vostok ice core, Antarctica. *Nature* 399:429–36.

Petite-Maire, Nicole, L. Beufort and N. Page. 1994. Holocene climate change and man in the present-day Sahara Desert. In *Third Millennium* BC *Climate Change and Old World Collapse*. Edited by H. Nüzhet Dalfes, George Kukla, and Harvey Weiss.

NATO ASI Series I. Global Environmental Change. Vol. 49. Berlin, New York: Springer-Verlag.

Pierce, Jennifer L., Grant A. Meyer, and A. J. Timothy Jull. 2004. Fire-induced erosion and millennial-scale climate change in northern ponderosa pine forests. *Nature* 432:87–90.

Pierrehumbert, Raymond T. 1999. Huascarán ^{18}O as an indicator of tropical climate during the last glacial maximum. *Geophysical Research Letters* 26:1345–48.

———. 2000. Climate change and the tropical Pacific: The sleeping dragon wakes. *Proceedings of the National Academy of Sciences of the United States of America* 97:1355–58.

Plass, Gilbert N. 1956a. Effect of carbon dioxide variations on climate. *American Journal of Physics* 24:376–87.

———. 1956b. The carbon dioxide theory of climatic change. *Tellus* 8:140–54.

———. 1959. Carbon dioxide and climate. *Scientific American,* July, 41–47.

Possehl, Gregory L. 1994. The drying up of the Sarasvati: Environmental disruption in South Asian prehistory. In *Third Millennium BC Climate Change and Old World Collapse*. Edited by H. Nüzhet Dalfes, George Kukla, and Harvey Weiss. NATO ASI Series I. Global Environmental Change. Vol. 49. Berlin, New York: Springer-Verlag.

Post (Athens, Ohio). 1997. OSU Student Falls Ill on Expedition. Athens: Ohio University, 27 October. http://thepost.baker.ohiou.edu/archives/102797/briefly.html.

Postel, Sandra. 1999. *Pillar of Sand: Can the Irrigation Miracle Last?* New York: W. W. Norton.

Potts, Richard. 1996. Evolution and climate variability. *Science* 273:922–23.

Powers, Thomas. 1993. *Heisenberg's War: The Secret History of the German Bomb*. New York: Knopf.

Quarterly Journal of the Royal Meteorological Society. 1965. Mr. G. S. Callendar. Obituary. vol. 91:38.

Quinn, William H. 1992. A study of Southern Oscillation–related climatic activity of A.D. 622–1900 incorporating Nile River flood data. In *El Niño. Historical and Paleoclimatic Aspects of the Southern Oscillation*. Edited by Henry F. Diaz and Vera Markgraf. Cambridge: Cambridge University Press.

Ramachandran, R. 2001. The receding Gangotri. *Frontline* (Madras) 18, no. 7, 31 March. http://www.flonnet.com/fl1807/18070690.htm.

Rasool, S. Ichtiaque, and Stephen H. Schneider. 1971. Atmospheric carbon dioxide and aerosols: Effects of large increases on global climate. *Science* 173:138–41.

Reader, John. 1982. *Kilimanjaro*. New York: Universe Books.

Regalado, Antonio. 2003a. Billionaire opens deep pockets for climate-theory research. *Wall Street Journal,* 17 July.

———. 2003b. Debating global warming: Global warming skeptics are facing storm clouds. *Wall Street Journal,* 31 July.

———. 2004. When a plant emerges from melting glacier, is it global warming? *Wall Street Journal,* 22 October.

Revelle, Roger Randall Dougan, Biography. Scripps Institution of Oceanography Archives. http://scilib.ucsd.edu/sio/archives/siohstry/revelle-biog.html.

———. 1988. Preparation for a scientific career. Based on an interview by Sarah L. Sharp. 1984. Berkeley: University of California Regional Oral History Office (copies at Bowling Green State University, Ohio, and Scripps Institution of Oceanography), 12a, 52.

Revelle, Roger, and Hans E. Suess. 1957. Carbon dioxide exchange between atmosphere and ocean and the question of an increase of atmospheric CO_2 during the past decades. *Tellus* 9:18–27.

Revkin, Andrew C. 2001. Glacier loss seen as clear sign of human role in global warming. *New York Times,* 19 February.

———. 2003. Politics reasserts itself in the debate over climate change and its hazards. *New York Times,* 5 August.

———. 2004a. Bush vs. the laureates: how science became a partisan issue. *New York Times,* 19 October.

———. 2004b. Climate debate gets its icon: Mt. Kilimanjaro. *New York Times,* 23 March.

———. 2004c. Election over, McCain criticizes Bush on climate change. *New York Times,* 16 November.

———. 2004d. The sky is falling! say Hollywood, and, yes, the Pentagon. *New York Times,* 29 February.

Rhodes, Richard. 1986. *The Making of the Atomic Bomb.* New York: Simon and Schuster.

———. 1995. *Dark Sun: The Making of the Hydrogen Bomb.* New York: Simon and Schuster.

Ricker, John E. 1977. *Yuraq Janka: Cordilleras Blanca and Rosko.* Banff: Alpine Club of Canada and New York: American Alpine Club.

Rignot, Eric, Andres Rivera, and Gino Casassa. 2003. Contribution of the Patagonia ice fields of South America to sea level rise. *Science* 302:434–37.

Rignot, Eric, and Robert H. Thomas. 2002. Mass balance of polar ice sheets. *Science* 297:1502–6.

Rind, David. 1999. Complexity and climate. *Science* 284:105–7.

Rohter, Larry. 2005. Antarctica, warming, looks ever more vulnerable. *New York Times,* 25 January.

Sarewitz, Daniel, and Roger Pielke Jr. 2000. Breaking the global-warming gridlock. *Atlantic Monthly,* July, 54–64.

Schiermeier, Quirin. 2004. Greenland's climate: A rising tide. *Nature* 428:114–15.

Schneider, David. 1996. Rain forest crunch. *Scientific American,* March.

———. 1998. The rising seas. *Scientific American,* August.

Schneider, Stephen H., and Clifford Mass. 1975. Volcanic dust, sunspots, and temperature trends. *Science* 190:741–46.

Schrag, Daniel P. 2000. Of ice and elephants. *Nature* 404:23–24.

Schrag, Daniel P., Gretchen Hampt, and David W. Murray. 1996. Pore fluid constraints on the temperature and oxygen isotopic composition of the glacial ocean. *Science* 272:1930–32.

Schwartz, Peter, and Doug Randall. 2003. An abrupt climate change scenario and its implications for United States national security. Report prepared for the Department of Defense, October. San Francisco: Global Business Network. www.gbn.org/ArticleDisplayServlet.srv?aid=26231.

Science. 2003a. In the eye of the storm. In Random Samples. Vol. 301:914c.

———. 2003b. Plankton in the tank. In Random Samples. Vol. 302:387c.

Seager, Richard, D. S. Battisti, J. Yin, N. Gordon, N. Naik, A. C. Clement, and M. A. Cane. 2002. Is the Gulf Stream responsible for Europe's mild winters? *Quarterly Journal of the Royal Meteorological Society* 128:2563–86.

Seltzer, Geoffrey O., D. T. Rodbell, P. A. Baker, S. C. Fritz, P. M. Tapia, H. D. Rowe, and R. B. Dunbar. 2002. Early warming of tropical South America at the last glacial-interglacial transition. *Science* 296:1685–86.

Service, Robert F. 2004. As the West goes dry. *Science* 303:1124–27.

Shackleton, Nicholas J. 2000. The 100,000-year ice-age cycle identified and found to lag temperature, carbon dioxide, and orbital eccentricity. *Science* 289:1897–1902.

Shackleton, Nicholas J., J. Backman, H. Zimmerman, D. V. Kent, M. A. Hall, D. G. Roberts, D. Schnitker, J. G. Baldauf, A. Despairies, R. Homrighausen, P. Huddlestun, J. B. Keene, A. J. Kaltenback, K. A. O. Krumsiek, A. C. Morton, J. W. Murray, and J. Westberg-Smith. 1984. Oxygen isotope calibration of the onset of ice-rafting and history of glaciation in the North Atlantic region. *Nature* 307:620–23.

Shepherd, Andrew, Duncan Wingham, Tony Payne, and Pedro Skvarca. 2003. Larsen Ice Shelf has progressively thinned. *Science* 302:856–59.

Shimada, Izumi. 1990. Cultural continuities and discontinuities on the northern north coast, middle-late horizon. In *The Northern Dynasties: Kingship and Statecraft in Chimor: A Symposium at Dumbarton Oaks, 12th and 13th October 1985*. Edited by Michael E. Moseley and Alana Cordy-Collins, 297–392. Washington, D.C.: Dumbarton Oaks Research Library and Collection.

———. 1994. *Pampa Grande and the Mochica Culture*. Austin: University of Texas Press.

Shimada, Izumi, Crystal B. Schaaf, Lonnie G. Thompson, and Ellen Mosley-Thompson. 1991. Cultural impacts of severe droughts in the prehistoric Andes: Application of a 1,500-year ice core precipitation record. *World Archaeology: Archaeology and Arid Environments* 22:247–70.

Shouse, Ben. 2001. For ice man, the band plays on. *Science* 293:2373.

Solomon, Susan, and Charles R. Stearns. 1999. On the role of the weather in the deaths of R. F. Scott and his companions. *Proceedings of the National Academy of Sciences of the United States of America* 96:13012–16.

Soon, Willie, and Sallie Baliunas. 2003. Proxy climatic and environmental changes of the past 1000 years. *Climate Research* 23:89–110.

Soon, Willie, Sallie Baliunas, Craig Idso, Sherwood B. Idso, and David R. Legates. 2003. Reconstructing climatic and environmental changes of the past 1000 years: A reappraisal. *Energy and Environment* (Brentwood Essex, UK) 14:233–96.

Sowers, Todd, and Michael Bender. 1995. Climate records covering the last deglaciation. *Science* 269:210–14.

Stainforth, David A., T. Aina, C. Christensen, M. Collins, N. Faull, D. J. Frame, J. A. Kettleborough, S. Knight, A. Martin, J. M. Murphy, C. Piani, D. Sexton, L. A. Smith, R. A. Spicer, A. J. Thorp, and M. R. Allen. 2005. Uncertainty in predictions of the climate response to rising levels of greenhouse gases. *Nature* 433:403–6.

Stauffer, Bernhard. 1999. Climate change: Cornucopia of ice core results. *Nature* 399:412–13.

Steele, Peter. 1998. *Eric Shipton: Everest and Beyond*. London: Constable.

Steig, Eric J. 2001. No two latitudes alike. *Science* 293:2015–16.

Steig, Eric J., and Richard B. Alley. 2002. Phase relationships between Antarctic and Greenland climate records. *Annals of Glaciology* 35:45–56.

Steig, Eric J., E. J. Brook, J. W. C. White, C. M. Sucher, M. L. Bender, S. J. Lehman, D. L. Morse, E. D. Waddington, and G. D. Clow. 1998. Synchronous climate changes in Antarctica and the North Atlantic. *Science* 282:92–95.

Steig Eric J., D. L. Morse, E. D. Waddington, M. Stuiver, P. M. Grootes, P. M. Mayewski, S. L. Whitlow, and M. S. Twickler. 2000. Wisconsinan and Holocene climate history from an ice core at Taylor Dome, Western Ross Embayment, Antarctica. *Geografiska Annaler* 82a:213–35.

Stevens, William K. 1996. Yes, you can do something about the weather. *New York Times,* 31 December.

———. 1998a. As Alaska melts, scientists consider the reasons why. *New York Times,* 18 August.

———. 1998b. Climatology guru is part curmudgeon, part imp. *New York Times,* 17 March.

———. 1999. *The Change in the Weather: People, Weather, and the Science of Climate.* New York: Delacorte.

———. 2000. Researchers find ocean temperature rising, even in the depths. *New York Times,* 24 March.

Stine, Scott. 1994. Extreme and persistent drought in California and Patagonia during medieval time. *Nature* 369:546–49.

———. 2001. The great droughts of Y1K. *Sierra Nature Notes.* El Portal, California: Yosemite Association. vol. 1, May. http://www.yosemite.org/naturenotes/paleo drought1.htm.

Stipp, David. 2004. The Pentagon's weather nightmare. *Fortune.* 9 February.

Stocker, Thomas F. 2003. South dials north. *Nature* 424:496–99.

Stone, Richard. 1999. Lake Vostok probe faces delays. *Science* 286:36–37.

Stute, Martin, M. Forster, H. Frischkorn, A. Serejo, J. F. Clark, P. Schlosser, W. S. Broecker, and G. Bonani. 1995. Cooling of tropical Brazil (5°C) during the last glacial maximum. *Science* 269:379–83.

Suess, Hans E. 1953. Natural radiocarbon and the rate of exchange of carbon dioxide between the atmosphere and the sea. In *Nuclear Processes in Geologic Settings,* 52–56. Washington, D.C.: National Research Council.

———. Papers (Background). MSS 0199. Mandeville Special Collections Library. Geisel Library. University of California, San Diego. http://orpheus.ucsd.edu/speccoll/testing/html/mss0199d.html.

Thompson, Lonnie G. 1974. Analysis of the concentration of micro-particles in ice cores from Camp Century, Greenland, and Byrd Station, Antarctica. *Antarctic Journal of the United States* 9:249–50.

———. 1992. Ice core evidence from Peru and China. Chap. 27 in *Climate Since A.D. 1500.* Edited by Raymond S. Bradley and Philip D. Jones. London: Routledge.

———. 1995. Late Holocene ice core records of climate and environment from the tropical Andes, Peru. *Bulletin de l'Institut Français d'Études Andines* 24:619–29.

———. 1996. Climatic changes for the last 2000 years inferred from ice-core evidence in tropical ice cores. In *Climatic Variations and Forcing Mechanisms of the Last 2000 Years.* Edited by Philip D. Jones, Raymond S. Bradley, and Jean Jouzel 281–95. NATO ASI Series I. Global Environmental Change. Vol. 41. Berlin: Springer-Verlag.

———. 2000. Ice core evidence for climate change in the tropics: Implications for our future. *Quaternary Science Reviews* 19:19–35.

Thompson, Lonnie G., and Willi Dansgaard. 1975. Oxygen isotope and micro-particle investigation of snow samples from the Quelccaya ice cap, Peru. *Antarctic Journal of the United States* 10:24–26.

Thompson, Lonnie G., Mary E. Davis, Ellen Mosley-Thompson, and Kam-Biu Liu. 1988. Pre-Incan agricultural activity recorded in dust layers in two tropical ice cores. *Nature* 336:763–65.

Thompson, Lonnie G., Mary E. Davis, Ellen Mosley-Thompson, Todd A. Sowers, Keith A. Henderson, Victor S. Zagorodnov, Ping-Nan Lin, Vladimir N. Mikhalenko, Richard K. Campen, John F. Bolzan, Jihong Cole-Dai, and Bernard Francou. 1998.

A 25,000-year tropical climate history from Bolivian ice cores. *Science* 282:1858–64.

Thompson, Lonnie G., Wayne L. Hamilton, and Colin Bull. 1973. Analysis of the concentration of microparticles in the long ice core from Byrd Station. *Antarctic Journal of the United States* 7:340–41.

Thompson, Lonnie G., Stefan Hastenrath, and Benjamín Morales Arnao. 1979. Climatic ice core records from the tropical Quelccaya ice cap. *Science* 203:1240–43.

Thompson, Lonnie G., and Ellen Mosley-Thompson. 1981. Microparticle concentration variations linked with climatic change: Evidence from polar ice cores. *Science* 212:812–15.

Thompson, Lonnie G., Ellen Mosley-Thompson, John F. Bolzan, and Bruce R. Koci. 1985. A 1,500 year record of tropical precipitation recorded in ice cores from the Quelccaya ice cap, Peru. *Science* 229:971–73.

Thompson, Lonnie G., Ellen Mosley-Thompson, Willi Dansgaard, and Pieter M. Grootes. 1986. The Little Ice Age as recorded in the stratigraphy of the tropical Quelccaya ice cap. *Science* 234:361–64.

Thompson, Lonnie G., Ellen Mosley-Thompson, Mary E. Davis, John F. Bolzan, Jihong Dai, Yao Tandong, N. Gundestrup, Wu Xiaoling, L. Klein, and Xie Zichu. 1989. Holocene–Late Pleistocene climatic ice core records from Qinghai-Tibetan Plateau. *Science* 246:474–77.

Thompson, Lonnie G., Ellen Mosley-Thompson, Mary E. Davis, Keith A. Henderson, Henry H. Brecher, Victor S. Zagorodnov, Tracy A. Mashiotta, Ping-Nan Lin, Vladimir N. Mikhalenko, Douglas R. Hardy, and Jürg Beer. 2002. Kilimanjaro ice core records: Evidence of Holocene climate change in tropical Africa. *Science* 298:589–93.

Thompson, Lonnie G., Ellen Mosley-Thompson; Mary E. Davis, Ping-Nan Lin, Keith A. Henderson, Jihong Cole-Dai, John F. Bolzan, Kam-Biu Liu. 1995. Late glacial stage and Holocene tropical ice core records from Huascarán, Peru. *Science* 269:46–50.

Thompson, Lonnie G., Ellen Mosley-Thompson, Mary E. Davis, Ping-Nan Lin, Keith Henderson, and Tracy A. Mashiotta. 2003. Tropical glacier and ice core evidence of climate change on annual to millennial time scales. *Climatic Change* 59:137–55.

Thompson, Lonnie G., Ellen Mosley-Thompson, Mary E. Davis, Ping-Nan Lin, Yao Tandong, Mark Dyurgerov, and Jihong Dai. 1993. "Recent warming": Ice core evidence from tropical ice cores with emphasis upon Central Asia. *Global and Planetary Change* 7:145–56.

Thompson, Lonnie G., Ellen Mosley-Thompson, and Keith A. Henderson. 2000. Ice core paleoclimate records in tropical South America since the last glacial maximum. *Journal of Quaternary Science* 15:377–94.

Thompson, Lonnie G., Ellen Mosley-Thompson, and Benjamín Morales Arnao. 1984. El Niño/Southern Oscillation events recorded in the stratigraphy of the tropical Quelccaya ice cap, Peru. *Science* 226:50–53.

Thompson, Lonnie G., Robert Mulvaney, and David A. Peel. 1989. A cooperative climatological-glaciological program in the Antarctic Peninsula. *Antarctic Journal of the United States* 24, no. 5:69–70.

Thompson, Lonnie G., David A. Peel, Ellen Mosley-Thompson, Robert Mulvaney, Jihong Dai, Ping-Nan Lin, Mary E. Davis, and Charles F. Raymond. 1994. Climate since A.D. 1510 on Dyer Plateau, Antarctic Peninsula: Evidence for recent climate change. *Annals of Glaciology* 20:420–26.

Thompson, Lonnie G., Yao Tandong, Mary E. Davis, Keith A. Henderson, Ellen

Mosley-Thompson, Ping-Nan Lin, Jürg Beer, H.-A. Synal, Jihong Cole-Dai, and John F. Bolzan. 1997. Tropical climate instability: The last glacial cycle from a Qinghai-Tibetan ice core. *Science* 276:1821–25.

Tillman, James E. Mars: Temperature overview. University of Washington. http://www-k12.atmos.washington.edu/k12/resources/mars_data-information/temperature_overview.html.

Tilman, Harold William. 1938. *Snow on the Equator*. New York: Macmillan.

Townsend, Mark, and Paul Harris. 2004. Now the Pentagon tells Bush: Climate change will destroy us. *Observer* (London), 22 February.

Union of Concerned Scientists. 2003. Global environment: Backgrounder: Responding to global warming skeptics: Prominent skeptics organizations. http://www.ucsusa.org/global_environment/global_warming/page.cfm?pageid=499.

United Nations Environmental Programme (UNEP). 2002. *Global Environment Outlook 3: Past, Present and Future Perspectives*. London: Earthscan. http://www.unep.org/geo/geo3/index.htm.

United States Senate Committee on Commerce, Science and Transportation. 1992. *Global Change Research: Indicators of Global Warming and Solar Variability: Hearing Before the Committee on Commerce, Science, and Transportation*. 102d Cong. 2d sess., 27 February.

University of Cambridge. Dept. of Physics. Charles Thomson Rees Wilson. http://www-outreach.phy.cam.ac.uk/camphy/physicists/physicists_wilson.htm.

Uppenbrink, Julia. 1996. Arrhenius and global warming. *Science* 272:1122.

Vaughan, David G., Gareth J. Marshall, William M. Connolley, John C. King, and Robert Mulvaney. 2001. Devil in the detail. *Science* 293:1777–79.

Vauxhall Society. Johnston, Sir Henry Hamilton (1858–1927). http://www.vauxhallsociety.org.uk/Johnston.html.

Visser, Katherine, Robert Thunell, and Lowell Stott. 2003. Magnitude and timing of temperature change in the Indo-Pacific warm pool during deglaciation. *Nature* 421:152–55.

Walker, Gabrielle. 2003. *Snowball Earth*. Bloomsbury: Crown.

———. 2004. Palaeoclimate: Frozen time. *Nature* 429:596–97.

Wang, W.-C., M. P. Dudek, X.-Z. Liang, and J. T. Kiehl. 1991. Inadequacy of CO_2 as a proxy in simulating the greenhouse effect of other radiatively active gases. *Nature* 350:573–77.

Wang, W.-C., Y. L. Yung, A. A. Lacis, T. Mo, and J. E. Hansen. 1976. Greenhouse effects due to man-made perturbation of trace gases. *Science* 194:685–90.

Watts, Jonathan. 2004. Highest icefields will not last 100 years, study finds. *Guardian* (Manchester, UK), 24 September. http://www.guardian.co.uk/print/0,3858,5023379-111400,00.html (accessed September 27, 2004).

Weart, Spencer R. 1997a. Global warming, Cold War, and the evolution of research plans. *Historical Studies in the Physical and Biological Sciences* 27:319–56.

———. 1997b. The discovery of the risk of global warming. *Physics Today,* January, 34–40.

———. 2003a. *The discovery of global warming*. Cambridge: Harvard University Press.

———. 2003b. Discovery of global warming Web site. College Park, Maryland: American Institute of Physics. http://www.aip.org/history/climate/.

Weaver, Andrew J., and Claude Hillaire-Marcel. 2004. Global warming and the next ice age. *Science* 304:400–402.

Weik, Martin H. 1961. The ENIAC story. *Ordnance* (Journal of the American Ordnance

Association, Washington D.C.), January–February. http://ftp.arl.mil/~mike/comphist/eniac-story.html.

Weiner, Jonathan. 1990. *The Next One Hundred Years: Shaping the Fate of Our Living Earth*. New York: Bantam.

Weiss, Harvey. 1994. Late Third Millennium abrupt climate change and social collapse in west Asia and Egypt. In *Third Millennium BC Climate Change and Old World Collapse*. Edited by H. Nüzhet Dalfes, George Kukla, and Harvey Weiss. NATO ASI Series I. Global Environmental Change. Vol. 49. Berlin, New York: Springer-Verlag.

———. 2001. Beyond the Younger Dryas: Collapse as adaptation to abrupt climate change in ancient West Asia and the eastern Mediterranean. In *Environmental Disaster and the Archeology of Human Response*. Edited by Garth Bawden and Richard M. Reycraft, 75–98. Albuquerque: University of New Mexico, Maxwell Museum of Anthropology.

Weiss, Harvey, and Raymond S. Bradley. 2001. What drives societal collapse? *Science* 291:609–10.

Weiss, Harvey, M.-A. Courty, W. Wetterstrom, F. Guichard, L. Senior, R. Meadow, and A. Curnow. 1993. The genesis and collapse of Third Millennium north Mesopotamian civilization. *Science* 261:995–1004.

White, James W. C., and Eric J. Steig. 1998. Timing is everything in a game of two hemispheres. *Nature* 394:717–18.

Whitfield, John. 2003. Alaska's climate: Too hot to handle. *Nature* 425:338–39.

Whitlock, Cathy. 2004. Land management: Forests, fires and climate. *Nature* 432:28–29.

Wright, James D. 2001. The Indonesian valve. *Nature* 411:142–43.

Wunsch, Carl. 2002. What is the thermohaline circulation? *Science* 298:1179–81.

Zwally, H. Jay, Waleed Abdalati, Tom Herring, Kristine Larson, Jack Saba, and Konrad Steffen. 2002. Surface melt-induced acceleration of Greenland ice-sheet flow. *Science* 297:218–22.

ACKNOWLEDGMENTS

The main thing for which I would like to thank Lonnie Thompson is for living the kind of life that is definitely worthy of a book. I hope this one does justice to his impeccability, his sagacity, his deep commitment, and his friendliness and decency. While this book, perforce, focuses on Lonnie's career, I have tried to make it clear that his many successes have been the result of a joint effort with his wife, Ellen Mosley-Thompson, a great woman who stands somewhat in the background here but has played a crucial and equal role, both in the orchestration of Lonnie's fieldwork and in the interpretation of the science that has arisen from it. I met Ellen before I met Lonnie, less than an hour after I learned that I would be going to Bolivia on rather short notice, and her first words were, "Don't panic now, we have everything figured out." Her grace under pressure, her commitment, her sharp intellect, and the friendliness and decency she shares with her husband have been clear from the outset.

They both took time from their exceedingly busy lives to tell me about their lives and their work, to show me around their hometown and their laboratories, and to share generously their knowledge and insight. I also thank Lonnie for being so gracious when I showed up ill-equipped on the summit of Sajama and for inviting me on the two Kilimanjaro expeditions. It is hard to imagine how I could have written this book had not he and Ellen both been so genuinely open and honest all along.

Other members of their group have been equally generous and open, particularly my friends Mary Davis, Bruce Koci, and Keith Mountain. Mary has been there to help, in her friendly and easygoing way, whenever I have asked. What can I say about Keith, except that he is some of the best company I have ever enjoyed; I remain astounded at his courageousness and optimism, and his insights into the big picture—both the science and, for lack of a better word, the culture of these unusual people—play a larger role here than it may appear. His and Bruce's photographs enhance this book tremendously. I thank Bruce not only for sharing his photographs and his time, but also for innumerable small acts of kindness: within a few hours of the moment I met him on the summit of Sajama, during the six weeks we spent together at the South Pole, and during our month together on and around Kilimanjaro.

Other members of the Thompsons' group have also given help and support. Vladimir Mikhalenko, Ping-Nan Lin, and Victor Zagorodnov were steady companions in the field; Henry Brecher and Zhongqui Li shared their time and their memories; and Dave Chadwell stepped out of the past to treat me with stories of the early days.

I feel privileged to have met some of the world's finest climatologists in course of writing this book. Ray Bradley, Wally Broecker, Colin Bull, Mark Cane, Peter deMenocal, Mark Dyurgerov, Jim Hansen, Dave Keeling, Alan Kolata (an anthropologist who has demonstrated the breadth of both his field and his mind by incorporating climatology into his work), Jim McCarthy, Michael McElroy, Michael Oppenheimer, Dan Schrag, and Jay Zwally all granted interviews, and I enjoyed a cordial and helpful exchange of e-mails with Michael Mann.

I would be remiss not to acknowledge the work of Spencer Weart. The history of greenhouse science that he has presented in a series of journal articles, on his extraordinarily comprehensive Web site, and in the book that has evolved from that Web site inform all historical sections of this book, particularly those in chapters 6 through 9. I may have a different take on one or two historical events but it was largely Dr. Weart who showed me where to look in the first place.

Special thanks go to Alison Osius, who sent me off on this adventure long ago. Even though we see each other only once or twice a decade, it sometimes feels as though I am the puppet and Alison is the benign entity pulling the strings.

Máire Crowe and Bruce Stutz, former editors at *Natural History,* provided the ground for this effort by hiring me to write the article about Sajama—on which I was privileged to work with George Steinmetz, one of the finest photographers working today. His photographs graced the article and now grace this book. The late Yossi Brain and his former girlfriend Ulli Schatz hosted me (perhaps a bit too enthusiastically) in La Paz. Lindsay Griffin, with his encyclopedic knowledge of climbers and climbing around the world, put me in touch with Yossi and other contacts in Bolivia and also set me straight about the high-altitude endurance records in the Andes that rival those of Lonnie and his crew. Doug Hardy, who has kept in touch ever since we first met on Sajama, continues to keep me posted on the goings-on in Africa and the Andes. Bernard Francou helped me get to Sajama and answered questions afterward. Todd Sowers provided my introduction to climatology in that spectacular setting and also helped afterward. Julias Minja was a friend and a coolheaded and sagacious companion on Kilimanjaro, and he honored me with a rare view into the life of his Chagga people. Phil Ndesamburo took time out of his busy schedule to show me around Old Moshi one day, teaching me about the history of the area. Garry McKenzie taught me about environmental geology. Jacques Leslie erased the few doubts I had that, yes, the skirmishes and conflagrations we hear about daily in the news are indeed "quotidian" in relation to the looming freshwater crisis. Tom Davis of Zoot Software has written a superb application enabling writers to keep volumes of information at their fingertips and has given me personal assistance, amazingly, in solving the few minor problems I have had in using it. Cilla Borras transcribed hours of audiotape. Fernando Garrido told me what it was like to live on top of Aconcagua for sixty-two days. Peter Athans described the changes he's seen on Everest and shared insights about Lonnie's Dasuopu expedition. Lloyd Athearn, Topher Donahue, Elizabeth Hawley in Kathmandu, Chris Holland, Cèsar Morales Arnao and his son and translator Gunter in Peru, Jim Nowak, Mark Richey, and Gerry Roach all told me about the changes they have seen in the high, white, but now darkening mountain world. Teresa Richey put me in touch with her uncle Ceŝar and cousin Gunter and helped me out with Peruvian names. Don Cutler, my former agent, made contact with Henry Holt and saw me through the process of signing my first book contract. And I would especially like to thank two mentors: Professor George Benedek of MIT, who has helped on this particular effort in one specific way and has stood as a shining example to me for three decades; and my dear friend Doug Scott, whose invitations to join him in some of the earth's wildest places provided the material for the magazine articles that opened the gate to this project, and whose company and example have enriched my life in ways that I am still coming to understand.

Writing this book involved a strange combination of field experience in some of the most breathtaking and expansive places on earth and foraging around in the physically confining but mentally expansive stacks of numerous libraries looking for books and journal articles. Now that it is done, I am moved to proclaim: Long live public libraries! My hometown is served by what must be one of the best library systems in existence, the Minuteman Library Network of eastern Massachusetts; the network and especially the helpful staff in my local branch have been invaluable. Twice, for example, they tracked down pristine first editions of Hans Meyer's 1891 book *Across East African Glaciers*, worth a few thousand dollars on the rare book market, and let me keep them for months at a time. I would also like to thank Joe Hankins of MIT's Lindgren Library not only for his professional help but also for making me feel that I was a member of the MIT community again.

Three readers gave helpful suggestions on the penultimate draft: Bill Atkinson, Ed Feldmann, and my mother, an English teacher, Leigh Bowen, whose lessons about paragraphs and subordination (in the grammatical sense, that is) I hope I have learned. I thank both my parents for their support.

My deepest gratitude goes to Craigen Bowen, the mother of our children, who took care of them while I was away for months at a time. As a member of the staff at Harvard, she was also in the vulnerable position of having access to one of the greatest library systems on the planet and was generous enough to spend many hours finding rare books and old journal articles for me, some of them virtually impossible to obtain elsewhere. She supported me in numerous ways, all through this effort.

My children Andrew and Anna have spent a significant portion of their childhoods seeing only my back as I hunched at my desk, and they have waited patiently for hundreds of dinners served long past the promised hour.

My profoundest thanks to Wendy for her understanding, her unflagging support, and her love, all of which have been sorely tested over the years that I have been involved in this selfish task.

The staff at Henry Holt has led me expertly through the enormous job of producing a work that is up to their standards. I would like to thank Kelly Too, the book's designer; Chris O'Connell, the production editor; and especially my taskmaster Supurna Banerjee, who has held it all together. I have especially appreciated Supurna's warmth, her attention to detail, and her unfailing good nature.

Saving the best for last, I feel very lucky to have worked with as wise and understanding an editor as Jack Macrae. He set the bar high and guided me firmly, gently, and with infinite patience until I finally struggled over it.

INDEX

Entries in *italics* refer to illustrations.

ABOUT THE AUTHOR

MARK BOWEN holds a Ph.D. in physics from MIT. He lives with his two children near Boston, Massachusetts.